Recent Trends in Geospatial AI

Dina Darwish
Ahram Canadian University, Egypt

Houssem Chemingui
Brest Business School, France

Vice President of Editorial	Melissa Wagner
Managing Editor of Acquisitions	Mikaela Felty
Managing Editor of Book Development	Jocelynn Hessler
Production Manager	Mike Brehm
Cover Design	Phillip Shickler

Published in the United States of America by
IGI Global Scientific Publishing
701 East Chocolate Avenue
Hershey, PA, 17033, USA
Tel: 717-533-8845
Fax: 717-533-8661
E-mail: cust@igi-global.com
Website: https://www.igi-global.com

Copyright © 2025 by IGI Global Scientific Publishing. All rights reserved. No part of this publication may be reproduced, stored or distributed in any form or by any means, electronic or mechanical, including photocopying, without written permission from the publisher.
Product or company names used in this set are for identification purposes only. Inclusion of the names of the products or companies does not indicate a claim of ownership by IGI Global Scientific Publishing of the trademark or registered trademark.

Library of Congress Cataloging-in-Publication Data

CIP Pending
ISBN: 979-8-3693-8054-3
EISBN: 979-8-3693-8056-7

Vice President of Editorial: Melissa Wagner
Managing Editor of Acquisitions: Mikaela Felty
Managing Editor of Book Development: Jocelynn Hessler
Production Manager: Mike Brehm
Cover Design: Phillip Shickler

British Cataloguing in Publication Data
A Cataloguing in Publication record for this book is available from the British Library.

All work contributed to this book is new, previously-unpublished material.
The views expressed in this book are those of the authors, but not necessarily of the publisher.
This book contains information sourced from authentic and highly regarded references, with reasonable efforts made to ensure the reliability of the data and information presented. The authors, editors, and publisher believe the information in this book to be accurate and true as of the date of publication. Every effort has been made to trace and credit the copyright holders of all materials included. However, the authors, editors, and publisher cannot assume responsibility for the validity of all materials or the consequences of their use. Should any copyright material be found unacknowledged, please inform the publisher so that corrections may be made in future reprints.

Editorial Advisory Board

Pankaj Bhambri, *Guru Nanak Dev Engineering College, India*
Monica Gupta, *Bharati Vidyapeeth's College of Engineering, India*
Arbia Hlali, *Taibah University, Saudi Arabia*
Vijaya Kittu Manda, *PBMEIT, India*
Rajesh Kanna Rajendran, *Christ University, Bangalore, India*
Theodore Tarnanidis, *International Hellenic University, Greece*

Table of Contents

Preface xiii

Chapter 1
Geospatial AI Concepts and Fundamentals 1
    Dina Darwish, Ahram Canadian University, Egypt

Chapter 2
Harnessing AI and Machine Learning for Enhanced Geospatial Analysis 27
    Rachna Rana, Ludhiana Group of Colleges, Ludhiana, India
    Pankaj Bhambri, Guru Nanak Dev Engineering College, Ludhiana, India

Chapter 3
Harnessing AI in Geospatial Technology for Environmental Monitoring and Management: Applications of AI for Geospatial Data Processing 73
    Monica Gupta, Bharati Vidyapeeth's College of Engineering, India
    Rupanshi Bhatnagar, Ernst & Young, India

Chapter 4
Geospatial Technologies for Smart Cities 103
    Amit Sai Jitta, Indiana University, Bloomington, USA
    Vijaya Kittu Manda, PBMEIT, India
    Theodore Tarnanidis, International Hellenic University, Greece

Chapter 5
PATCH.AI: Forest Cover Virtualization on Digital Maps Using Satellite Imagery (Google Maps API) 137
    R. Parvathi, Vellore Institute of Technology, India
    V Pattabiraman, Vellore Institute of Technology, India
    B. Shakthi, Vellore Institute of Technology, India

Chapter 6
Optimizing Flood Risk Management Through Geospatial AI and Remote Sensing 155
> G. Prabhanjana, Christ University, India
> Yashas Shetty, Christ University, India
> Daksh Vats, Christ University, India
> Rajesh Kanna Rajendran, Christ University, India

Chapter 7
The Role of IoT in Shaping the Future of Geospatial AI 177
> Rachna Rana, Ludhiana Group of Colleges, Ludhiana, India
> Pankaj Bhambri, Guru Nanak Dev Engineering College, Ludhiana, India

Chapter 8
Traffic Flow Optimization Using AI 217
> Rajesh Kanna Rajendran, Christ University, India
> N. R. Wilfred Blessing, University of Technology and Applied Sciences, Ibri, Oman
> T. Mohana Priya, Christ University, India

Chapter 9
Current Trends, Opportunities, and Futures Research Directions in Geospatial Technologies for Smart Cities 239
> Vijaya Kittu Manda, PBMEIT, India
> Veena Christy, SRM Institute of Science and Technology, India
> Arbia Hlali, Taibah University, Saudi Arabia

Chapter 10
Benefits and Opportunities of Geospatial AI Adoption 271
> Sucheta Yambal, Dr. Babasaheb Ambedkar Marathwada University, India
> Yashwant Arjunrao Waykar, Dr. Babasaheb Ambedkar Marathwada University, India

Chapter 11
Geospatial AI Future Perspectives 297
    Dina Darwish, Ahram Canadian University, Egypt

Compilation of References 325

About the Contributors 363

Index 369

Detailed Table of Contents

Preface xiii

Chapter 1
Geospatial AI Concepts and Fundamentals 1
    Dina Darwish, Ahram Canadian University, Egypt

In terms of its progress and the ways in which it might be applied in our everyday lives, geospatial artificial intelligence (Geo-AI) is an intriguing issue. When it comes to the economic and social growth of a region or country, one of the factors that contributes to this development is spatial planning. The primary focus of this investigation is based on spatial data, with the objective of maximizing the efficiency of land use of spatial data on the region as upstream data and developing GIS-based urban planning apps that automatically present the findings of analysis and predictions of urban objects. Also, the plans for urban areas for both spatial and land use can be made, and geospatial AI generates a significant amount of spatial data based on the characteristics of the environment and the geography. This is done in order to maximize the revenue that is generated by tourists. In the context of urban planning and the tourism industry, this Geo-AI can offer many advantages and benefits. This chapter discusses several topics related to geospatial AI, including geo-computation, geospatial AI applications, challenges facing it, and its future prospects.

Chapter 2

Harnessing AI and Machine Learning for Enhanced Geospatial Analysis 27
    Rachna Rana, Ludhiana Group of Colleges, Ludhiana, India
    Pankaj Bhambri, Guru Nanak Dev Engineering College, Ludhiana, India

In recent years, artificial intelligence (AI) and machine learning (ML) have changed geospatial analysis, allowing for more accurate, efficient, and scalable processing of massive volumes of geographical data. Traditionally, geospatial analysis depended on human-driven approaches and rule-based systems, which were frequently time-consuming and restricted in their capacity to handle large datasets. The combination of AI and ML has resulted in the development of revolutionary approaches like as deep learning, neural networks, and automated feature extraction, which have transformed the use of geographic information systems (GIS). This chapter investigates the critical role of artificial intelligence and machine learning in developing geospatial analysis in a variety of disciplines, including environmental monitoring, urban planning, disaster management, and agriculture. AI-powered models can now do predictive analytics, real-time data processing, and pattern identification in satellite images, LiDAR, and sensor networks.

Chapter 3

Harnessing AI in Geospatial Technology for Environmental Monitoring and Management: Applications of AI for Geospatial Data Processing 73
    Monica Gupta, Bharati Vidyapeeth's College of Engineering, India
    Rupanshi Bhatnagar, Ernst & Young, India

The chapter explores the role of Geospatial Technology in processing, and analyzing geographically-referenced data from various sources, such as satellite imagery, sensor networks, and climate models. Geospatial Technology tools like GPS, remote sensing, GIS, and LiDAR have proven invaluable in areas like urban planning, disaster management, and environmental conservation. These technologies provide real-time, accurate geographic data, enabling organizations to make informed decisions. Geospatial Technology is widely used in urban development for optimizing infrastructure, tracking deforestation, and monitoring biodiversity. In disaster management, it supports early warning systems and enhances coordination during crisis response. Additionally, it helps manage natural resources and monitor agricultural productivity. The chapter highlights the evolution of Geospatial Technology, emphasizing its growing importance in environmental monitoring and resource management, while showcasing how its applications continue to expand across sectors such as defense, urban planning, and agriculture.

Chapter 4
Geospatial Technologies for Smart Cities 103
    Amit Sai Jitta, Indiana University, Bloomington, USA
    Vijaya Kittu Manda, PBMEIT, India
    Theodore Tarnanidis, International Hellenic University, Greece

Geospatial technologies, both traditional and modern technologies, have changed the way spatial data is collected, stored, processed, analyzed, and visualized for decision-making in Smart cities. Popular geospatial technologies are geospatial information systems (GIS), remote sensing, and global positioning systems (GPS). Computing technologies have undergone rapid development in recent times. Artificial Intelligence (AI), the Internet of Things (IoT), Big Data, and others are used alongside geospatial technologies for improved decision-making by city planners and administrators. These technologies help Smart cities offer various services to the citizens, promote a circular economy, and be sustainable. Real-time data processing and predictive analytics help proactively manage infrastructure, optimize resource allocation, and enhance overall urban resilience. The chapter is novel in its comprehensive overview of geospatial technologies and data in smart cities, integrating both traditional and modern technologies and emphasizing the significance of geospatial data visualization.

Chapter 5
PATCH.AI: Forest Cover Virtualization on Digital Maps Using Satellite
Imagery (Google Maps API) 137
    R. Parvathi, Vellore Institute of Technology, India
    V Pattabiraman, Vellore Institute of Technology, India
    B. Shakthi, Vellore Institute of Technology, India

In a situation where the severe threats of down environmental challenges are prevalent, there is no doubt that there has never been a time when innovative tools for the forest ecosystems monitoring and management are relevant. PATCH.AI stands out as a pioneering technology that enables the integration of satellite imagery and AI algorithms to execute a through coverage status survey of forests. PATCH.AI provides analytical solutions based on the state-of-the-art Google Maps API. This technology makes use of high-definition satellite photographs that allow for the precise visualization and analysis of the given area. The core functions of PATCH.AI are in getting accurate details on areas that are forested or not through the artificial intelligence technology. Through the use of the latest in image processing technology, the system automatically recognizes and colors any forest regions in green, and thereafter, the background is plainly white, indicating the absence of forest.

Chapter 6
Optimizing Flood Risk Management Through Geospatial AI and Remote Sensing 155
    G. Prabhanjana, Christ University, India
    Yashas Shetty, Christ University, India
    Daksh Vats, Christ University, India
    Rajesh Kanna Rajendran, Christ University, India

Flooding presents an increasing threat to communities globally, intensified by climate change and urban expansion. Effective flood risk management requires precise and timely information to guide decision-making and planning. This chapter explores the use of geospatial artificial intelligence (AI) in combination with remote sensing to enhance flood risk management. Advanced AI techniques are applied to analyze satellite and aerial images, enabling more accurate identification of flood-prone areas and prediction of potential flood events.Machine learning is utilized to integrate historical and real-time data, improving flood prediction models and evaluating the effectiveness of various mitigation strategies. A decision-support system is developed to leverage this technology, providing valuable insights for policymakers, emergency responders, and urban planners. This chapter demonstrate that the integration of geospatial AI and remote sensing can significantly advance flood risk management, offering a more proactive and resilient approach to addressing this critical issue.

Chapter 7
The Role of IoT in Shaping the Future of Geospatial AI 177
    Rachna Rana, Ludhiana Group of Colleges, Ludhiana, India
    Pankaj Bhambri, Guru Nanak Dev Engineering College, Ludhiana, India

This chapter shows about the new expertise for instance AI, ML, and IoTs which has altered the geospatial sector, which involves the collecting, analysis, and display of geographical data. These technologies drive industrial innovation and growth by allowing for more precise and efficient data collecting, analysis, and decision-making. AI & ML are especially significant in the geospatial business because they enable the analysis of enormous quantity of facts that would be too time-consuming for people to handle manually. AI and ML methods can examine and understand geographical data such as satellite imaging, aerial photography, and LiDAR scans, revealing patterns and trends that humans may miss. The IoTs is also propelling the geospatial sector forward by allowing for the capture of real-time data from sensors implanted in actual things. Weather sensors, traffic sensors, and GPS trackers are examples of IoTs devices that may provide useful geographical data for decision-making in a kind of business, including farming, transportation, and town development.

Chapter 8
Traffic Flow Optimization Using AI 217
    Rajesh Kanna Rajendran, Christ University, India
    N. R. Wilfred Blessing, University of Technology and Applied Sciences,
      Ibri, Oman
    T. Mohana Priya, Christ University, India

Traffic flow optimization is a critical challenge in urban planning and transportation management, aimed at reducing congestion, improving travel times, and enhancing overall roadway efficiency. This paper explores the application of artificial intelligence (AI) techniques to address these challenges. Leveraging machine learning algorithms, neural networks, and advanced data analytics, AI-driven systems can dynamically adjust traffic signals, predict traffic patterns, and optimize routing in real-time. This approach utilizes historical traffic data, real-time sensors, and predictive modeling to make data-driven decisions that enhance traffic flow and reduce delays. Integrating AI with existing traffic management infrastructure, cities can achieve more responsive and adaptive traffic control and improved quality of life for commuters. This Chapter presents a review of current AI applications in traffic optimization, evaluates their effectiveness through case studies, and discusses potential future developments in this evolving field.

Chapter 9
Current Trends, Opportunities, and Futures Research Directions in
Geospatial Technologies for Smart Cities 239
    Vijaya Kittu Manda, PBMEIT, India
    Veena Christy, SRM Institute of Science and Technology, India
    Arbia Hlali, Taibah University, Saudi Arabia

Studying current trends and opportunities in geospatial technologies for Smart cities helps city planners and administrators understand technology advancements and aids in better implementation practice. Similarly, understanding future research directions enables researchers and policymakers to harness these technologies fully and anticipate upcoming developments. Overall, this approach supports the creation of more livable, sustainable, and equitable cities for all. This chapter explores the current trends, emerging opportunities, and future research directions. Geospatial AI is supported by several cutting-edge technologies such as IoT, digital twins, and 3D/4D urban modeling. Future geospatial research should use advanced AI models, real-time analytics, and privacy-preserving technologies. Understanding the technologies' ethical and inclusive implementation is essential to support long-term urban sustainability and citizen well-being. Such Smart cities can ensure more sustainable, resilient, and equitable urban environments for future generations.

Chapter 10

Benefits and Opportunities of Geospatial AI Adoption 271
    Sucheta Yambal, Dr. Babasaheb Ambedkar Marathwada University, India
    Yashwant Arjunrao Waykar, Dr. Babasaheb Ambedkar Marathwada University, India

The convergence of artificial intelligence (AI) and geospatial intelligence (GEOINT) is transforming industries such as agriculture, disaster management, urban planning, environmental conservation, and defense. AI's data processing power, combined with GEOINT's geographic insights, is enhancing decision-making and predictive models. Urban planners are using this technology to create smarter cities and optimize infrastructure amid growing urbanization. In agriculture, AI-powered precision farming improves crop yields and food security while reducing environmental impact. In disaster management, the integration enables faster evaluations and better coordination of relief efforts. Despite its potential, challenges like AI biases, ethical concerns, and data privacy issues must be addressed for responsible deployment. Together, AI and GEOINT offer innovative solutions for a resilient, sustainable future.

Chapter 11

Geospatial AI Future Perspectives 297
    Dina Darwish, Ahram Canadian University, Egypt

Geocomputation and geospatial artificial intelligence (GeoAI) play crucial roles in propelling geographic information science (GIS) and Earth observation into a new era. GeoAI has transformed conventional geospatial analysis and mapping, changing the approaches for comprehending and overseeing intricate human–natural systems. Nonetheless, challenges persist in multiple facets of geospatial applications concerning natural, built, and social environments, as well as in the integration of distinctive geospatial features into GeoAI models. At the same time, geospatial and Earth data play essential roles in geocomputation and GeoAI studies, as they can efficiently uncover geospatial patterns, factors, relationships, and decision-making processes. This chapter focuses on several topics related to geospatial AI, including advancements in this field and future perspectives of Geospatial AI.

Compilation of References 325

About the Contributors 363

Index 369

Preface

In the topic of geo-spatial artificial intelligence, often known as GeoAI, techniques and methods from engineering, computer science, statistics, and space science are combined in order to analyze and model spatial and temporal events using artificial intelligence (AI) approaches. GeoAI is an interdisciplinary field. Due to the fact that it focuses on problems that actually exist in the world, this topic has a significant impact on both society and the economy. In the field of geospatial data, such as satellite photographs, aerial photography, and other types of photographs, artificial intelligence is rapidly acquiring the ability to automatically recognize attributes. With the goal of assisting machines in comprehending their environment through the analysis of satellite data, the computer community has a strong interest in satellite photography. The application of this kind of treatment provides the ability to test broad areas at a significant cost. For both civilian and military purposes, the collecting, analysis, and processing of global observation data is made possible through the use of remote sensing and geographic data capabilities. In this article, an overview of GeoAI methodologies in urban planning is presented. This overview includes the definition of GeoSpatial Artificial Intelligence as well as the differences between GeoAI and traditional AI. There are also depictions of several kinds of satellites that collect geographical data. Both the incorporation of Artificial Intelligence (AI) into Geographic Information Systems (GIS) and the utilization of GeoAI tools and techniques are essential elements in the process of successfully analyzing geographic data.

Smith 1984, Couclelis 1986, Openshaw 1992, Openshaw and Openshaw 1997, and Janowicz et al. 2020 are some of the historical sources that describe the historical foundations of the junction of artificial intelligence and geographic studies. This connection is not wholly new. Before the recent explosion of deep learning studies by LeCun, et al. 2015, major AI developments included theoretical speculations in the 1950s and 1960s; artificial neural networks (ANN), heuristic search, knowledge-based expert systems, neurocomputing and artificial life (e.g., cellular automata) in the 1980s; genetic programming, fuzzy logics, and development of hybrid intelligent systems in the 1990s; and ontology and web semantics for geographic information retrieval (GIR) in the 2000s. Every one of these advancements has made a contribution to the study topics that GeoAI is focusing on. Why spatial is unique in artificial intelligence is a fundamental question that drives contributions in GeoAI. One possible explanation could be because geographic location is frequently the most important factor in integrating disparate data sets that have been extensively utilized for the purpose of training advanced artificial intelligence models. There is a question about the most important geographical issues that we are now able to answer more effectively with the use of AI as opposed to more conventional methods, and How can artificial intelligence be used to solve problems that have not yet been solved. When it comes to the construction of models and data pipelines in geographic information systems (GIS), there is a need to develop novel theories or intelligent methodologies. In recent publications, geographers and computer scientists have made significant contributions to these topics. For example, a special issue on artificial intelligence techniques for geographic knowledge discovery was published in the International Journal of Geographical Information Science. Additionally, the ACM SIGSPATIAL GeoAI workshops (2017, 2018, 2019) and discussions in the American Association of Geographers (AAG) GeoAI and Deep Learning symposiums (2018, 2019, 2020) have been held. Currently, artificial intelligence (AI) is introducing geospatial research with a plethora of new prospects and challenges that are of significant importance. Its rapid progress is driven by theoretical advancements, large amounts of data, computer hardware (such as the graphics processing unit, or GPU), and high-performance computing platforms that enable the development, training, and deployment of artificial intelligence models in a reasonable amount of time.

During the past several years, there have been major advancements in geospatial artificial intelligence (GeoAI), which is the combination of geospatial studies with artificial intelligence, particularly machine learning and deep learning techniques, as well as the most recent AI technologies at both academic institutions and private businesses. For the purpose of advancing our knowledge and finding solutions to problems in human environmental systems and their interactions, with a particular emphasis on spatial contexts and roots in geography or geographic information science (GIScience), GeoAI can be considered a study subject that aims to develop intelligent computer programs that mimic the processes of human perception, spatial reasoning, and discovery about geographical phenomena and dynamics. Therefore, in order to be competent in GeoAI research, it would be necessary to have understanding of artificial intelligence theory, programming, and computation procedures, in addition to having knowledge of the geographic domain. In the fields of geographic information science (GIScience), remote sensing, the physical environment, and human civilization, there have already been an increasing number of combined geoAI studies. At this point in time, it would be beneficial to give a critical reference list for educators, students, researchers, and practitioners who are interested in keeping up with the most recent GeoAI research issues. This bibliographic entry will begin by providing a brief overview of the historical origins of artificial intelligence in the fields of geography and geographic information science (GIScience). Subsequently, it will provide a list of up to ten selected recent works, along with annotations that provide a concise explanation of the significance of these works for each topic of interest in the GeoAI landscape. These topics include fundamental spatial representation learning, spatial predictions, and various advancements in cartography, earth observation, social sensing, and geospatial semantics.

By improving the capacity to process, analyze, and extract insights from spatial data, artificial intelligence (AI) plays a significant role in the field of spatial analysis. The use of artificial intelligence algorithms may handle big and complicated spatial datasets more efficiently than traditional approaches, which can lead to improvements in decision-making across a variety of organizations.

Through the application of machine learning algorithms, artificial intelligence is utilized in spatial analysis in a significant way. These algorithms are able to understand patterns and relationships in geographical data, which enables them to perform tasks such as image categorization, object detection, and prediction modelling. Systems that are powered by artificial intelligence, for instance, are able to analyze satellite pictures in order to recognize patterns of land use, monitor changes in the environment, or forecast agricultural harvests.

Over the course of its history, geospatial technology has seen a series of key developments that have distinguished its progress. In its early stages, geospatial technology was essentially comprised of maps and charts that were utilized for the purposes of land surveying and navigation. The development of computers and satellite technology, on the other hand, has resulted in the evolution of geospatial technology into a complex system that is capable of capturing, storing, analyzing, and displaying geographic information.

These technologies make it possible to perform automated analysis on enormous geographic datasets, which ultimately results in decisions that are more precise and effective in a variety of fields, including agriculture, urban planning, and disaster management, among others.

WHAT THE IMPORTANCE OF GEOSPATIAL AI IS

A discussion on the significance of geospatial artificial intelligence and the ways in which it may assist in revolutionizing business has followed:

1. *Achievable Accuracy:*

The information that geospatial artificial intelligence delivers is extremely accurate and exact, which enables firms to make informed decisions in a variety of industries. It encompasses a wide range of fields, including agriculture, urban planning, and disaster management.

2. *Observation in Real Time:*

The real-time monitoring capabilities offered by GeoAI are particularly useful in dynamic scenarios such as the flow of traffic in urban areas or natural disasters. Because of this, prompt action and intervention can be taken whenever it is required.

3. *Effectiveness and mechanization:*

In the past, tasks that required time and labor from humans might be automated by geospatial artificial intelligence. By way of illustration, it is able to rapidly analyze satellite photographs in order to monitor alterations in land use or automatically recognize irregularities in infrastructure.

4. *Applications That Cut Across Domains:*

Applications of geospatial artificial intelligence can be found in a variety of fields, including defense, transportation, healthcare, and environmental protection for example. Because of its adaptability, it is a useful instrument for resolving a wide variety of problems.

5. *Observations and Forecasts:*

The approaches of geospatial artificial intelligence assist find deeper insights and forecasts. Business organizations are able to discover patterns, correlations, and trends in geographic data by utilizing them. They make it possible to foresee events like as traffic patterns, agricultural output, and other aspects, which is a result of their characteristics.

THE WORKINGS OF GEOSPATIAL ARTIFICIAL INTELLIGENCE

The processing, analysis, and interpretation of spatial data in order to derive meaningful insights is the method by which geospatial artificial intelligence, also known as Geographic Information Systems (GIS) mixed with artificial intelligence (AI), operates. The operation is as follows:

- *Data Collection:* The first step in the process of conducting geospatial artificial intelligence is to collect spatial data from a variety of sources, including satellites, drones, GPS devices, and ground sensors. This data consists of geographical information such as images, coordinates, and other location-based details.
- *Preprocessing of the Data:* The data that has been obtained is prior to being processed in order to eliminate noise, rectify errors, and standardize formats. This guarantees that the data are free of any flaws and are prepared for analysis.
- Data Integration: It is the responsibility of geospatial artificial intelligence to combine the preprocessed data with other pertinent datasets, such as maps of infrastructure, demographic information, and land use data. With this integration, it is possible to conduct an in-depth investigation of the surrounding spatial environment.
- *Data Analysis:* In order to analyze the integrated data, geospatial artificial intelligence makes use of several AI algorithms, such as machine learning and deep learning. On the basis of spatial data, these algorithms are able to recognize patterns, put features into categories, and forecast trends.
- *Geospatial Modelling:* Geospatial artificial intelligence is responsible for the creation of spatial models that depict the connections between various spatial features. Simulating events, predicting outcomes, and making decisions based on accurate information are all possible with these models.

With the help of maps, charts, and several other graphical representations, geospatial artificial intelligence is able to visualize the data that has been analyzed. It is easier for consumers to comprehend the spatial patterns and trends that are present in the data as a result of this.

- *Decision Support:* Geospatial artificial intelligence offers decision support capabilities that assist users in making well-informed decisions upon the basis of the data that has been analyzed. Among these tools, you'll find everything from straightforward advice to intricate optimization techniques.

APPLICATIONS OF ARTIFICIAL INTELLIGENCE IN GEOSPATIAL

Geospatial artificial intelligence, which is the combination of artificial intelligence (AI) and geographic information systems (GIS), offers a wide range of applications that may be found in a variety of different industries. Among the most important applications are:

- *Urban planning:* Geospatial artificial intelligence is utilized to perform analyses of urban areas, optimize the development of infrastructure, and enhance traffic management. The ideal places for new buildings, parks, and transportation routes can be determined with the assistance of this function.
- *Agriculture:* Geospatial artificial intelligence is utilized in the agricultural sector for the purposes of precision farming, crop health monitoring, and yield prediction. It assists farmers in making the most of their resources and rising their overall productivity.
- *Monitoring the Environment:* Geospatial artificial intelligence is utilized for the purpose of monitoring environmental changes, including pollution, deforestation, and degradation of land. The assessment of the impact that human activities have on the environment and the implementation of mitigation measures are both aided by this technology.
- *Disaster response:* The use of geospatial artificial intelligence in disaster response is extremely important since it allows for the mapping of impacted areas in real time, the evaluation of damage, and the facilitation of rescue efforts. It has a positive impact on the coordination of relief activities and the reduction of the effects of disasters.
- *The management of natural resources:* including water, forests, and minerals, is one of the functions that geospatial artificial intelligence can do. Monitoring the exploitation of resources, conservation initiatives, and sustainable management methods are all aided by this capability.
- *Management of Infrastructure:* Geospatial artificial intelligence is utilized in the management of infrastructure assets, which include roads, bridges, and utilities. It assists in the tracking of assets, the planning of maintenance, and the evaluation of risks.

For the purpose of tracking disease outbreaks, monitoring health trends, and planning healthcare services, geospatial artificial intelligence is utilized in the field of public health. The identification of high-risk locations and the implementation of preventative measures are both aided by this.

- *Climate change*: Geospatial artificial intelligence is utilized to investigate the effects of climate change on the environment and to devise solutions for adaptation. Changes in temperature, precipitation, and sea levels can be monitored more effectively with its assistance.

THE ADVANTAGES OF GEOSPATIAL ARTIFICIAL INTELLIGENCE FOR BUSINESSES

- *Enhancement of Decision-Making Capabilities:* Geospatial artificial intelligence offers organizations significant insights that are gained from spatial data analysis, thereby assisting them in making decisions that are better informed. The decision-making processes are improved by geographic artificial intelligence, whether it be for the purpose of optimizing store sites, calculating delivery routes, or spotting market trends.
- *Cost Reductions:* Businesses are able to decrease expenses by optimizing their operations based on the insights provided by geospatial data. As an illustration, optimizing the routes that delivery fleets take can result in cost savings for gasoline, while optimizing the placements of stores can lead to an increase in foot traffic and sales.

Geospatial artificial intelligence enables organizations to better understand the requirements and preferences of their customers based on location data, which results in an enhanced customer experience. Therefore, personalized marketing efforts, enhanced service offerings, and focused client involvement are all made possible as a result of this.

- *Risk Mitigation:* Geospatial artificial intelligence assists businesses in identifying and mitigating risks connected with natural disasters, disruptions in supply chains, and other causes. Businesses are able to proactively manage risks and assure the continuity of their operations by collecting and analyzing geographical data.
- *Efficiency in Operations:* Geospatial artificial intelligence helps to streamline operations by optimizing resource allocation, enhancing logistics, and increasing overall efficiency. This results in workflows that are more streamlined and increased productivity.
- *Competitive Advantage:* Businesses that make use of geospatial artificial intelligence are able to achieve a competitive advantage by employing advanced analytics to better understand market dynamics, customer behavior,

and industry trends. Because of this, they are able to develop and adapt more swiftly than their corresponding competitors.
- *Environmental Sustainability:* Geospatial artificial intelligence may assist organizations in accomplishing their environmental sustainability objectives by maximizing the efficiency of energy usage, lowering carbon emissions, and encouraging environmentally responsible behaviors.

OBSTACLES AND RESTRICTIONS IN GEOSPATIAL ARTIFICIAL INTELLIGENCE

Even though geospatial artificial intelligence has numerous advantages, it also has a number of disadvantages and restrictions:

- *Data Quality and Availability:* The quality of geographic data and its availability are extremely important to the success of geospatial artificial intelligence. There is a possibility that errors in analysis and decision-making could be caused by problems such as erroneous or missing data.
- *Problems Regarding Privacy:* Geospatial artificial intelligence frequently interacts with sensitive location data, which raises problems regarding privacy and data security. This data presents a big issue in terms of ensuring that it is used in an ethical and responsible manner.
- *Ethical Considerations:* The application of artificial intelligence algorithms in geospatial analysis might give rise to ethical considerations, such as the possibility of discriminatory outcomes and the existence of bias within the algorithms themselves.
- *The difficulty of analysis:* Geospatial artificial intelligence frequently necessitates the utilization of intricate analytical methods, such as machine learning and spatial modelling, which can be difficult to effectively execute and comprehend.

The implementation of geospatial artificial intelligence systems can be expensive and resource-intensive because it requires specialized software, hardware, and experienced professionals. It is possible that this will be a hurdle for less significant organizations or underdeveloped nations.

- *Interoperability:* If there is a lack of standardization and interoperability concerns, it can be difficult to integrate geographic data that comes from a variety of sources and formats.

- ***Legal and Regulatory Obstacles:*** Geospatial artificial intelligence is subject to a variety of legal and regulatory frameworks, which can differ dramatically between countries and regions. It can be difficult to ensure that these frameworks are maintained in compliance.
- ***Impact on the Environment:*** The gathering and processing of geospatial data, in particular satellite imagery, can have an impact on the environment, including the consumption of energy and the development of electronic waste.
- ***Human Factors:*** Because individuals are responsible for the design and implementation of geospatial artificial intelligence systems, these systems are susceptible to being influenced by human biases, errors, and limits.

INNOVATIONS AND EMERGING TRENDS IN THE FIELD OF GEOSPATIAL ARTIFICIAL INTELLIGENCE

There are a number of emerging technologies that are taking spatial intelligence to new heights. These technologies include enhanced satellite photography and drones that are able to analyze terrain. There may appear more advanced algorithms and machine learning models that are capable of processing spatial data in a manner that has never been seen before, thereby making geospatial analysis easier.

THE INFLUENCE OF ARTIFICIAL INTELLIGENCE IN GEOSPATIAL APPLICATIONS ON FUTURE INDUSTRIES

A wide range of industries, including transportation and agriculture, as well as urban planning and healthcare, are just beginning to feel the effects of spatial intelligence. Geographic artificial intelligence has made it possible to optimize logistical routes, achieve precision through the utilization of machine learning algorithms and statistical methods, artificial intelligence can provide assistance in the study of complex spatial data. The data can be analyzed to reveal previously unknown links, correlations, and spatial patterns, which can assist in the generation of insights and the formulation of forecasts. That knowledge is especially helpful in areas such as the planning of transportation, logistics, and the management of natural resources.

With the help of geospatial information, a number of leading transport companies across the world have been able to cut their operational expenses by fifteen percent. Generally speaking, artificial intelligence has the potential to transform geospatial analysis by automating tasks, extracting meaningful information from vast datasets, and providing valuable insights for a wide range of applications in fields such as urban planning, agriculture, disaster management, and environmental monitoring farming, and carry out smart city planning.

This book includes several topics related to Geospatial AI and how it is used in different industries. Also, this book is targeting industry experts, researchers, students, practitioners and higher education institutions interested in Geospatial AI and its influence on real life and industry.

Dina Darwish
Ahram Canadian University, Egypt

Houssem Chemingui
Brest Business School, France

Chapter 1
Geospatial AI Concepts and Fundamentals

Dina Darwish
Ahram Canadian University, Egypt

ABSTRACT

In terms of its progress and the ways in which it might be applied in our everyday lives, geospatial artificial intelligence (Geo-AI) is an intriguing issue. When it comes to the economic and social growth of a region or country, one of the factors that contributes to this development is spatial planning. The primary focus of this investigation is based on spatial data, with the objective of maximizing the efficiency of land use of spatial data on the region as upstream data and developing GIS-based urban planning apps that automatically present the findings of analysis and predictions of urban objects. Also, the plans for urban areas for both spatial and land use can be made, and geospatial AI generates a significant amount of spatial data based on the characteristics of the environment and the geography. This is done in order to maximize the revenue that is generated by tourists. In the context of urban planning and the tourism industry, this Geo-AI can offer many advantages and benefits. This chapter discusses several topics related to geospatial AI, including geo-computation, geospatial AI applications, challenges facing it, and its future prospects.

INTRODUCTION

Geospatial artificial intelligence, also known as GeoAI, is a combination of deep learning techniques and spatial machine learning, also known as geographic information systems (GIS). Its purpose is to assist in the resolution of difficult problems and the acquisition of more profound insights in a manner that is both powerful and inventive. The most recent developments in GeoAI, which includes geospatial data

DOI: 10.4018/979-8-3693-8054-3.ch001

collecting (sensors, Internet of Things, high-resolution satellite imaging), spatial analysis techniques, artificial intelligence, and high-performance spatial computing, have made it possible to investigate the influence of location on health in greater depth. When it comes to the health of people as well as populations, location is a very important factor. What is known as a Geographical Information System (GIS) is a computer-based system that is capable of collecting, storing, representing, and manipulating spatial data. At both the individual and the population level, it has developed into a powerful instrument for studies that investigate health outcomes and the provision of healthcare. The Geographic Information System (GIS) provides assistance to the field of public health in a variety of ways, including the mapping, monitoring, and modelling of infectious and chronic diseases, disaster preparedness, disease surveillance, and the planning of health promotion efforts. Innovative sources of big spatial data, such as personal sensing, satellite remote sensing, and social media, are being utilized in conjunction with GeoAI in order to provide answers to research questions pertaining to a variety of fields, such as epidemiology, social and behavioral sciences, genetics, infectious diseases, and environmental health.

As an example, GeoAI has been utilized to model the surrounding environment and link locations to prospective consequences. This has resulted in the provision of useful insights into the ways in which environmental, social, and other exposures may have an impact on people. As an additional benefit, GeoAI has resulted in the development of hypotheses, the establishment of new data linkages, and the forecasting of the emergence of environmental problems and diseases. These innovative ideas will make a significant contribution through the integration of technology and will establish the groundwork for further research into related subjects. Using Geographic Information Systems (GIS), geographical analysis, artificial intelligence, and large amounts of spatial data, it intends to advance research on different sectors. Research from a variety of interdisciplinary perspectives will provide insights into the challenges of integrating these technologies in different fields. Additionally, researches will bring together specialists from different fields to share their ideas, experiences, and insights with one another. This chapter discusses the geospatial AI and geo-computation, applications of geospatial AI in different domains, as well as challenges and future directions in this domain.

BACKGROUND

It is also possible to describe GeoAI as a new discipline that combines innovations in spatial science, artificial intelligence (AI) methods such as Machine Learning (ML) and Deep Learning (DL), data mining (data mining), and high-performance computing (high-performance computing). According to Gartner, GeoAI is the utilization of artificial intelligence (AI) methods, including ML and DL (Pierdicca & Paolanti, 2022), to generate knowledge through spatial and image data analysis. There have been a number of factors that have led to the growing significance and potential of GeoAI, these factors include the development of artificial intelligence, the availability of massive processing powers, and the growing availability of geographic data. Machine learning is a sub-discipline of artificial intelligence that is used to extract knowledge from geographic data. This notion is fed into the larger framework of artificial intelligence. Implementation of Machine Learning On Urban Planning using Geo-AI now has an important role to play in advancing traditional AI technologies and innovating new ways to solve specific problems posed by the massive, complex, diverse, and ever-increasing nature of geospatial data, which is considered geo-referenced data containing geotagging locations or position markers. However, geo-AI is not yet fully developed. The use of geospatial data is widespread across a wide range of scientific disciplines and applications, such as smart cities, transportation, business, public health, public safety, resilience to natural disasters, climate change, and many more. According to Cugurullo (Cugurullo, 2020), the objective of addressing this issue is to enhance the standard of living of the ever-increasing urban population around the world. This interdisciplinary field is shaped by a number of different disciplines, including computer science, geography, geographic information systems (GIS), and urban studies, among others. This study aims to present an overview of the primary concepts underlying the burgeoning subject of GeoAI, clarify the differences between GeoAI and more general AI, and integrate AI with GIS, with the goal of making visualization and software that contain AI characteristics with the most important aspects of the field. Researchers (Abelha, Fernandes, Mesquita, Seabra, & Ferreira-Oliveira, 2020) conduct study on an overview of GeoAI technology. This research includes the definition of GeoAI as well as the distinctions between GeoAI and traditional AI. By combining artificial intelligence with geographic information systems (GIS) and making use of GeoAI tools and technologies, one can successfully analyze geographic data. Additionally, it demonstrates the primary areas of application and models in GeoAI, as well as the difficulties associated with implementing GeoAI methods and technologies and the advantages that they offer. Research conducted by authors (Alastal & Shaqfa, 2022) is based on a variety of different sources of imagery or structured data, including as satellite and drone photography, street views, and geoscientific data, and their application to a wide range of image analysis and machine vision tasks. The primary advantages of GeoAI research are summarized here. These advantages include the following: (1) the ability to perform large-scale analytics; (2) automation; (3) high accuracy; (4) sensitivity in detecting minor changes; and (5) noise tolerance. Although different applications tend to use different types of data and models, authors summarized these advantages. This research on GeoAI for spatial knowledge discovery, which was conducted by Janowicz et al. (Janowicz, Gao, McKenzie, Hu, & Bhaduri, 2020), illustrates how changes in data are driving the rapid expansion of GeoAI and points out future research prospects. In addition, it discusses the creation of spatially explicit models and the dissemination of high-quality geospatial datasets for the purpose of advancing research in the field of geoinformation technology (GeoAI) that can be replicated in subsequent study. The research is based on two practical examples: how geospatial products can be generated in the proposed architecture, how these products can be used in machine learning for tactical planning, and how learned action courses and intelligence products may be delivered to planners in decision support. Although there are some research performed on topics related to geospatial AI, there are still many areas in the geospatial AI that need more investigation and research. Figure 1 illustrates the intersection between AI, GIS and Spatial data.

Figure 1. Intersection between AI, GIS, spatial data

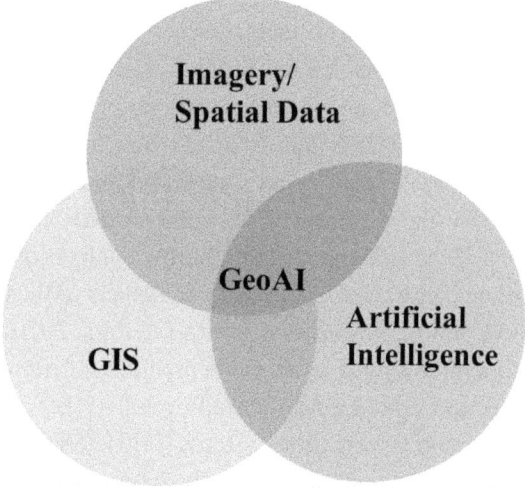

MAIN FOCUS OF THE CHAPTER

Geo-Spatial Artificial Intelligence and Geo-Computation

Geo-computation and Geospatial Artificial Intelligence (GeoAI) are two examples of novel approaches that are aimed at improving Geographic Information Systems (GIS) and Earth Observation. The utilization of computational methods and tools to investigate geographical data and earth data in order to develop new knowledge is one of the benefits of geo-computation, according to Janowicz et al.'s research from 2020 (Janowicz, Gao, McKenzie, Hu, & Bhaduri, 2020). In the meantime, GeoAI offers learning algorithms and techniques, such as machine learning, deep learning (Li, 2021), and knowledge transfer, with the goal of developing solutions that are both effective and innovative for geospatial and earth problems (PS Chauhan & Shekhar, 2021), (Pierdicca & Paolanti, 2022), (Liu & Biljecki, 2022). In geographic information systems (GIS) and earth observation, mapping is an essential component that contributes to a better understanding of both the natural and constructed environments. Maps are typically created by the use of spatial analysis, which is founded on the idea of spatial statistical inference. The following are some of the categories that can be used to classify difficulties pertaining to geographical analysis: recognizing spatial patterns (Xu, 2021), examining spatial factors (Liu & Biljecki, 2022), spatial simulation (Wen & Li, 2022), and geographic decision-making (Chadzynski et al., 2021). Geo-computation and GeoAI have considerably expanded methods of geospatial analysis and mapping in recent years. These advancements have the potential to enhance the way we understand and manage the complex interactions that occur between human and natural systems. Despite variations in scope and focus, geo-computation and GeoAI have made tremendous progress in these areas. In order to tackle difficult geospatial and earth-related problems, geo-computation and geo-artificial intelligence have developed more sophisticated methods. More prospects for creative applications of geospatial artificial intelligence (GeoAI) and earth observation are made available as a result of the integration of modern computing tools. Among the advanced computational tools that have been developed, such as big data analysis (Song, Kalacska, Gašparović, Yao, & Najibi, 2023), cloud computing (including Google Earth Engine) (Teja, Liu, & Chopra, 2023), and graph knowledge (Nizzoli, Avvenuti, Tesconi, & Cresci, 2020), GeoAI has emerged as a significant driving force in the advancement of geospatial data utilization. In spite of the fact that the application of geo-computation and geo-artificial intelligence in mapping is gradually expanding, there is still a requirement to enhance application development from a variety of perspectives. First, it is becoming increasingly vital to grasp the geographical consequences of the methods and results provided by geo-computation and GeoAI at the present time, from the point of view of an algorithm or model. The application of computational methods and direct learning to geospatial data is a significant part of both geo-computation and geo-artificial intelligence. The incorporation of geographical attributes and spatial relationships in models is consequently simplified as a result of this. Traditional methods of spatial analysis make use of a variety of geospatial characteristics, including spatial autocorrelation, which is used to measure the degree of similarity between observations (Arundel, Li, & A Wang, 2020), spatial heterogeneity, which is used to describe variations in geospatial data across space, spatial singularity and spatial anomalies, which are used to identify observations that are unusual or rare, and to measure the degree of similarity and complexity of geospatial data based on their respective geographic configurations. Although there have been some recent studies that have characterized spatial dependence by exploiting the link between data and their correlation (Wen & Li, 2022), there is still a need for improvement in the incorporation of these geospatial aspects. In addition, geospatial data is both complicated and varied, with sources and types as varied as satellite imagery, aerial photography, photogrammetric data, geographical data, and location data from social media. These data are treated as samples or images in the same manner as other fields, independent of the geospatial characteristics of the data. Unique in nature. It is common knowledge that geospatial data can accurately express geospatial information with a variety of spatial kinds (such as points, polylines, areas, and grids) and at a variety of scales, in addition to the place itself, such as longitude and latitude. Because of this, it is necessary to incorporate these one-of-a-kind geographical characteristics into the algorithms and models that are used by GeoAI in order to make full use of its skills in resolving issues that are associated with geospatial and geodata. Figure 2 illustrates the integration of communication means to transport spatial data.

Figure 2. Integration of communication means

Geo-Spatial Data Modelling Techniques

The capacity to infer the behavior of various variables in the spatial dimension is essential for gaining an understanding of the many occurrences and events that take place, as well as the interrelationships that exist between them. S. Law et al. (Law et al., 2017) trained a Convolutional Neural Network (CNN) that was able to predict the probability of having an active frontage on every single street segment in London. This provided important insights for urban design and security in public spaces. The CNN was trained using images from Google Street View and an abstract 3D model of a city that was created with Esri City Engine. In their study, K. Elgarroussi and colleagues (Elgarroussi et al., 2018) developed a framework for spatiotemporal emotion change analysis. This framework is designed to monitor and summarize the change of positive and negative emotions across time and location. An emotional score was assigned to each tweet by the authors through the use of geolocated tweets that were gathered in the state of New York in June of 2014. Additionally, a contour-based spatial clustering algorithm and VADER, a tool for sentiment analysis, were utilized. A fresh method for influencing people's feelings through storytelling is presented in this study. The authors of the contribution made by G. Xi and S. Mei (Xi and S. Mei, 2009) used a geolocated dataset of influenza-like illnesses (ILI) activities in Shenzhen City, China, in order to train a CNN with residual learning that was capable of predicting influenza trends by combining the spatial-temporal features of influenza at an intra-urban scale. According to the results of the studies, the model performed better than the other four baseline models when it came to making predictions one week ahead and two weeks ahead. By training a CNN with Google Street View images and a dataset of Amsterdam's buildings, S. Srivastava et al. (Srivastava et al., 2018) were able to predict numerous co-occurring building function classes for each building. The model that was developed takes into account many zoom levels of the pictures that were captured, and it performed better than comparable baseline CNN models. In their article (Vopham et al., 2018) T. Vopham and colleagues provide an overview of the application of GeoAI in environmental epidemiology. According to the authors, the utilization of spatial and temporal big data in conjunction with high performance computing, data mining, deep learning, and big data infrastructures can assist in the prediction of the quantity of an environmental factor at a specific time and location, as well as the production of high-resolution exposure models that are representative of a specific environmental variable. They highlight the potential applications of GeoAI in the field of tackling issues that are connected to human health. In this manner, research (Boulos et al., 2019) discusses the significance of geography in terms of both population and individual health, as well as the growing significance of GeoAI models and new data sources for the purpose of enhancing human health. However, the use of mobility trajectory datasets that are publicly available is restricted due to actual privacy concerns, despite the fact that these datasets are essential for the training and evaluation of algorithms. In order to solve this issue, V. Kulkarni (Kulkarni, 2017) trained a Long Short-term Memory (LSTM) recurrent neural network (RNN) that extracts substantive behavioral patterns of users from actual mobility traces datasets. The dataset consisted of 191 users who had their mobility traces collected by Nokia Mobile over a period of two years in Switzerland. The purpose of this study is to generate new datasets that are greater in size and that imitate the actual features of users with respect to a specific dataset.

Geospatial Analysis Tools

A broad variety of software programs, libraries, and platforms that are utilized for the purpose of processing, analyzing, and visualizing geographic data are together referred to as Geospatial Analysis Tools. Geospatial professionals and academics are able to conduct spatial analysis, produce maps, and extract important insights from geospatial datasets with the assistance of these technologies.

Among the geospatial analytic tools that are often utilized are:

- *ArcGIS* is a sophisticated geographic information system (GIS) software package that was developed by Esri. It provides a wide variety of features for geospatial analysis and mapping.
- *The QGIS program* is an open-source geographic information system (GIS) application that offers strong geospatial analysis and mapping features with an intuitive user interface.

Known for its sophisticated capabilities in spatial analysis and modelling, GRASS GIS is a Geographic Information System (GIS) program that is open-source.

- *PostGIS* is an extension for PostgreSQL that adds spatial database characteristics to a relational database. This makes it possible to do extensive geographic analysis within the database.
- *Python geographic Libraries* are libraries that offer geographic data processing and analysis capabilities in the Python programming language. Some examples of these libraries are Geopandas, Fiona, Shapely, and Pyproj.
- *R Spatial Packages*: R packages such as sf, sp, and raster that provide functions for the manipulation of spatial data, visualization of spatial data, and statistical analysis of spatial data.
- *The Google Earth Engine:* is a cloud-based platform that allows users to analyze and visualize data gathered from Earth observations. It is especially helpful for comprehensive environmental monitoring on a broad scale.
- *Tools for processing and analyzing satellite and aerial pictures* are examples of remote sensing software. Some examples of these tools are ENVI, Erdas Imagine, and SNAP (Sentinel Application Platform).
- *Frameworks for online Mapping* This category includes frameworks like Leaflet and Mapbox, which are used for the creation of interactive online maps and applications.
- *Software applications* such as ArcGIS Network Analyst and GraphHopper are examples of network analysis tools. These tools are used to solve complicated network routing issues.

- According to Wikipedia, ***"OpenEV*** is a library, and reference application for viewing and analyzing raster and vector geospatial data." Even if you only need to digitize anything for a short period of time, OpenEV can help you get up and running quickly. Despite the fact that it does not have a large number of sophisticated editing tools, it is capable of reading in hundreds of different raster and vector data types, which you can then use as a foundation for drawing your own forms. The Python programming environment is included, and it comes with a wide variety of picture improvement tools.
- ***Quantum Geographic Information System (QGIS)*** "Quantum GIS (QGIS) is a Geographic Information System that is compatible with Windows, Linux Mac OS X, and Unix." Additionally, QGIS is compatible with raster, vector, and database formats. PostGIS databases, in addition to dozens of additional vector and raster formats, are among the formats that it is able to access. A wonderful user community is available, and it allows for the labelling of features. Extensibility is made available by means of an environment supporting plugins.

These geospatial analytic tools are designed to accommodate a diverse variety of applications, including urban planning and environmental monitoring, as well as disaster response and transportation logistics. In order to select the appropriate tools, it is necessary to take into consideration the unique needs of a project, the kind of geographical data that is being analyzed, and the user's expertise.

Application of Geospatial AI technology in a Variety of Domains

According to Li and Hsu (Li & Hsu, 2022), geospatial artificial intelligence (GeoAI) is a study topic that is both growing and promising. It is a field that merges artificial intelligence with geospatial science in order to solve problems and difficulties that are of a geographic character. The development of GeoAI brings the benefits of traditional artificial intelligence studies in computer science to the field of geographic research. This is accomplished by empowering its quantitative methods with revolutionary technologies such as machine and deep learning, high-performance computing power, and big data mining (Liu and Biljecki, 2022; Liu et al., 2022). According to Leszczynski and Crampton (Leszczynski & Crampton, 2016), as a result of the fact that more than 80 percent of big data contains spatial information, this rising trend in research that is focused on artificial intelligence is especially significant for geographic studies in this era of big data. GeoAI enables researchers to better monitor human behaviors and the surrounding environment, which are frequently spatially dependent and autocorrelated. This is accomplished through the advocacy of Tobler's first law of geography, which states that "everything is related to everything else, but near things are more related than distant things" (Miller, 2004). A recent breakthrough in GeoAI, and more specifically deep learning, has made it possible for a new research paradigm to emerge. This paradigm integrates data science and geography in order to analyze, mine, and visualize large volumes of spatiotemporal data. Additionally, it enables researchers to better capture the human-environment relationship, which is a complex, multifaceted, and non-linear relationship (Li, 2022).

According to Hoggart (Hoggart, 2002), human geography is the subfield of geography that investigates the spatial links that exist between human groups, cultures, and economies, as well as the ways in which these entities interact with the surrounding environment. In contrast to the field of physical geography, which focuses on the spatial and environmental processes that shape the natural world and tends to draw on the natural and physical sciences for its scientific underpinnings and methods of investigation, the field of human geography focuses on the spatial organization and processes that shape the lives and activities of people, as well as their interactions with places and nature (Gregory et al., 2011). The field of human geography is comprised of a number of sub-disciplinary domains that concentrate on various aspects of human activity and organization (Gregory et al., 2011). These sub-domains primarily include cultural, economic, political, historical, urban, population, social, health, rural, regional, tourism, behavioral, environmental, and transport geography. The application of a set of fundamental geographical concepts and notions, such as the idea that the world operates spatially and temporally, and the idea that social relations are thoroughly grounded in and through of place and environment, are what differentiate human geography from other related disciplines, such as development, economics, politics, and sociology. This is where the implementation of GeoAI is well positioned and urgently needed. Although a range of review papers give attention to various related topics such as the application of deep learning in geography (Li and Hsu, 2022), GeoAI applications in urban planning and development (Alastal and Shaqfa, 2022) and urban geography (Liu and Biljecki, 2022, Liu et al., 2022), GeoAI approaches for complex geomatics data (Pierdicca and Paolanti, 2022), unsupervised machine learning in urban studies (Wang and Biljecki, 2022), and more broadly GeoAI in social science from a scoping review perspective (Li, 2022), what is lacking from the current scholarship is a holistic, comprehensive, and systematic understanding of GeoAI application in various domains of human geography. Figure 3 shows the different types of maps showing elevation, land usage, parcels, streets and political/administrative boundaries. Figure 4 illustrates applications of GeoAI in different domains, like, public health and environment, disaster response and community resilience, ecosystems and biodiversity.

A systematic review on the implementation of GeoAI in quantitative human geography, including the subdomains listed above, with the goals of 1) providing a comprehensive picture of the state-of-the-art GeoAI techniques and applications that have been used in human geography, as well as the data sources that were used to support GeoAI, 2) outlining the future directions for geographers to grasp the AI-oriented opportunities, while at the same time highlighting the future challenges and risks that require us to think critically and tackle them specifically. Those that fail to do so run the risk of incurring significant costs and falling behind in the mainstream of scientific research as other researchers learn from the ever-increasing data flood. The modelling tasks that have been implemented by GeoAI applications in Human Geography studies are classified into broad and secondary categories.

Examples of these categories include: 1) classification (such as image-based, vector-based, sequential-data-based, number-based, and text-based classification); 2) prediction (such as linear and non-linear regression); 3) simulation (such as at the aggregated and individual level); 4) embedding (such as the reduction of data dimensions; feature extraction); and 5) geolocating (also known as geoparsing). Prediction (especially non-linear regression) and classification (especially image-based classification) have been widely utilized in various subdomains of human geography, regardless of the number of papers that have been published. Therefore, some common patterns of GeoAI modelling tasks may be observed across a variety of subdomains of human geography. A variety of simulation tasks were also carried out by GeoAI in the field of urban geography. In particular, the aggregated level simulation was carried out (for example, by utilizing cellular automata, deep neural networks, and deep enforcement learning models).

Figure 3. Types of maps

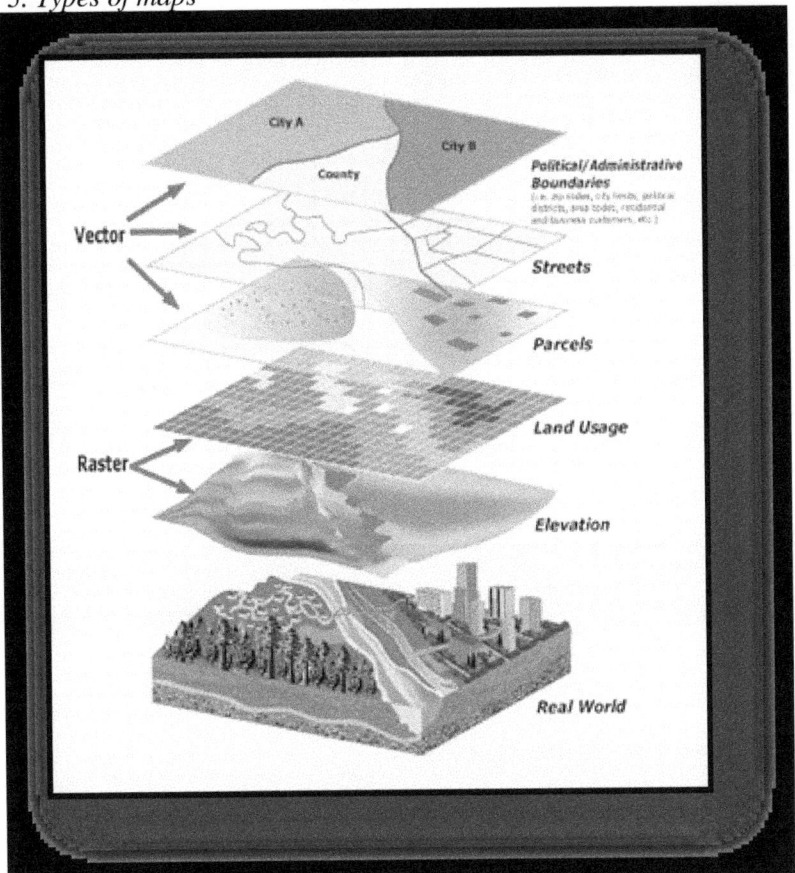

Figure 4. Different types of GeoAI applications

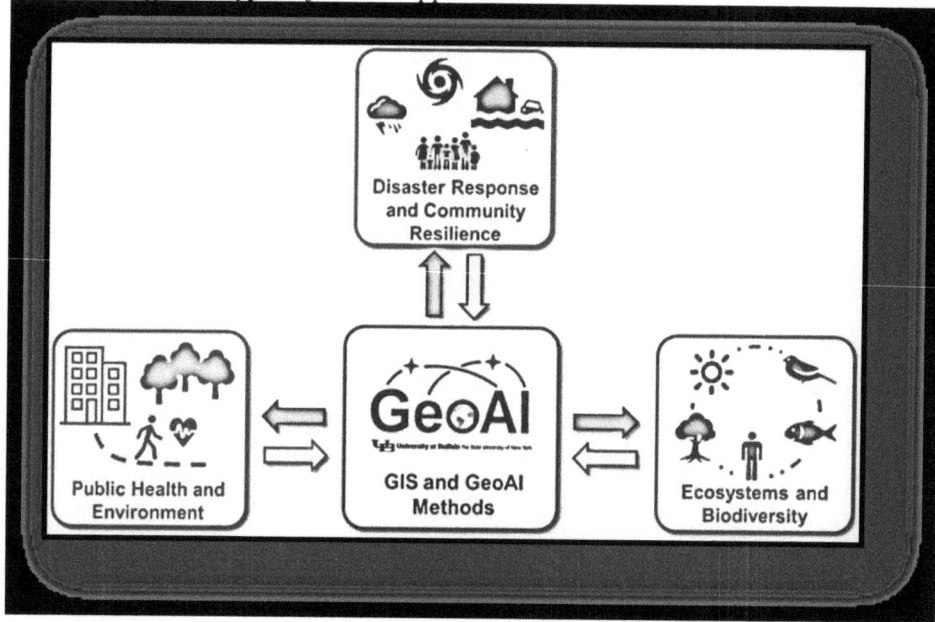

Case Studies for Using GeoAI

1. Navigation

Getting Around CNN was taught by Q. Li (Li, 2017) utilizing sampled pictures taken from the camera videos of smartphones when they were in interior areas. The Hidden Markov Model, the Viterbi algorithm, and a topological map were utilized by the authors in order to get position information by identifying landmarks within the confines of an indoor setting. The professionally trained CNN was able to identify the locations in the area accurately. In order to reduce the amount of distance gap that exists between the predictions and the ground-truth traces of an actual automobile journey A CNN that was trained by J. Murphy (Murphy, 2017) was able to reach human-level performance when it came to categorizing the noise level of the input data from the Global Positioning System (GPS) on a specific route. In order to determine the most accurate estimation of a driving path, the algorithm makes a conditional choice between utilizing the raw GPS data and the route that is matched to the map. The writers of research (Xu et al., 2018) conduct an analysis of a number of research publications on the localization of images captured by Unmanned Aerial Vehicles (UAV). Both classic visual localization systems that are based on picture retrieval and contemporary image localization systems that are based on deep learning methods are described by the authors. Visual localization systems are designed to generate location information in a reasonable amount of time while simultaneously locating the picture that is the most similar to the reference image database and the aerial image captured by the unmanned aerial vehicle (UAV). An improved road extraction quality was achieved by the training of a CNN model by T. Sun and colleagues (Sun et al., 2018), which integrates satellite images and GPS data. One hundred twenty satellite photos from Beijing, data on paved roads from OpenStreetMap as ground truth, and GPS data from sixty-five cabs in Beijing (a total of two hundred and twenty-four hours of driving) were utilized in the research, which demonstrated exceptional performance in comparison to other options. A study conducted by N. Pourebrahim and colleagues (Pourebrahim et al., 2018) examined the effectiveness of Neural Networks and Gravity Models in predicting people migration across cities by utilizing data from Twitter. For the purpose of gathering ground-truth data, the authors utilized the Origin Destination job Statistics (LODES) for New York City, which were provided by the United States Census Bureau. These statistics included workers' home and job locations, in addition to other characteristics such as age, incomes, industry distributions, and local workforce indicators. Additionally, greater than two million geolocated tweets that were posted in New York City between June 2015 and May 2016 were used for the research project. The findings demonstrated that the performance of both models was improved by the use of Twitter data.

2. Governance, as well as societal relations

It is the obligation of governments to gather and manage huge amounts of data; however, their primary responsibility is to improve the quality of life of their population by making decisions that are driven by data. In addition to the contribution that D. Jha et al. (Jha and Singh, 2018) made for the identification of swimming pools, it also proposes that it might be used for the control of vector-borne diseases. This is because the model can differentiate between pools that have been neglected and those that have not been neglected, hence avoiding the transmission of viruses that are carried by mosquitoes. On the other hand, governments may reap the benefits of this approach by using it to update the records of swimming pools for the purpose of administering taxes. Three-dimensional city models have emerged as an indispensable instrument for decision-making in recent years. In order to train a CNN that automatically creates segment polygons of roofs from a raster picture, D. Kudinov et al. (Kudinov et al., 2020) employed 200 square miles of aerial LiDAR with 213,000 roof segments that were manually digitized by human editors. These roof segments included kinds such as flat, gable, hip, shed, dome, vault, and mansard. The model makes a contribution to the improvement of human city models in three dimensions. Using the 2011 Tohoku Earthquake-Tsunami as a case study, T. O. (T. O, 2018) suggested a deep learning semantic segmentation method for post catastrophe damage mapping. The algorithm's goal was to speed up the process of operational disaster response practice. In addition to collecting ground truth data from damage inventory for the study region, high-resolution Worldview-2 photos were also gathered both before and after the earthquake. However, the scientists pointed out that the optical remote sensor has a disadvantage when it comes to recognizing damage that is below the roof, despite the fact that the model demonstrated a decent performance. In order to carry out this investigation, the AI for Earth grant program made available the resources that were available on Microsoft Azure. As can be seen from the applications that were discussed earlier, GeoAI models are typically supervised learning algorithms that are highly dependent on the availability of labelled data in order to achieve satisfactory levels of performance.

Challenges Facing Geospatial AI

According to the findings, the field of human geography has made significant strides in recent years thanks to the emergence of spatiotemporal big data. This data has made it possible for geographers to follow, monitor, and quantify complex human behaviors on either a large spatial or temporal scale. Furthermore, it demonstrates that artificial intelligence plays a significant role in human geography, while simultaneously drawing attention to the restricted interpretability of existing models, which is a topic that has been brought up in recent research. This issue, in addition to the possibility of bias in GeoAI models, is extremely important for additional research in the future. For example, models that predict human perceptions based on imagery, which are frequently built through extensive surveys, might not be suitable for the requirements of certain geographic area. More than that, these models often need for a large amount of data that is specific to a certain region, which creates difficulties in terms of data availability and applicability across a variety of domains. It is absolutely necessary, in order to make progress in the field of GeoAI, to address these data restrictions and biases. The data can include a wide range of categories and originates from a diverse range of sources, including authoritative data from governments as well as crowdsourced examples. The majority of the data that was utilized in the papers that were examined was derived from OpenStreetMap, which is a map of the world that can be edited without restriction. In certain regions, OpenStreetMap provides data that is acceptable for some of the use cases. On the other hand, the data continues to be quite diverse (Biljecki et al., 2023) and in certain regions, it might not be adequate or even hazardous for the analysis. It is important to paying attention to the quality of the data, particularly those that are obtained from a crowdsourced provenance. Furthermore, taking into consideration the impact that the propagation of errors has effect on the outcome of an analysis and this can affect the prospective study direction that could be pursued. Several aspects, including completeness, positional correctness, and thematic accuracy, are included in the concept of spatial data quality, which is a subject that is frequently disregarded in the field of human geography (Hou and Biljecki, 2022). Because the quality of each of these elements has a varied impact on various use cases in a variety of different ways, it would be necessary to conduct considerable research in order to comprehend the dependability of an analysis that is based on the input dataset.

Future Prospects of Geospatial AI

Human geography has undergone a revolution that has resulted in it being increasingly supported by spatiotemporal big data, more robust in research design to address the non-linear complex relationship between human society and its potential drivers, more diverse in empirical studies, and ultimately leading to the advancement of theoretical foundations. The subject of GeoAI offers a multitude of options for study that spans several disciplines, with the goal of establishing connections between human geography and other fields such as public health, environmental science, medical science, decision-making and policy-making, and industrial practices in general. In particular, human geography subdomains that were deeply rooted in social science in the past, such as cultural, historical, and political geography, have been empowered by GeoAI and spatiotemporal big data and crowdsourcing data (for example, data from social media) to broaden its research impact and the coverage of empirical contexts. This is something that can sometimes be accomplished by using small data (for example, questionnaires) and qualitative methods. In addition, there is a great deal of potential that can be realized by utilizing the power of GeoAI in conjunction with broader fields in social science such as psychology, sociology, and anthropology in order to analyze and predict human behavior. Furthermore, this can be done in order to improve research on human-environment interactions, as it was previously believed that human behavior was mediated through the environment in which they reside (for example, the natural environment and the built environment in urban areas) (Wang et al., 2023, Wang et al., 2021b). It is possible for public sectors and authorities to rely on the quantitative results simulated by GeoAI in order to optimize policy implementation and reduce social and financial costs. On the practical side, GeoAI helps to achieve decision-making and evaluate different scenarios of policies in the initiatives of smart and healthy cities. It also helps to facilitate citizen participation in urban planning and design. With the advent of new technologies like generative artificial intelligence, digital twins, knowledge graphs, 5G, and the Internet of Things (Zhang et al., 2022, Zhang and Zhao, 2022), the possibilities for cross-disciplinary research are virtually limitless, and the potential benefits are substantial. These technologies have the potential to assist researchers and policymakers in gaining a more in-depth understanding of complex urban systems and in making decisions that are more informed, thereby positively linking the academic outcomes to the real world.

GeoAI is essential for addressing environmental and sustainability concerns, since it allows for the effective processing of geographical data. For techniques that are consistent, transferable, and scalable, their solutions need to be repeatable, reproducible, and extensible. In order to achieve repeatability, GeoAI must deliver consistent findings using the same data and methodologies. This necessitates the existence of well documented algorithms and procedures that are open to scrutiny. Open-source platforms, standardized data, and shared code all contribute to the enhancement of collaboration and development, which is done through reproducibility, which is accomplished by the use of a variety of data sets or contexts. In order to ensure that GeoAI systems continue to be adaptable and applicable, expandability enables adaption to larger or new data sets as well as evolving queries. This is accomplished through the use of modular architectures, cloud computing, and improved algorithms for big data. Furthermore, the repeatability, reproducibility, and expandability of GeoAI approaches are all impacted by the complexity of the procedures and the surroundings in which they are carried out. Developing and executing GeoAI models can be accomplished in three primary ways: 1) by utilizing pre-existing GIS or analysis software such as Geoda, ArcGIS, and other similar programs, which is user-friendly but less reproducible and scalable; 2) by developing and executing complex models through code, which improves reproducibility and scalability, particularly with Jupyter Notebooks. However, Trisovic et al. (Trisovic et al., 2022) conduct a large-scale study on research code quality and execution, and found that 74% of data science research code failed execution tests (3) utilizing visual programming tools like ArcGIS's Model Builder, QGIS, Knime, Orange3, and Alteryx to develop and execute GeoAI models in an executable workflow, reducing the burden on researchers and improving understandability, it provides a promising idea to improve repeatability, reproducibility, and expandability of GeoAI modes. On the other hand, there are still a relatively small number of applications for this at the moment. To ensure that solutions are repeatable, replicable, and extendable, sustainable geoscientific artificial intelligence necessitates taking a holistic approach. By concentrating on these qualities, the GeoAI community has the potential to cultivate work that is collaborative, long-term, and effective, so contributing to a future that is more sustainable.

Significant progress has been made in artificial intelligence (AI) with the introduction of Artificial General Intelligence (AGI), which refers to AI systems that are capable of executing intellectual tasks better than a person and sometimes even surpassing human intelligence (for example, ChatGPT). The utilization of artificial general intelligence (AGI) in human-centered geoAI has the potential to bring significant benefits, such as the enhancement of decision-making processes, the optimization of resource management, and the improvement of disaster response and recovery capabilities. It is envisaged that the active participation of stakeholders in the design and development of artificial intelligence solutions would increase confidence and acceptability among the community. This will ensure that the solutions that are developed are tailored to meet the particular interests and requirements of the stakeholders. This strategy encourages inclusiveness and makes it possible for stakeholders to have a sense of ownership, which ultimately results in the creation of a collaborative environment that is an ideal environment for the development of effective AI solutions. Using artificial general intelligence (AGI) in geospatial research raises concerns about potential invasions of privacy and the strengthening of existing inequities. It is possible for geospatial data to violate individuals' rights to privacy and liberty if it is misused by either public or private groups. Using historical data, artificial general intelligence (AGI) could potentially perpetuate prejudices and discrimination, hence exacerbating existing socioeconomic inequities and marginalizing populations. Human-centered geoAI should place an emphasis on openness and accountability, hence providing stakeholders with full access to information pertaining to artificial general intelligence (AGI). To prevent reinforcing the biases that already present in geospatial analysis, it is essential that artificial general intelligence (AGI) be inclusive and equitable. Building trust and contributing to a more equitable and sustainable future through the implementation of AGI solutions that are open, responsible, and inclusive is necessary in order to accomplish this goal. Fair data practices and actively minimizing data biases are required to do this.

Conclusion

The field of human geography has experienced a significant transformation, with an increased reliance on spatiotemporal big data to improve study design and address the complex and non-linear linkages that exist between human society and the potential drivers of that society. During this history, there has been a wider diversity in empirical studies, which has contributed to the advancement of theoretical underpinnings. The incorporation of developing spatiotemporal big data has considerably driven human geography research, allowing geographers to follow, monitor, and quantify complex human behaviors on a wide spatial and temporal scale. This has been a tremendous accomplishment for the field of human geography. In the near future, the convergence of GeoAI and quantum computing is going to bring about a further revolution in the field of human geography research. This will be accomplished by giving sophisticated tools that can simulate spatial phenomena and improve forecasts concerning the environment and population dynamics. Through the utilization of this integration, researchers will be able to process and analyze enormous datasets at a speed that has never been seen before, which will allow for a more in-depth investigation of geographical correlations. Nevertheless, it is of the utmost importance to approach the development and use of GeoAI in a responsible and ethical manner, taking into consideration the potential social and environmental implications that its adoption may have. Through the concerted measures that are being taken, the roadmap of GeoAI in human geography will be enriched, and its application will be extended to encompass a wider range of geographic paradigms. There are possibilities for research and make use of the insights that may be gained from the flood of new data and artificial intelligence.

REFERENCES

Abelha, M., Fernandes, S., Mesquita, D., Seabra, F., & Ferreira-Oliveira, A. T. (2020). Graduate employability and competence development in higher education—A systematic literature review using PRISMA. *Sustainability (Basel)*, 12(15), 5900. DOI: 10.3390/su12155900

Alastal, A. I., & Shaqfa, A. H. (2022). Geoai technologies and their application areas in urban planning and development: Concepts, opportunities and challenges in the smart city (Kuwait, study case). *Journal of Data Analysis and Information Processing*, 10(2), 110–126. DOI: 10.4236/jdaip.2022.102007

Arundel, S. T., Li, W., & Wang, S. (2020). GeoNat v1. 0: A dataset for natural feature mapping with artificial intelligence and supervised learning. *Transactions in GIS*, 24(3), 556–572. DOI: 10.1111/tgis.12633

Biljecki, F., Chow, Y. S., & Lee, K. (2023). Quality of crowdsourced geospatial building information: A global assessment of OpenStreetMap attributes. *Building and Environment*, 237, 110295. DOI: 10.1016/j.buildenv.2023.110295

Chadzynski, A., Krdzavac, N., Farazi, F., Lim, M. Q., Li, S., Grisiute, A., Herthogs, P., von Richthofen, A., Cairns, S., & Kraft, M. (2021). Semantic 3D City Database—An enabler for a dynamic geospatial knowledge graph. *Energy and AI*, 6, 100106. DOI: 10.1016/j.egyai.2021.100106

Chauhan, P. S. Lokendra, & Shekhar, Shashi. (2021). GeoAI–Accelerating a Virtuous Cycle between AI and Geo. 2021 Thirteenth International Conference on Contemporary Computing (IC3-2021), 355–370.

Cugurullo, F. (2020). Urban artificial intelligence: From automation to autonomy in the smart city. *Frontiers in Sustainable Cities*, 2, 38. DOI: 10.3389/frsc.2020.00038

Elgarroussi, K., Wang, S., Banerjee, R., & Eick, C. F. (2018). "Aconcagua: A Novel Spatiotemporal Emotion Change Analysis Framework," *Proc. 2Nd ACM SIGSPATIAL Int. Work. AI Geogr. Kwl. Discov.*, no. Ccdm. DOI: 10.1145/3281548.3281552

Gregory, D., Johnston, R., Pratt, G., Watts, M., & Whatmore, S. (2011). *The dictionary of human geography*. John Wiley & Sons.

Hoggart, K., (2002). Researching human geography.

Hou, Y., & Biljecki, F. (2022). A comprehensive framework for evaluating the quality of street view imagery. *International Journal of Applied Earth Observation and Geoinformation*, 115, 103094. DOI: 10.1016/j.jag.2022.103094

Janowicz, K., Gao, S., McKenzie, G., Hu, Y., & Bhaduri, B. (2020). GeoAI: Spatially explicit artificial intelligence techniques for geographic knowledge discovery and beyond. [Taylor & Francis.]. *International Journal of Geographical Information Science*, 34(4), 625–636. DOI: 10.1080/13658816.2019.1684500

Jha, D., & Singh, R. (2018). "Swimming pool detection and classification using deep learning." [Online]. Available: https://medium.com/geoai/swimming-pool-detection-and-classification-using-deep-learning-aaf4a3a5e652

Kamel Boulos, M. N., Peng, G., & Vopham, T. (2019). An overview of GeoAI applications in health and healthcare. *International Journal of Health Geographics*, 18(1), 1–9. DOI: 10.1186/s12942-019-0171-2 PMID: 31043176

Kudinov, D., Hedges, D., & Maher, O. (2020). "Reconstructing 3D buildings from aerial LiDAR with AI: details." [Online]. Available: https://medium.com/geoai/reconstructing-3d-buildings-from-aerial-lidar-with-ai-details-6a81cb3079c0

Kulkarni, V. (2017). *Generating Synthetic Mobility Traffic Using RNNs*. DOI: 10.1145/3149808.3149809

Law, S., Shen, Y., & Seresinhe, C. (2017). "An application of convolutional neural network in street image classification," *Proc. 1st Work. Artif. Intell. Deep Learn. Geogr. Knowl. Discov. -. GeoAI*, 17, 5–9. DOI: 10.1145/3149808.3149810

Leszczynski, A., & Crampton, J. (2016). Introduction: Spatial big data and everyday life. *Big Data & Society*, 3(2), 3. DOI: 10.1177/2053951716661366

Q. Li, (2017). "Visual Landmark Sequence-based Indoor Localization," no. 1.

Li, W. (2021). *GeoAI and deep learning*. The International Encyclopedia of Geography. DOI: 10.1002/9781118786352.wbieg2083

Li, W. (2022). GeoAI in social science. Handbook of Spatial Analysis in the Social Sciences, 291-304.

Li, W., & Hsu, C.-Y. (2022). GeoAI for large-scale image analysis and machine vision: Recent progress of artificial intelligence in geography. *ISPRS International Journal of Geo-Information*, 11(7), 385. DOI: 10.3390/ijgi11070385

Liu, K., Chen, J., Li, R., Peng, T., Ji, K., & Gao, Y. (2022). Nonlinear effects of community built environment on car usage behavior: A machine learning approach. *Sustainability (Basel)*, 14(11), 6722. DOI: 10.3390/su14116722

Liu, P., & Biljecki, F. (2022). A review of spatially-explicit GeoAI applications in Urban Geography. *International Journal of Applied Earth Observation and Geoinformation*, 112, 102936. DOI: 10.1016/j.jag.2022.102936

Miller, H. J. (2004). Tobler's first law and spatial analysis. *Annals of the Association of American Geographers*, 94(2), 284–289. DOI: 10.1111/j.1467-8306.2004.09402005.x

J. Murphy, (2017). "Image-based Classification of GPS Noise Level using Convolutional Neural Networks for Accurate Distance Estimation."

Nizzoli, L., Avvenuti, M., Tesconi, M., & Cresci, S. (2020). Geo semantic-parsing: AI-powered geoparsing by traversing semantic knowledge graphs. *Decision Support Systems*, 136, 113346. DOI: 10.1016/j.dss.2020.113346

Pierdicca, R., & Paolanti, M. (2022). GeoAI: A review of artificial intelligence approaches for the interpretation of complex geomatics data. *Geoscientific Instrumentation, Methods and Data Systems*, 11(1), 195–218. DOI: 10.5194/gi-11-195-2022

Pourebrahim, N., Thill, J.-C., Sultana, S., & Mohanty, S. (2018). "Enhancing trip distribution prediction with twitter data: Comparison of neural network and gravity models," *Proc. 2nd ACM SIGSPATIAL Int. Work. AI Geogr. Knowl. Discov. GeoAI 2018*. DOI: 10.1145/3281548.3281555

Song, Y., Kalacska, M., Gašparović, M., Yao, J., & Najibi, N. (2023). Advances in geocomputation and geospatial artificial intelligence (GeoAI) for mapping. [Elsevier.]. *International Journal of Applied Earth Observation and Geoinformation*, 120, 103300. DOI: 10.1016/j.jag.2023.103300

Srivastava, S., Vargas-Muñoz, J. E., Swinkels, D., & Tuia, D. (2018). "Multilabel Building Functions Classification from Ground Pictures using Convolutional Neural Networks," *Proc. 2nd ACM SIGSPATIAL Int. Work. AI Geogr. Knowl. Discov.*, pp. 43–46. DOI: 10.1145/3281548.3281559

Sun, T., Di, Z., & Wang, Y. (2018)... *Combining Satellite Imagery and GPS Data for Road Extraction*, 3281550(c), 4–7.

T. O. (2018). Satellite-based and D. U. U. Convolutional. "Towards Operational Satellite-Based Damage-Mapping Using U-Net Convolutional Network: A Case Study of 2011 Tohoku".

Teja, K. R., Liu, C. M., & Chopra, S. R. (2023). Water Assessment Using Geospatial and Data Science Tools. *2023 International Conference on Emerging Smart Computing and Informatics (ESCI)*, 1–6. IEEE. DOI: 10.1109/ESCI56872.2023.10099538

Trisovic, A., Lau, M. K., Pasquier, T., & Crosas, M. (2022). A large-scale study on research code quality and execution. *Scientific Data*, 9(1), 60. DOI: 10.1038/s41597-022-01143-6 PMID: 35190569

Vopham, T., Hart, J. E., Laden, F., & Chiang, Y. Y. (2018). "Emerging trends in geospatial artificial intelligence (geoAI): Potential applications for environmental epidemiology," *Environ. Heal. A Glob.Access Sci. Source*, 17(1), 1–6.

Wang, J., & Biljecki, F. (2022). Unsupervised machine learning in urban studies: A systematic review of applications. *Cities (London, England)*, 129, 103925. DOI: 10.1016/j.cities.2022.103925

Wang, S., Cai, W., Tao, Y., Sun, Q. C., Wong, P. P. Y., Thongking, W., & Huang, X. (2023). Nexus of heat-vulnerable chronic diseases and heatwave mediated through tri-environmental interactions: A nationwide fine-grained study in Australia. *Journal of Environmental Management*, 325, 116663. DOI: 10.1016/j.jenvman.2022.116663 PMID: 36343399

Wang, S., Liu, Y., Lam, J., & Kwan, M.-P. (2021). The effects of the built environment on the general health, physical activity and obesity of adults in Queensland, Australia. *Spatial and Spatio-temporal Epidemiology*, 39, 100456. DOI: 10.1016/j.sste.2021.100456 PMID: 34774262

Wen, R., & Li, S. (2022). Spatial Decision Support Systems with Automated Machine Learning: A Review. *ISPRS International Journal of Geo-Information*, 12(1), 12. DOI: 10.3390/ijgi12010012

G. Xi and S. Mei,(2009). "A Deep Residual Network Integrating Spatial-temporal Properties to Predict Influenza Trends at an Intra-urban Scale".

Xu, Q. (2021). Evaluation of Rural Tourism Spatial Pattern Based on Multifactor-Weighted Neural Network Algorithm Model in Big Data Era. *Scientific Programming*, 2021, 1–11. DOI: 10.1155/2021/8108287

Xu, Y., Pan, L., Du, C., Li, J., Jing, N., & Wu, J. (2018). "Vision-based UAVs Aerial Image Localization: A Survey," *Proc. 2nd ACM SIGSPATIAL Int. Work. AI Geogr. Knowl. Discov.*, pp. 9–18, 2018. DOI: 10.1145/3281548.3281556

Zhang, X., & Zhao, X. (2022). Machine learning approach for spatial modeling of ride sourcing demand. *Journal of Transport Geography*, 100, 103310. DOI: 10.1016/j.jtrangeo.2022.103310

Zhang, Z., Wen, F., Sun, Z., Guo, X., He, T., & Lee, C. (2022). Artificial intelligence-enabled sensing technologies in the 5G/internet of things era: From virtual reality/augmented reality to the digital twin. *Advanced Intelligent Systems*, 4(7), 2100228. DOI: 10.1002/aisy.202100228

KEY TERMS AND DEFINITIONS

Artificial general intelligence (AGI): is a type of artificial intelligence (AI) that falls within the lower and upper limits of human cognitive capabilities across a wide range of cognitive tasks.

Artificial intelligence (AI): is technology that enables computers and machines to simulate human learning, comprehension, problem solving, decision making, creativity and autonomy.

Deep Learning (DL): DL is a subset of machine learning. With this model, an algorithm can determine whether or not a prediction is accurate through a neural network without human intervention.

GeoAI: the integration of artificial intelligence (AI) with spatial data, science, and geospatial technology to increase understanding and solve spatial problems.

Geographic Information Systems (GIS): is a computer system for capturing, storing, checking, and displaying data related to positions on Earth's surface.

Machine Learning (ML): is a branch of artificial intelligence (AI) and computer science that focuses on the using data and algorithms to enable AI to imitate the way that humans learn, gradually improving its accuracy.

OpenStreetMap: is a free, open geographic database updated and maintained by a community of volunteers via open collaboration.

Chapter 2
Harnessing AI and Machine Learning for Enhanced Geospatial Analysis

Rachna Rana
Ludhiana Group of Colleges, Ludhiana, India

Pankaj Bhambri
https://orcid.org/0000-0003-4437-4103
Guru Nanak Dev Engineering College, Ludhiana, India

ABSTRACT

In recent years, artificial intelligence (AI) and machine learning (ML) have changed geospatial analysis, allowing for more accurate, efficient, and scalable processing of massive volumes of geographical data. Traditionally, geospatial analysis depended on human-driven approaches and rule-based systems, which were frequently time-consuming and restricted in their capacity to handle large datasets. The combination of AI and ML has resulted in the development of revolutionary approaches like as deep learning, neural networks, and automated feature extraction, which have transformed the use of geographic information systems (GIS). This chapter investigates the critical role of artificial intelligence and machine learning in developing geospatial analysis in a variety of disciplines, including environmental monitoring, urban planning, disaster management, and agriculture. AI-powered models can now do predictive analytics, real-time data processing, and pattern identification in satellite images, LiDAR, and sensor networks.

DOI: 10.4018/979-8-3693-8054-3.ch002

INTRODUCTION

Geospatial analysis, or the collection, processing, and interpretation of geographical data, is important in a variety of fields such as urban planning, environmental monitoring, agriculture, and disaster management. Historically, geographic information systems (GIS) were the major tool for geographical data analysis, with a heavy reliance on manual input, predetermined models, and human interpretation. While successful, these traditional approaches frequently fail to keep up with the fast-expanding volume, complexity, and diversity of geospatial data produced by current technologies such as satellite photography, LiDAR, and sensor networks (Vigneshwari et al., 2024).

The advent of artificial intelligence (AI) and machine learning (ML) has created new opportunity to improve the speed, accuracy, and depth of geospatial research. AI, with its ability to replicate human-like thinking, and machine learning, with its ability to learn from data patterns and improve over time, have revolutionized geographical data analysis and use. These technologies enable automated feature extraction, predictive modeling, real-time data processing, and the detection of previously indistinguishable spatial patterns. This integration of AI and machine learning into geospatial research signifies a paradigm change, transitioning from manual, labor-intensive techniques to more efficient, automated systems capable of managing enormous datasets. Deep learning models, a type of AI, have performed exceptionally well in image identification tests, making them ideal for processing satellite images and other remote sensing data. Similarly, ML algorithms improve spatial forecasts and anomaly identification, allowing for better decision-making in crucial areas such as disaster preparedness and response, land use planning, and climate change study.

However, using AI and machine learning to geospatial research is not without hurdles. Issues like as the requirement for big, high-quality training datasets, assuring AI model openness, and eliminating data biases continue to be key concerns. Furthermore, AI's ethical ramifications, particularly in terms of data protection and misuse, must be carefully explored. This introduction presents an overview of how AI and machine learning are altering geospatial analysis, laying the groundwork for a more in-depth investigation of the methodologies, advantages, and issues connected with their use. As these technologies advance, their potential to transform how we comprehend and interact with geographical data grows exponentially, providing tremendous tools for addressing complex global concerns in an increasingly data-driven society (Ahnaf et al., 2023; Anand et al., 2024; El Baba et al., 2020; Eseosa Halima & Hiroaki, 2022). Figure 1 is related to Tools for Geospatial Analysis.

Figure 1. Tools for Geospatial Analysis

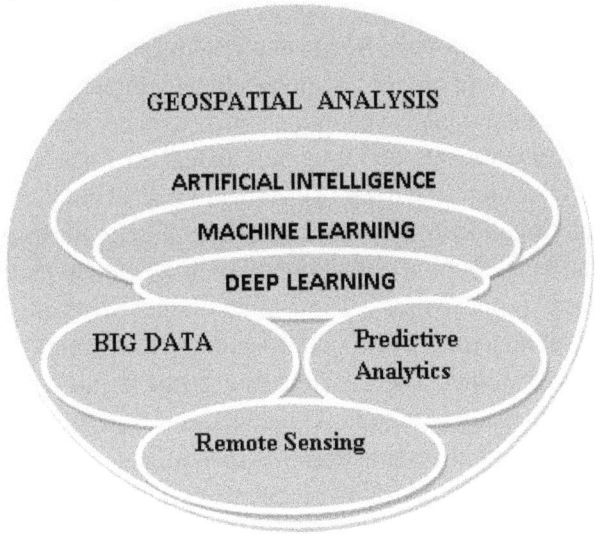

Artificial Intelligence

Geospatial analysis is the collecting, processing, and interpretation of data that has a spatial or geographical component. It is crucial in many fields, including urban planning, environmental protection, transportation, disaster management, and natural resource exploration. Traditional geospatial analytic approaches are useful but increasingly challenged by the enormous expansion in both the number and complexity of spatial data. The introduction of large data sources such as satellite imaging, drones, sensor networks, and social media geotagging has outperformed traditional analytical approaches. AI, with its ability to learn from enormous datasets and identify complicated patterns, has emerged as a game changer in geospatial analysis. AI includes a variety of sophisticated algorithms that include ML, deep learning, and computer vision, which allow machines to solve problems and make decisions in the same way that humans do. These technologies have radically revolutionized the processing, analysis, and application of geospatial data. AI-powered geospatial analysis can automate time-consuming tasks like picture categorization, feature extraction, and anomaly detection. More crucially, it can manage large datasets, enabling real-time processing and analysis, so boosting decision-making in both government and industry.

In remote sensing, for example, AI can analyze high-resolution satellite pictures to detect changes in land cover, forecast environmental threats, and monitor natural resources. Deep learning models can outperform standard approaches in spotting patterns in complicated datasets, detecting tiny changes in urban landscapes, woods, and water bodies with more accuracy (Thirumalaiyammal et al., 2024). Similarly, AI-powered prediction models are being utilized in disaster management to assess susceptibility, anticipate natural catastrophes such as floods and wildfires, and improve emergency response times. These technologies are changing the way we prepare for and respond to real-world situations. However, using AI into geospatial research is not without hurdles. There are technical obstacles, such as the requirement for huge, well-labeled datasets to train AI models, and worries regarding the interpretability of these models. AI is transforming several sectors, including geospatial analysis. Geospatial analysis, which involves collecting, processing, and analyzing geographical data from sources such as satellite images, sensor networks, and GIS, has typically used manual or semi-automated methods (Sharma & Bhambri, 2024). While useful in some situations, these old approaches have substantial limits when dealing with the enormous volume, complexity, and diversity of data produced by current technology. The emergence of AI has transformed the sector, providing fresh techniques to automate and improve spatial activities.

AI, particularly ML and deep learning (DL), is ideally suited to addressing many of the issues of geospatial analysis. These technologies enable powerful pattern recognition, feature extraction, and predictive modeling that significantly outperform human skills. Deep learning algorithms, for example, can automatically categorize objects in satellite photos, detect changes in land use over time, and anticipate environmental trends. This has far-reaching consequences for businesses like urban planning, agriculture, environmental monitoring, disaster management, and military. AI's capacity to analyze massive volumes of data rapidly and reliably is a game changer in geospatial analytics. Traditional approaches frequently need substantial human effort and time, especially for tasks like picture interpretation and data categorization. In contrast, AI-powered systems may automate these operations, decreasing human labor while enhancing accuracy. The integration of AI with cloud computing and big data infrastructures expands its capabilities, allowing for real-time analysis of enormous geographical information at scale. Despite its benefits, the use of AI in geospatial research presents various problems. These include the necessity for high-quality training data, assuring algorithm openness and interpretability, and reducing biases that might lead to incorrect or immoral results.

Furthermore, as AI systems become more widely utilized for vital decision-making in domains such as disaster response and resource management, it is critical to assure their dependability and equity(Gupta & Pandya, 2024; Ighile et al., 2022; Jurišić et al., 2022; Kibona et al., 2011; Li & Bao, 2021; Mkumbo et al., 2022)

Machine Learning

ML has quickly emerged as a vital tool in geospatial analysis, transforming spatial data collection, processing, and interpretation. Geospatial analysis, which examines geographical data from a variety of sources such as satellite imagery, geographic information systems (GIS), LiDAR, and sensor networks, has traditionally relied on rule-based systems and human expertise for tasks such as classification, pattern recognition, and prediction (Rana & Bhambri, 2024a). While useful to some extent, both manual and semi-automated procedures have limits, especially when dealing with the massive volumes of complicated geographical data created today. Machine learning is especially useful in this situation.

At its heart, machine learning is intended to allow systems to learn from data, spot patterns, and make predictions or judgments with minimum human involvement. In geospatial analysis, ML models can analyze large-scale information more effectively, detect subtle patterns in geographic characteristics, and make high-accuracy predictions that traditional approaches would struggle to achieve. For example, land cover categorization, urban expansion monitoring, and catastrophe risk assessment using satellite imaging and remote sensing data are increasingly popular applications. Despite its transformational promise, machine learning in geospatial analysis poses problems. These include the necessity for high-quality training datasets, the danger of overfitting or underfitting models, and assuring transparency and interpretability in the outputs produced by complicated machine learning algorithms. Furthermore, while ML is excellent at spotting patterns in data, it is critical to contextualize these patterns with domain expertise to prevent drawing incorrect inferences (Mohammadi, 2022; N & J, 2023 Nayak et al., 2022; "Predictive analysis of campus placement of student using machine learning algorithms," 2024).

Geospatial Analysis

Geospatial analysis is the act of gathering, analyzing, and interpreting data about the Earth's surface and its features, utilizing the spatial connections between distinct geographic entities. This type of study is essential for a variety of applications, including urban planning, environmental management, disaster response, transportation, agriculture, and national security. Geospatial analysis, which examines spatial patterns, correlations, and trends in geographical data, provides for a better understanding of physical surroundings, human activities, and their interactions.

Historically, geospatial analysis depended on conventional approaches such as cartography, GIS, and remote sensing technologies to map and evaluate geographical events. These technologies allow for the presentation and administration of spatial data, which can include everything from satellite images and sensor readings to population information and land use maps. Over time, advances in processing power, data gathering techniques, and the development of big data have dramatically enhanced geospatial analysis' possibilities. Geospatial analysis relies heavily on geographical data, which can be organized or unstructured. This data, which is frequently obtained via satellites, drones, LiDAR (Light Detection and Ranging), and GPS technology, gives insights into anything from landscape changes to infrastructure development. Geospatial analysis enables enterprises and governments to analyze and map data, discover patterns, anticipate trends, and make data-driven choices that are vital for resource management, environmental conservation, disaster preparedness, and economic growth. The rising availability of high-resolution data, along with developments in AI and ML, is reshaping the geospatial analytics environment. These technologies allow for the automation of complicated processes like image categorization, feature extraction, and predictive modeling, resulting in quicker, more accurate insights than ever before. Furthermore, cloud computing and the growth of open data platforms are making geospatial tools more accessible, allowing a broader spectrum of people to participate with and profit from spatial analysis.

Despite its numerous uses and increasing complexity, geospatial analysis poses obstacles. To successfully handle and understand geographical data, significant technologies and knowledge are required due to its sheer amount and complexity. Furthermore, data privacy, the quality of accessible data, and the ethical implications of surveillance and monitoring are all key considerations that must be addressed as the sector evolves.

Geospatial analysis is a critical tool for comprehending and controlling the physical world, allowing stakeholders to make informed decisions on anything from urban growth to environmental preservation. As technology advances, the possibility for more precise, real-time, and predictive geospatial analysis develops, creating new opportunities to address some of the world's most critical issues ("Preface: The 3rd annual advanced technology, applied science and engineering conference 2021 (ATASEC 2021)," 2023; Radočaj & Jurišić, 2022; Rana, 2022; Rana & Sharma, 2020). Recent advances in geospatial analysis reflect technological breakthroughs, increasing data accessibility, and the incorporation of artificial intelligence and machine learning. These trends improve the scope, precision, and applicability of geospatial analysis:

Combining AI and Machine Learning: ML and DL are revolutionizing geographic data analysis. AI algorithms automate feature identification, categorization, and predictive modeling, decreasing human labor and increasing accuracy. For example, ML is commonly employed in automated land use categorization, urban planning, and disaster response.

Big Data & Cloud Computing: The growing availability of large-scale geospatial data from satellites, IoT devices, and mobile sources has made big data critical for geospatial research. Cloud systems such as Google Earth Engine and AWS provide scalable infrastructure, making it simpler to handle and analyze large datasets, resulting in faster insights and cooperation on worldwide projects.

Real-time Data Analysis: Real-time data from sources such as drones, satellites, and sensors enables quick analysis, allowing for the monitoring of events as they occur. This is especially useful in disaster response, traffic management, and environmental monitoring, where fast information may lead to more effective decision making.

Increased Utilization of UAVs (Drones) for Data Collection: Unmanned Aerial Vehicles (UAVs), sometimes known as drones, offer high-resolution and adaptable data collecting capabilities, making them perfect for thorough surveys in agriculture, construction, mining, and environmental monitoring. Drones are frequently selected in circumstances requiring regular, site-specific data gathering, as they provide a cost-effective alternative to traditional surveying techniques.

Geographic Information System

A GIS is a sophisticated tool for collecting, analyzing, visualizing, and interpreting spatial and geographical data. At its heart, GIS combines data with location-based information, allowing users to see correlations, patterns, and trends in a geographical context. This makes GIS a vital tool for a wide range of applications, including urban planning and environmental management, as well as disaster response, transportation, and public health. A GIS' fundamental role is to merge different forms of data, such as satellite images, census data, topographic maps, and sensor readings, with geographic coordinates or spatial references. By doing so, GIS generates dynamic, data-driven maps and models that may be used to examine and comprehend spatial relationships in a variety of sectors. For example, urban planners use GIS to find the optimum places for new infrastructure, environmental experts evaluate deforestation trends, and emergency responders map out disaster-affected regions to properly distribute resources. GIS is more than simply a tool for making maps. It supports complicated geographical analysis by merging layers of data such as topography, population density, climate, and land use. Users may query the data, do geographical analysis, and create visualizations that highlight insights that are not readily apparent in standard tabular data. These features make GIS a crucial decision-making tool, enabling enterprises and governments to make educated decisions based on spatial linkages and geographic patterns in their data. GIS are frameworks for collecting, organizing, analyzing, and displaying geographic data. It combines many forms of data, enabling users to build maps and evaluate patterns, correlations, and trends in geographical contexts. Urban planning, environmental management, transportation, healthcare, and other areas make use of GIS technology. (Rinaldi & Nielsen, 2024; Rustamov et al., 2023). The following key components of GIS: Hardware includes computers, servers, and mobile devices that run GIS software. Specialized applications for processing and analyzing geographic data include Arc GIS, QGIS, and others. Data is the foundation of every GIS system, consisting of spatial data (places) and attribute data (descriptive information about those locations. The method of gathering, processing, and analyzing is geographic data. The Skilled experts such as GIS analysts, developers, and researchers oversee and study the system.

> **Functions of GIS:** Gathering geographic information from a variety of sources, including satellite photos, surveys, and GPS devices. Data storage is the organization of geographical data in databases to make it easier to access and retrieve information. Using geographical studies such as overlay, proximity analysis, and buffering to identify patterns and trends. Visualization is the process of creating maps, 3D models, and interactive apps to graphically display geographic data.
>
> **Applications of GIS:** Urban planning entails developing infrastructure, zoning, and land-use plans. Environmental management entails monitoring deforestation, animal habitats, and climate change effects. Transportation includes route planning, traffic management, and logistics. Public health: Disease outbreak mapping and distribution of healthcare resources. Disaster management includes assessing risk regions, arranging evacuations, and coordinating recovery operations.

1. DEEP LEARNING

Deep Learning in Geospatial Analysis combines AI, specifically deep neural networks, with geographic data to improve spatial comprehension and decision-making (Rana & Bhambri, 2024b). Deep learning has considerably enhanced the capabilities of classical geospatial analysis by allowing for the processing and interpretation of vast amounts of geographical data with greater accuracy, speed, and automation Rinaldi & Nielsen, 2024). Deep learning is a form of machine learning that use artificial neural networks, specifically deep neural networks with several layers, to model and solve difficult problems. These networks can automatically learn patterns and features from enormous datasets without the need for manual feature engineering, making them ideal for applications such as image recognition, natural language processing, and, more recently, geospatial analysis (Rustamov et al., 2023; Sudarmadji & Santoso, 2023; (Taromideh et al., 2022).

1.1 Remote Sensing and Image Analysis

Deep learning models can categorize many forms of land cover (for example, forests, urban areas, and water bodies) in satellite photos. Object detection is the process of identifying items in aerial photography, such as buildings, roads, and cars, using convolutional neural networks. Change detection is the process of analyzing satellite data over time to monitor environmental changes such as deforestation, urban expansion, and agricultural growth.

1.2 Autonomous Mapping and Navigation

Deep learning can increase mapping accuracy for autonomous vehicles (e.g., drones, self-driving automobiles) by utilizing real-time spatial data to comprehend their surroundings and navigate safely.

1.3 Geospatial Data Fusion

Integrating data from many sources (satellite, aerial, IoT sensors, etc.) to generate a comprehensive perspective of a geographic region, generally utilizing deep learning models to manage the complicated relationships between diverse data types.

1.4 Disaster Prediction and Management

Deep learning can anticipate natural disasters like as floods, earthquakes, and hurricanes by studying spatial patterns in past data. Deep learning algorithms can analyse satellite data to automate post-disaster damage assessment.

1.5 Urban Planning and Infrastructure Development

Deep learning uses geographical data to evaluate and anticipate urban growth, estimate infrastructure needs, and improve city planning.

1.6 Key Techniques of Deep Learning for Geospatial Analysis

Figure 2 displays the key techniques of Deep Learning for Geospatial Analysis.

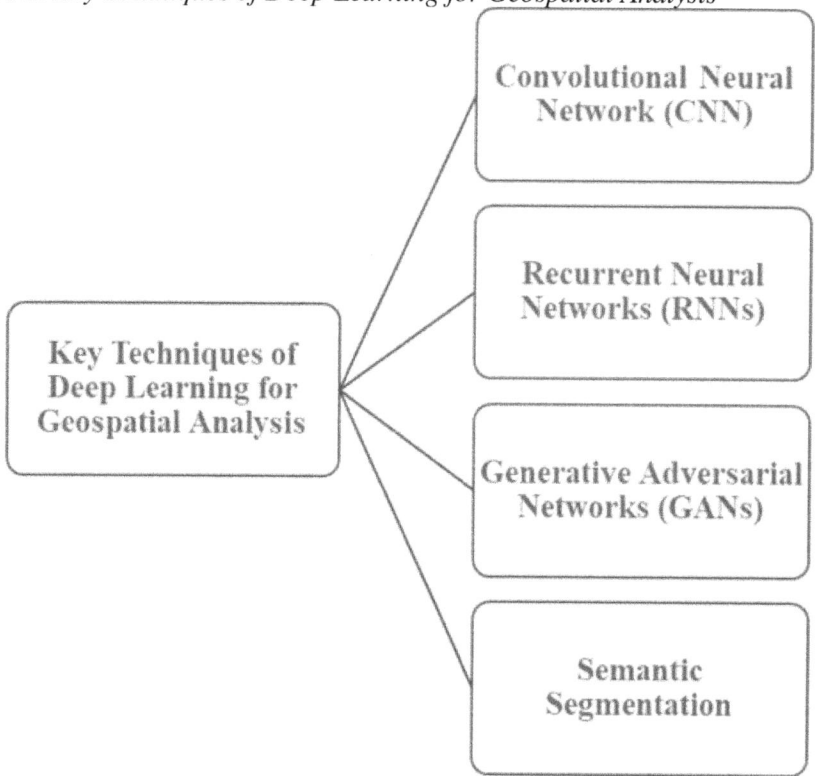

Figure 2. Key Techniques of Deep Learning for Geospatial Analysis

- **Convolution Neural Network (CNN):** CNNs are commonly utilized for geospatial image analysis applications such as land-use categorization, object recognition, and satellite imagery segmentation.
- **Current Neural Networks (RNNs)** are excellent for assessing time-series geographical data, such as weather patterns, tracking changes over time, and forecasting future environmental conditions.
- **Generative Adversarial Networks (GANs):** GANs may produce realistic geographic data or fill gaps in incomplete satellite images, hence enhancing data quality.

- **Semantic segmentation:** It is a technique for classifying geographical information at the pixel level, allowing for detailed mapping of land cover, plant kinds, and other features.

1.7 Benefits of Deep Learning for Geospatial Analysis

- **Automation:** Deep learning allows for the automatic processing and interpretation of huge geographic datasets, eliminating the need for manual intervention.
- **Accuracy:** It considerably enhances spatial data prediction and classification accuracy, particularly for applications like as item recognition and change detection.
- **Scalability:** Deep learning's capacity to analyze huge datasets from sources such as satellites and UAVs (drones) makes it a suitable tool for national and global geospatial analysis.

Deep learning models, when combined with real-time data streams (e.g., from sensors or drones), allow for fast insights and decision-making (Ruby et al., 2024).

1.8 Challenges

- **Data Quality:** The accuracy of deep learning models is dependent on the quality of the training data, which might be noisy or incomplete.
- **Computer Resources:** Deep learning necessitates tremendous computer resources, particularly when working with large-scale geographical datasets.
- **Interpretability:** Deep neural networks sometimes function as "black boxes," making it difficult to explain how a model reached a specific judgment.

2. REMOTE SENSING

Remote Sensing in Geospatial Analysis refers to the combination of remote sensing technology and data with geographic information systems (GIS) and spatial analysis methods. This combination improves our knowledge of the Earth's surface and environment by delivering high-resolution, timely, and complete spatial data. Remote sensing is an important technology for gathering and evaluating geographic information, allowing for more informed decision-making in a variety of disciplines (Tohidi & B. Rustamov, 2020). sensing is the process of gathering information about an item or region from a distance, usually using satellites or airplanes outfitted with sensors. These sensors detect and quantify electromagnetic radiation reflected or emitted by the Earth's surface. The information gathered may be utilized to generate comprehensive maps and evaluate spatial patterns (Wei et al., 2021).

2.1 The characteristics of Remote Sensing in Geospatial Analysis(Wei et al., 2021)

- Data collection at different scales and resolutions: Remote sensing offers data gathering at various geographical, temporal, and spectral resolutions, enabling detailed or broad-scale analysis. This versatility facilitates data collecting for unique research purposes, such as monitoring local vegetation or assessing global climate patterns.
- Multispectral and hyperspectral imaging: Remote sensing, which captures many wavelengths (visible, infrared, thermal, etc.), offers information about surface material qualities, plant health, water bodies, and other topics. Hyper spectral imaging, in particular, allows for accurate material identification, which is useful in agriculture, environmental monitoring, and mineral prospecting.
- Monitoring and Detecting Changes over Time: Remote sensing systems collect data over time, allowing users to track and detect changes like deforestation, urbanization, and catastrophe impacts. This skill is critical for monitoring environmental trends, natural catastrophes, and human growth.
- Worldwide Coverage and Accessibility:Satellites give worldwide data coverage, reaching locations that would otherwise be inaccessible. This function is important for researching remote, huge, or hazardous places (e.g., arctic regions, deserts, or dense woods) and is easily accessible to all users.
- Terrain and Elevation Analysis (LiDAR and Radar): Active remote sensing systems, such as LiDAR and radar, provide three-dimensional representations of the Earth's surface. These are required for topographic mapping, analyzing landforms, comprehending geomorphology, urban development, and hydrological modeling.
    Integration with geographical information systems (GIS): Remote sensing data is routinely combined with GIS, which improves spatial analytic skills and allows more complicated spatial modeling. Together, these techniques offer insights into urban planning, environmental protection, agriculture, and disaster management.
- Data collection and analysis in real time: Some remote sensing platforms (such as drones or modern satellite systems) give near-real-time data, making them useful for disaster management, emergency response, and military operations that need quick action.

- Image Processing Techniques Using Automation and Machine Learning: Advances in artificial intelligence have enabled the automation of huge dataset processing, improving classification, segmentation, and feature detection accuracy. This efficiency enables large-scale geospatial initiatives with complicated data requirements.

2.2 Applications of Remote Sensing in Geospatial Analysis

Figure 3 is displaying the applications of Remote Sensing in Geospatial Analysis.

Figure 3. Applications of Remote Sensing in Geospatial Analysis

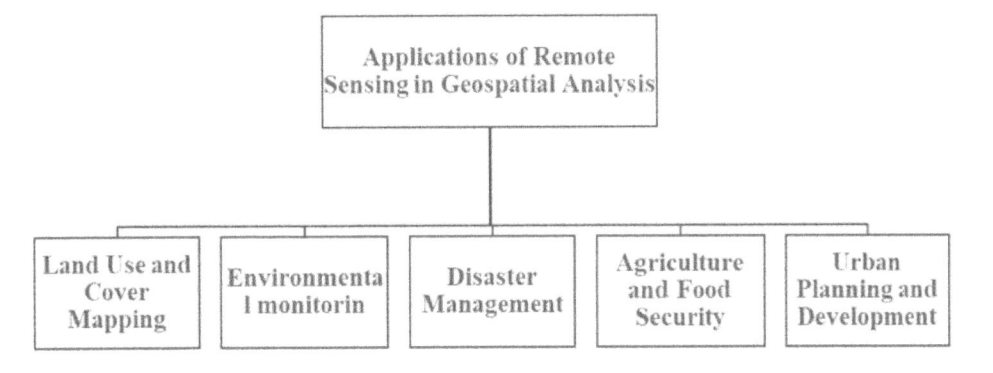

- **Land Use and Cover Mapping:** Remote sensing analyzes spectral fingerprints to provide accurate, up-to-date maps of land use and land cover types (urban, agricultural, and wooded sectors).
- **Environmental monitoring** allows for the evaluation of environmental changes such as deforestation, desertification, and habitat loss, which provides crucial information for conservation initiatives.
- **Disaster Management:** Remote sensing data aids in disaster planning and response by monitoring natural hazards (floods, earthquakes, hurricanes) and analyzing damage after the event.
- **Agriculture and Food Security:** Remote sensing technologies make precision agriculture possible by monitoring crop health, determining soil moisture levels, and optimizing irrigation operations.

- **Urban Planning and Development:** Remote sensing helps urban planners analyze growth trends, infrastructure development, and transportation networks, resulting in improved urban management.

2.3 Advantages of Using Remote Sensing in Geospatial Analysis

- **High spatial resolution:** Remote sensing produces precise pictures capable of revealing fine-scale features on the Earth's surface.
- **Temporal Analysis:** The capacity to gather data across time enables the tracking of changes and trends, making it useful for long-term investigations.
- **Cost-Effectiveness:** Remote sensing can cover huge regions quickly and efficiently, eliminating the need for lengthy ground surveys.
- **Comprehensive Coverage:** It may collect data from remote or inaccessible regions, offering a more complete picture of geographic phenomena

2.4 Challenges

- **Data interpretation:** Interpreting remote sensing data needs skill and can be difficult owing to factors such as atmospheric conditions and sensor calibration.
- **Resolution Trade-offs:** There is usually a choice between spatial resolution (detail) and temporal resolution (frequency of data collection); high-resolution data may be obtained less frequently (Mohana Sundari et at., 2024).
- **Data Management:** Handling vast amounts of remote sensing data requires reliable storage, processing, and analytical tools.

3. PREDICTIVE ANALYTICS

Predictive analytics in geospatial analysis is the use of statistical algorithms, machine learning techniques, and historical data to estimate future occurrences or trends based on geographic and geographical information. Organizations may make more informed decisions, optimize resource allocation, and improve planning across several domains by combining predictive analytics with geospatial data (Yang, 2024). Predictive analytics is a collection of strategies that use historical and present data to forecast future results. This involves using statistical approaches, data mining, and machine learning to discover patterns and correlations in data. When applied to geographic data, predictive analytics provides insights into spatial trends and events, making it a useful tool in a variety of industries.

3.1 The Characteristics of Predictive Analytics in Geospatial Analysis

- **Data Collection:** Collecting historical and real-time geographic data from a variety of sources, such as satellite images, sensor networks, social media, and demographic data.
- **Data processing** is cleaning, converting, and organizing data for analysis, which frequently includes combining multiple data types (e.g., raster and vector data) into a coherent dataset.
- **Statistical and machine learning techniques** include using regression analysis, decision trees, neural networks, and clustering algorithms to model and predict geographical data outcomes.
- **Geospatial Visualization** is the presentation of prediction model findings using maps and visuals to effectively communicate insights.

3.2 Applications of Predictive Analytics in Geospatial Analysis

- **Urban Planning and Development:** Predictive analytics may assist estimate urban growth trends, analyze housing demand, and identify regions ideal for infrastructure development, allowing for more effective city planning.
- **Disaster Management:** By evaluating past data on natural catastrophes, predictive models can anticipate the possibility of future incidents, allowing emergency services to better prepare and allocate resources.
    - **Environmental Monitoring:** Predictive analytics may be used to simulate environmental changes such as land cover change or the effects of climate change, offering insights into future ecological conditions.
- **Transportation and Traffic Management:** Analyzing traffic patterns and demographic data can aid in congestion prediction, public transportation route optimization, and infrastructure development (Geetha et al., 2024).
- **Predictive models** can anticipate disease outbreaks or infection transmission by examining geographical and temporal data, which helps with public health response and resource allocation.
- **Retail and Marketing:** Geospatial predictive analytics may help businesses identify customer behavior, improve shop locations, and customize marketing techniques to local patterns.

3.3 Benefits of Predictive Analytics for Geospatial Analysis

- **Enhanced Decision-Making:** Provides data-driven insights to assist companies in making educated decisions based on projected outcomes.Resource optimization enables more effective resource allocation by predicting future requirements and obstacles.
- **Risk management** is the process of identifying possible hazards and implementing proactive mitigation measures, particularly in disaster-prone locations.
- **Personalization:** Predictive analytics in retail and marketing may personalize services and products to specific geographic areas and client preferences.

3.4 Challenges

- **Data Quality and Availability:** The accuracy of forecasts is dependent on the quality and completeness of the data, which might be a problem in particular areas.
- **Model Complexity:** Creating predictive models may be difficult, necessitating competence in statistical analysis and machine learning, as well as a knowledge of geographic data.
- **Interpretability:** Predictive models, particularly those based on machine learning, can behave like "black boxes," making it difficult to grasp how predictions are formed.

4. BIG DATA

Big Data in geographic Analysis refers to the utilization of huge, complex datasets derived from a variety of geographic sources, including satellite imaging, GPS sensors, social media, and mobile devices. This combination of big data technology with geospatial analysis allows for the extraction of useful insights from massive volumes of spatial data, resulting in enhanced decision-making in a variety of disciplines such as urban planning, environmental monitoring, transportation, and public health (Yohana et al., 2023). Big Data is defined by the "three Vs."

4.1 Volume

The amount of data created is massive and increasing exponentially, frequently exceeding terabytes or petabytes.

4.2 Velocity

Data is created and gathered at unprecedented rates, necessitating real-time processing and analysis.

4.3 Variety

Data arrives in many formats (structured, unstructured, and semi-structured) from numerous sources.

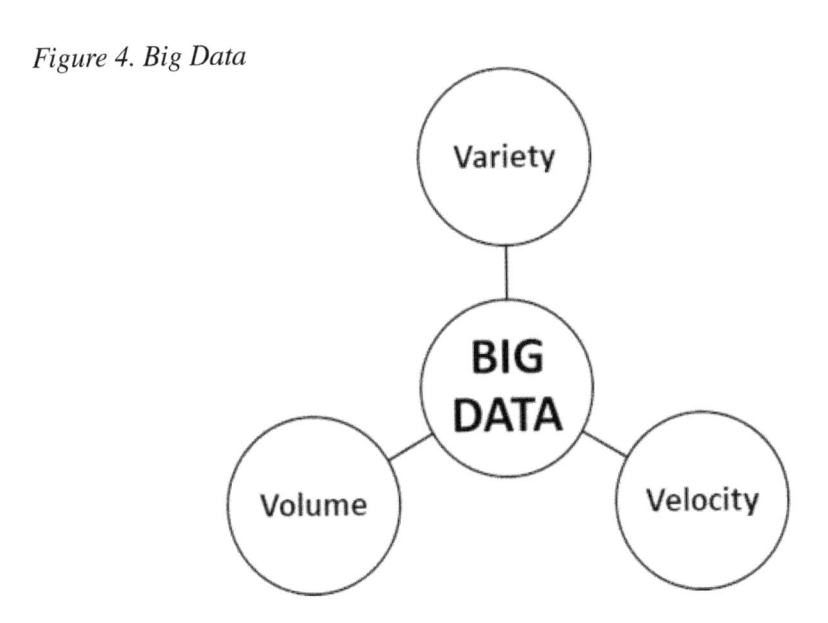

Figure 4. Big Data

Relevant components of the Big Data are shown in Figure 4.

4.3.1 Major Big Data Components for Geospatial Analysis

Big data for geospatial analysis comes from a variety of sources, including:

- **Remote sensing** uses satellite and aerial photos to provide high-resolution data on the Earth's surface.
- **Social media:** Geotagged posts and user-generated material that give real-time information on human activities and behavior.
- **IoT devices** are sensor networks that collect data on weather, traffic, and environmental variables.

- **Surveys and Demographic Data:** Large-scale surveys and census data provide information on population characteristics.
    Big data necessitates modern storage solutions, such as distributed databases and cloud storage systems, in order to efficiently store and handle big datasets.
- **Big data analytics** is the use of modern technology and algorithms to process and analyze big datasets, such as:
- **Distributed computing:** Frameworks such as Apache Hadoop and Spark allow for concurrent processing of huge datasets across numerous computers.
- **Machine Learning and Artificial Intelligence:** Techniques for detecting patterns, trends, and correlations in data.
- **Geospatial Analytics:** The combination of standard GIS methodologies and big data analytics allows for the analysis of geographical correlations and patterns, revealing previously unobtainable insights.

4.4 Application of Big Data in Geospatial Analysis

- **Urban planning** is analyzing massive datasets to better understand urban growth trends, traffic flow, and public service requirements, resulting in more effective city planning.
- **Environmental monitoring** involves using big data to track land use changes, monitor pollutant levels, and estimate the effects of climate change on ecosystems.
- **Transit and Logistics:** Real-time traffic data analysis optimizes routes, reduces congestion, and improves public transit systems.
- **Big data analytics** may improve disaster preparedness by evaluating past trends and forecasting the effects of natural catastrophes.
- **Public Health:** Using geographic data from health records, social media, and environmental sensors to anticipate epidemics and distribute resources efficiently.
- **Market Analysis:** Retailers may utilize geographical big data to better analyze customer behavior, improve shop locations, and customize marketing strategies to local patterns.

4.5 Benefits of Big Data for Geospatial Analysis

- **Enhanced insights:** The combination of big data with geospatial analysis allows for deeper insights into previously unknown spatial patterns and linkages.

- **Access to real-time geographic information** improves data-driven decision-making and makes it more robust and timely.
- **Cost efficiency:** Optimizing resource allocation and service delivery saves money for governments and enterprises.
- **Predictive Capabilities:** Analyzing past big data can assist in projecting future trends and results based on geographical patterns.

4.6 Applications of Big Data for Geospatial Analysis

- **Urban planning** is analyzing massive datasets to better understand urban growth trends, traffic flow, and public service requirements, resulting in more effective city planning.
- **Environmental monitoring** involves using big data to track land use changes, monitor pollutant levels, and estimate the effects of climate change on ecosystems.
- **Transit and Logistics:** Real-time traffic data analysis optimizes routes, reduces congestion, and improves public transit systems.
- **Big data analytics** may improve disaster preparedness by evaluating past trends and forecasting the effects of natural catastrophes.
- **Public Health:** Using geographic data from health records, social media, and environmental sensors to anticipate epidemics and distribute resources efficiently.
- **Market Analysis:** Retailers may utilize geographical big data to better analyze customer behavior, improve shop locations, and customize marketing strategies to local patterns.

4.7 Challenges

- **Data Management:** Handling, processing, and analyzing large volumes of data may be difficult and resource-intensive.
- **Data Quality:** Ensuring the correctness and dependability of data from many sources may be difficult, especially for unstructured data.
- **Interoperability:** It might be challenging to integrate data from diverse sources and formats into a single analytical framework.
- **Privacy Concerns:** The use of personal geographic data involves privacy and ethical concerns, needing caution and adherence to legislation.

5. THE CRITICAL ROLE OF AI AND ML IN GEOSPATIAL ANALYSIS

AI and ML are changing geospatial analysis in a variety of fields by improving the capacity to handle, analyze, and understand massive volumes of geographical data (Chithra & Bhambri, 2024). Following are graphics highlighting the crucial functions of AI and ML in geospatial analysis (York & Bamberger, 2024). The critical role of artificial intelligence and machine learning in developing geospatial analysis in a variety of disciplines is shown in Figure 5.

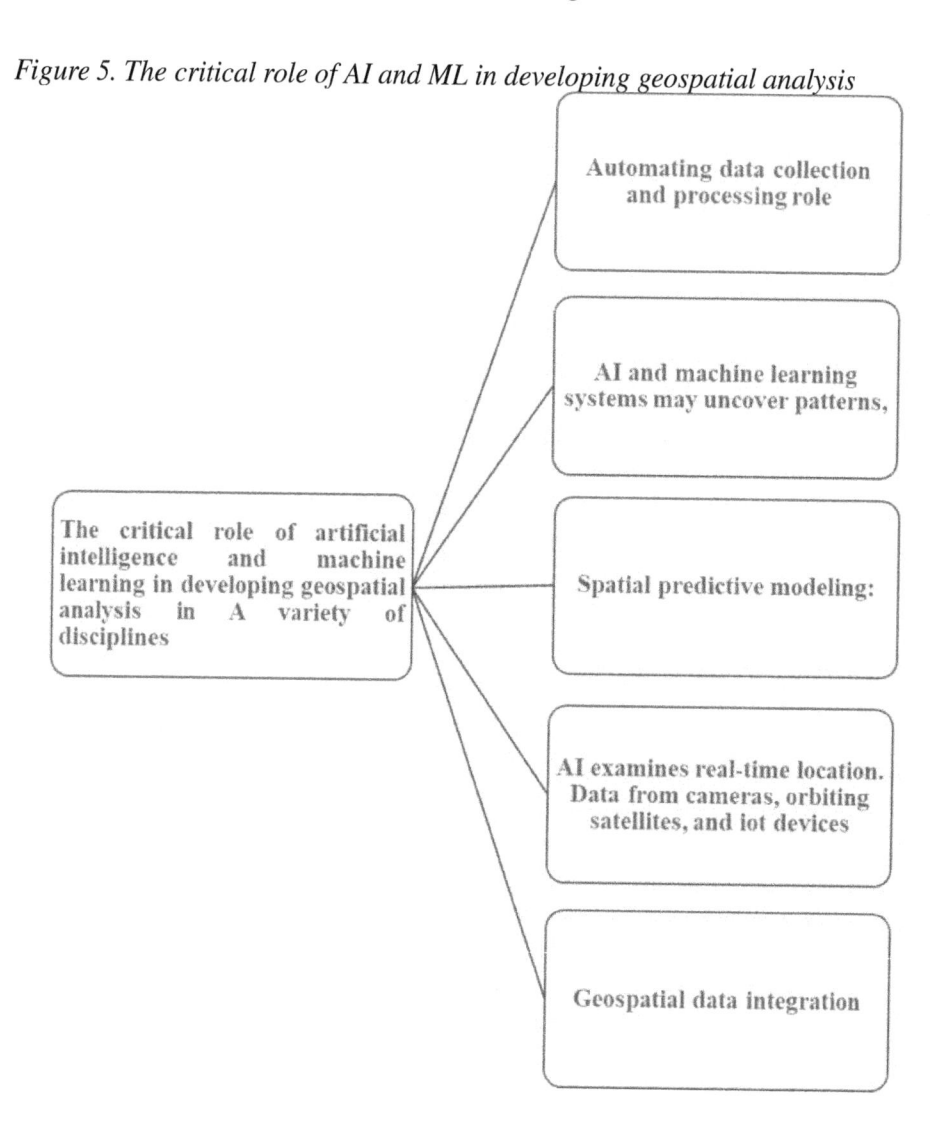

Figure 5. The critical role of AI and ML in developing geospatial analysis

- The critical role of artificial intelligence and machine learning in developing geospatial analysis in A variety of disciplines
  - Automating data collection and processing role
  - AI and machine learning systems may uncover patterns,
  - Spatial predictive modeling:
  - AI examines real-time location. Data from cameras, orbiting satellites, and iot devices
  - Geospatial data integration

5.1 Automating Data Collection and Processing Role

AI-powered automation facilitates the collecting of geospatial data from satellite photos, drones, LiDAR, and other remote sensing technologies. Machine learning algorithms filter and sanitize this data for future analysis, eliminating human error and manual work.

Applications include Monitoring environmental changes (such as deforestation and urbanization), Infrastructure Development Tracking, and Agricultural yield monitoring.

5.2 AI and Machine Learning Systems May Uncover Patterns

Characteristics and abnormalities in geographical data that is difficult to notice manually. CNNs are commonly used for image categorization, including land-use mapping and item recognition. It's Applications include identifying highways, buildings, and other features in satellite photography. Predicting natural dangers (such as landslides and floods). Identifying patterns in human mobility

5.3 Spatial Predictive Modeling

ML models anticipate future spatial occurrences using previous data. By combining temporal and geographical data, machine learning enhances model predictive capabilities, allowing for reliable forecasting of numerous occurrences. The applications in which spatial predictive modeling is used like Forecasting the Spread of Forest Fires, Urban growth projection, and Risk Assessment for Disaster Management etc.

5.4 AI Examines Real-Time Location

Data from cameras, orbiting satellites, and IoT devices. This enables faster decision-making in a variety of disciplines, including transportation, emergency response, and urban planning. The applications by which AI examines real time location such as Traffic flow monitoring, Smart City Infrastructure, Monitoring weather trends for real-time forecasting.

5.5 Geospatial Data Integration

AI integrates several geospatial datasets from various sensors or sources to provide a comprehensive view of an area of interest. This is especially useful for producing high-precision maps and undertaking multidisciplinary research. Combining topography, demographic, and climate data, Multi-layer analysis for urban planning, and Ecosystem modeling and conservation planning applications which are used for geospatial data integration.

6. REVOLUTIONARY APPROACH OF AI AND ML

The combination of AI and machine learning has resulted in significant advances, notably in the collection, processing, and analysis of geographical data. These technologies enable the automation of data extraction from a variety of sources, including satellites, drones, and sensors, considerably lowering manual work and improving accuracy (Bhambri & Bajdor, 2024a). Here are some important ways AI and machine learning have changed geospatial analysis ("Flood prediction in Nigeria using ensemble machine learning techniques," 2023).

6.1 Automatic Data Collection

AI-powered systems can automatically collect data from remote sensing platforms like satellites and drones. These systems can capture massive volumes of high-resolution data in real time, providing a steady stream of information for processing.

6.2 Data Preprocessing

Machine learning models may clean and preprocess raw geographic data by identifying and correcting abnormalities, reducing noise, and converting it into a useful format. This procedure considerably increases the data's quality and dependability.

6.3 Pattern Recognition

AI and machine learning systems excel in recognizing complex patterns in geographical data. For example, they may discern between land use categories, such as urban regions and woods, or detect bodies of water in satellite data.

6.4 Predictive Modeling

By analyzing past geographical data, ML models may estimate future spatial phenomena such as urban expansion, deforestation, and the impact of natural catastrophes. This promotes proactive decision-making and planning.

6.5 Real-Time Analysis

The combination of AI with IoT sensors allows for real-time monitoring of urban infrastructure, traffic patterns, environmental changes, and catastrophe management. AI systems can continually evaluate incoming data and deliver real-time insights, which is essential for applications such as traffic management and disaster response.

7. NEXT TRANSFORMATIONAL STAGE OF GEOSPATIAL ANALYSIS

The combination of AI and ML has immense potential to catapult geospatial analysis into the next transformational stage. This new age will provide more sophistication, accuracy, and scalability, allowing for deeper insights into geographical data. Here's how artificial intelligence and machine learning are impacting the future of geospatial analysis

7.1 Hyper-accurate spatial data interpretation

AI and machine learning can dramatically increase the accuracy of geographical data interpretation. With powerful pattern recognition algorithms, these technologies can recognize minor land characteristics, detect changes over time, and accurately categorize regions, even in difficult terrain or urban contexts. This will allow for more accurate land-use mapping, environmental monitoring, and catastrophe risk assessment.

7.2 Scalable Real-Time Data Processing

Real-time data from IoT devices, satellites, and drones will power the next phase of geospatial analysis. AI/ML models excel at analyzing enormous volumes of data in real time, delivering fast, actionable insights into urban planning, resource management, and traffic monitoring. This skill is critical for smart city programs, which require real-time decision-making.

7.3 Automatic Data Annotation and Feature Extraction

Traditionally, geospatial data analysis required significant human labor for activities such as annotating satellite photos and extracting geographic characteristics. AI and machine learning are transforming this by automating these processes.

Advanced models can now automatically detect buildings, roads, woods, and bodies of water, saving time and money.

7.4 Predictive spatial modeling

ML models may use previous geographical data to predict future spatial occurrences with increasing accuracy. For example, they can anticipate urban development, flood-prone locations, and the effects of climate change on certain regions. This predictive capacity enables more proactive planning and efficient resource allocation.

7.5 Advanced Integration of Remote Sensing Technologies

Artificial intelligence and machine learning are boosting remote sensing technology, enabling for seamless integration with high-resolution satellite imaging, LiDAR, and multispectral data. These systems can evaluate several data sources simultaneously, resulting in a more complete picture of an area's spatial dynamics.

7.6 Geospatial AI to achieve environmental and sustainability goals

AI-driven geospatial technologies will be critical in addressing environmental issues. AI-enhanced geospatial analysis may assist governments and organizations in meeting sustainability goals by offering data-driven insights for informed policy choices, ranging from monitoring deforestation and illicit mining to projecting climate change implications.

7.7 The democratization of geospatial insights

Non-experts will have easier access to geospatial analysis thanks to AI and ML. Automated systems will enable enterprises, governments, and researchers to get insights from geographical data without requiring extensive technical skills. This democratization will enable new applications in areas like as agriculture, logistics, urban development, and environmental conservation.

8. PROMOTION OF ENTREPRENEURSHIP WITH AI AND ML

The combination of AI and ML is promoting entrepreneurship in both the public and commercial sectors in profound ways. Here are a few reasons why.

8.1 Innovation & New Business Models

AI and machine learning empower entrepreneurs to build previously unimaginable goods and services. For example, in healthcare, AI-powered diagnostic tools and predictive analytics enable entrepreneurs to provide creative solutions, whilst in finance, AI-driven fintech companies are transforming the way individuals manage their money.

8.2 The Automation of Jobs With AI

This enables entrepreneurs to concentrate on important growth areas. AI systems may now simplify tasks that formerly required human labor, such as data analysis, customer care, and marketing.

8.3 Cost Efficiency

Artificial intelligence saves operational costs in both the public and commercial sectors by automating repetitive work, streamlining supply chains, and improving decision-making processes. This allows young firms to operate at a lesser cost, while older enterprises may develop with fewer resources.

Governments are also embracing AI and ML to improve public services, fostering an environment in which companies may provide AI-driven solutions to government concerns, particularly in infrastructure, health, and public safety.

8.4 Data-Driven Decision Making

ML algorithms examine large volumes of data to identify patterns and trends, guiding strategic decisions. Entrepreneurs and government organizations may use these information to improve anything from consumer experiences to operational efficiency.

8.5 Personalization at Scale

In the private sector, AI-powered customization enables startups and major corporations to provide individualized products, services, and marketing campaigns. ML models can anticipate user preferences, allowing organizations to provide more tailored experiences that boost customer happiness and loyalty. AI may be used by public sector institutions to customize citizen services, increase public involvement, and provide more focused, efficient services.

8.6 Democratisation of Technology

Open-source AI platforms and cloud-based machine learning tools have increased access to advanced technologies. Entrepreneurs no longer need to make large initial commitments to use cutting-edge AI solutions. This allows startups to compete on a more equal footing with established enterprises.

Similarly, public sector bodies are increasingly working with private AI providers and startups to implement these technologies, promoting a mutually beneficial.

8.7 Public/Private Partnerships

Governments are increasingly encouraging entrepreneurship by offering AI-focused innovation initiatives, subsidies, and collaborations with firms. Public sector entities frequently aim to use AI and ML for infrastructure, governance, healthcare, and education, resulting in possibilities for AI entrepreneurs.

For example, the application of AI to public data analytics, urban planning, and e-governance systems has resulted in the development of various public-private partnerships.

8.8 Risk Mitigation

Both industries benefit from predictive analytics enabled by ML, which helps spot dangers before they occur. These technologies can help entrepreneurs develop solutions for efficiently managing financial, operational, and market risks.

9. SOLUTIONS FOR GLOBAL CHALLENGES THROUGH AI AND ML

The combination of AI and ML is enabling the creation of better, more resilient solutions to global challenges in a variety of areas. Here's how AI and ML help to address some of the most important global challenges:

9.1 Climate Change and Environmental Sustainability

Predictive Analytics in Climate Modeling: Artificial intelligence and machine learning are being used to improve climate models and anticipate environmental changes. By analyzing large datasets from satellites, weather stations, and sensors, these technologies can give more accurate forecasts of extreme weather occurrences, assisting governments and communities in preparing for natural catastrophes such as floods, storms, and droughts.

9.2 Healthcare

AI-powered technologies can analyze medical data to detect illnesses earlier, anticipate epidemics, tailor therapies, and improve diagnostic accuracy. Radiology, genetics, and drug discovery are already utilizing machine learning models to accelerate breakthroughs and enhance patient outcomes.

9.3 Food Security

Artificial intelligence aids in agricultural activity monitoring, crop production prediction, and supply chain optimization. Machine learning algorithms can predict insect infestations, weather fluctuations, and agricultural illnesses, allowing farmers to act more proactively and decrease food wastage.

9.4 Disaster Response and Humanitarian Aid

AI and machine learning models can foresee disasters, allowing for faster and more efficient responses. They also assist organizations with logistics and distribution during catastrophes, ensuring that resources reach people in most need promptly.

9.5 Economic and social inequality

Artificial intelligence and machine learning can detect and evaluate patterns of inequality, such as biases in financial systems or access to education. This allows politicians to create more egalitarian programs and treatments that address structural challenges.

9.6 Cybersecurity and Fraud Detection

Artificial intelligence systems can identify anomalous patterns of behavior or weaknesses in real time, preventing cyberattacks, identity theft, and financial fraud. Machine learning models develop over time, becoming more robust to new and changing threats.

9.7 Sustainable Development

Artificial intelligence may improve efficiency in sectors such as transportation, urban planning, and energy usage. Self-driving cars, smart cities, and intelligent resource management systems may all help to decrease waste and maximize resources, hence advancing sustainable development goals.

10. ONGOING RESEARCH EFFORTS TO ADDRESS THESE CHALLENGES OF AI-POWERED GEOSPATIAL TOOLS

Ongoing research on AI-powered geospatial technologies aims to address global issues by improving the accuracy, scalability, and accessibility of geospatial data for a variety of applications. Some significant fields of research are (Taromideh et al., 2022).

10.1 Climate Monitoring and Environmental Protection

- **Satellite and Drone Data Analysis:** Researchers are working on improving AI models to analyze satellite and drone imagery for monitoring deforestation, tracking wildlife, and detecting illegal activities like poaching or mining. AI-powered geospatial tools help in understanding environmental changes over time by processing vast amounts of remote sensing data.
- **Predictive Climate Models:** AI and machine learning algorithms are being developed to predict climate-related events like hurricanes, floods, and droughts. By incorporating geospatial data, these models can help governments and organizations prepare for and mitigate the impact of climate
  ### 11.2 AI-Driven Crop Management
  Research aims to enhance farming methods with AI and GIS tools. By integrating satellite images, meteorological data, and machine learning models, AI may help monitor soil health, estimate agricultural yields, and improve irrigation system efficiency.
- **Pest and Disease Detection:** AI-powered geospatial technologies are also used to track and forecast the spread of pests and plant diseases. Early identification using satellite data assists farmers in mitigating hazards and improving food security.

10.2 Farming processes

- **AI-Driven Crop Management:** Researchers want to improve farming processes using AI and GIS tools. AI can assist monitor soil health, predict agricultural yields, and increase irrigation system efficiency by combining satellite imagery, meteorological data, and machine learning models.
- **Pest and Disease Detection:** AI-powered geospatial technologies are also utilized to monitor and anticipate pest and plant disease outbreaks. Early detection utilizing satellite data helps farmers mitigate risks and improve food security.

  11.4 AI-Driven Crop Management: Researchers want to improve farming processes using AI and GIS tools. AI can assist monitor soil health, predict agricultural yields, and increase irrigation system efficiency by combining satellite imagery, meteorological data, and machine learning models.

  11.5 Pest and Disease Detection: AI-powered geospatial technologies are also utilized to monitor and anticipate pest and plant disease outbreaks. Early detection utilizing satellite data helps farmers mitigate risks and improve food security.

  11.6 Urban Planning and Smart Cities
- **Traffic and Mobility Optimization:** Researchers are utilizing AI-powered geospatial technologies to improve urban mobility and alleviate traffic congestion. Real-time data from sensors, GPS, and satellite images are evaluated to improve public transportation systems, route planning, and emissions reduction in cities.
- **Infrastructure Monitoring:** AI algorithms use geographical data to monitor the state of roads, bridges, and buildings. These techniques aid in anticipating infrastructure wear and tear and recognizing repair needs, hence boosting urban resilience.

10.3 Disaster Response and Risk Management

- **AI for Disaster Prediction:** Current research focuses on enhancing AI models that anticipate and respond to natural catastrophes such as earthquakes, wildfires, and floods using geospatial data. AI-powered technologies are being utilized to develop early warning systems that leverage satellite images, seismic data, and machine learning to enhance reaction times and resource allocation.
- **Post-Disaster Damage Assessment:** Researchers are using AI-powered geospatial tools to swiftly estimate damage following a disaster. These tech-

niques use satellite and drone photos to offer real-time information on the degree of damage, enabling for more effective humanitarian help and resource allocation.

10.4 Biodiversity and Conservation Animal

- **Tracking and Habitat Mapping:** AI-powered geospatial techniques are used to monitor animal movements and ecosystem changes. This research focuses on developing AI models capable of processing satellite data to detect ecosystems, migratory patterns, and possible biodiversity concerns.
- **Ecosystem Service Valuation:** AI is being used to assess ecosystem services (such as carbon sequestration and water purification) by analyzing geographical data. These methods aid in determining the worth of natural resources and directing conservation activities.

10.5 Global Health and Epidemiology

- **Disease Surveillance and Outbreak Prediction:** AI-powered geospatial technologies are being created to monitor the spread of infectious illnesses and forecast future outbreaks. For example, satellite data on environmental elements such as temperature and humidity, paired with AI models, may be used to forecast the development of illnesses such as malaria and dengue fever.
- **Healthcare Resource distribution:** The research focuses on combining AI and geographical data to optimize the distribution of healthcare resources in neglected locations. AI models can assist governments in better planning for pandemics and distributing medical resources by assessing population density, infrastructure, and health information.

10.6 Water Resource Management

- **AI for Water Monitoring:** Ongoing research aims to improve AI models that use satellite images and sensor data to monitor water resources. These techniques aid in regulating water supplies, identifying pollution, and forecasting droughts. Researchers are developing more accurate and scalable models for real-time water management.
- **Flood Prediction and Mitigation:** Geospatial AI algorithms anticipate flood hazards based on rainfall patterns, geography, and river levels. The goal of research is to improve the precision of these models in order to offer early warnings and lessen the impact of floods on vulnerable populations.

10.7 Geospatial AI Ethics and Fairness

- **Bias Mitigation in Geospatial AI:** Geospatial AI models, like other AI applications, are susceptible to bias, especially when the data used to train them reflects past disparities. Researchers are working on approaches to detect and reduce biases in geospatial information in order to achieve more fair and equitable outcomes, particularly in resource allocation and disaster assistance.
- **Data Privacy and Security:** There are ongoing initiatives to resolve the privacy problems associated with the usage of sensitive geospatial data. Researchers are working on ethical frameworks and privacy-preserving AI algorithms that can evaluate geographical data while protecting individual privacy.
- **Sustainable Land Use Planning:** AI-powered geospatial technologies are being investigated to improve land-use planning that balances environmental conservation and human development objectives. These technologies assist with zoning choices, urban growth planning, and the development of sustainable infrastructure.
- **Monitoring Progress toward the SDGs:** Geospatial AI techniques are being used to track progress toward the United Nations' Sustainable Development Goals (SDGs), which include eliminating poverty and hunger and boosting access to clean water (Bhambri & Bajdor, 2024b). These applications use satellite and survey data to monitor development parameters in real time.

10.8 Key Research Challenges

- **Data Quality and Integration:** Cleaning and combining massive geographic information remains challenging. Researchers are working on strategies to combine data from many sources, such as satellite images, sensor networks, and social media, to construct more complete AI models.
- **Scalability with Real-Time Processing:** The sheer volume of data created makes it challenging to process and analyze geographical data in real time. Research is being performed to improve computing efficiency and scalability, allowing geospatial AI to be used in real-time disaster response and environmental monitoring.
- **Access to High-Quality Geospatial Data:** Many regions, particularly those with low incomes, lack access to high-quality geospatial data. Ongoing research aims to democratize access to these datasets through projects such as open-source geospatial platforms, which will enable more communities to benefit from AI-powered technologies. These ongoing research projects seek to increase the usefulness of AI-powered geospatial technologies in tackling

some of the world's most serious crises, as well as issues of data accessibility, prejudice, and ethics.

11. IMPORTANCE OF INTERDISCIPLINARY COLLABORATION OF AI-POWERED GEOSPATIAL TOOLS

The development and successful implementation of AI-powered geospatial applications rely largely on multidisciplinary cooperation. Interdisciplinary teams may handle the technological, social, and environmental difficulties of geospatial AI applications by bringing together experts from many sectors. Here's why this partnership is important (Sudarmadji & Santoso, 2023; Petrocchi et al., 2024).

11.1 Integration of Different Expertise

- **AI and Data Science:** AI and machine learning professionals create algorithms for analyzing large volumes of geographical data. Their experience is critical for improving models for tasks such as image recognition, predictive analytics, and automation.
- **Geographical Science:** Geographers, cartographers, and remote sensing professionals provide domain-specific expertise of spatial patterns, data gathering methodologies, and geospatial analytic approaches. Their insights are critical for ensuring that AI systems draw correct and relevant conclusions from geographical data.
- **Ecologists and environmental scientists** offer vital expertise about ecosystems, biodiversity, and natural resource management to geospatial AI applications used in environmental monitoring, climate change mitigation, and conservation activities.
- **Urban planners and civil engineers** give context for how AI-powered solutions might improve transit, land use, and resource allocation in cities. Their collaboration guarantees that AI models are feasible in real-world scenarios.
- **Healthcare and Epidemiology:** Interdisciplinary work with healthcare professionals and epidemiologists guarantees that AI-powered models handle real-world difficulties and enhance population health outcomes in applications such as public health, disease monitoring, and disaster response.

11.2 Improving Accuracy and Applicability

- **Contextual Knowledge:** AI models frequently require fine-tuning to reflect the characteristics of the region or task at hand. By working across disci-

plines, specialists may add domain knowledge into models, making them more accurate and appropriate to local situations. For example, including ecological insights aids in the training of AI models to better interpret environmental data and wildlife migration patterns.
- **Validating the AI Results:** Interdisciplinary collaboration enables the cross-validation of AI-driven discoveries. For example, environmental scientists can test a machine learning model that predicts deforestation patterns based on satellite images using on-the-ground data. This guarantees that the AI tool is both data-driven and educated by real-world observations.

11.3 Addressing Ethics and Social Implications (Arellano et al., 2023)

- **Ethical AI Use:** Working with social scientists, ethicists, and legal experts is critical for addressing the ethical implications of AI-powered geospatial products. These systems frequently gather and analyze sensitive data, raising questions about privacy, spying, and prejudice. Ethical frameworks and rules are crucial for ensuring that AI applications are utilized properly, especially in disaster response and public health.
- **Equitable Access and Use:** Interdisciplinary teams guarantee that geospatial AI technologies are inclusive and usable. Social scientists, anthropologists, and human rights activists should emphasize how these tools may affect disadvantaged or vulnerable populations, ensuring that their development and use benefit all rather than worsening existing disparities.

11.4 Resolving Complex Global Challenges

- **Climate change** is a multidimensional topic that requires the knowledge of meteorologists, environmental scientists, AI developers, and legislators. AI-powered geospatial tools for climate monitoring and catastrophe prediction become far more successful when they are guided by a shared knowledge of climate systems, environmental regulations, and computational models.
- **Disaster Response and Risk Management Interdisciplinary teamwork** facilitates more efficient catastrophe response. AI models that forecast and analyze the effect of natural catastrophes must collaborate closely with emergency response teams, geospatial analysts, and humanitarian groups. This guarantees that data is used efficiently, projections are correct, and resources are allocated on time, saving lives and reducing damage.

11.5 Innovation and Knowledge Share
New Ideas and Approaches

- **Collaboration across fields promotes creativity.** For example, an urban planner who collaborates with AI developers and geospatial scientists may create new tools for improving city infrastructure based on data-driven insights such as traffic patterns or energy use. Such multidisciplinary connection allows for the exchange of ideas, resulting in creative solutions to challenging issues.
- **Open Data and Shared Resources:** Satellites and drones, as well as social media platforms and public documents, are common sources of geospatial data collection. Interdisciplinary collaboration promotes the exchange of datasets and procedures, therefore standardizing data gathering, processing, and analysis. This collaborative atmosphere also encourages the creation of open-source geospatial platforms, allowing academics and practitioners from diverse domains to contribute and benefit from shared

11.6 Improving Public Policy and Decision Making

- **Data-Driven Policy:** Policymakers use accurate and trustworthy data to make educated decisions in urban planning, public health, and environmental protection. Interdisciplinary collaboration guarantees that AI-powered geospatial tools give actionable insights customized to decision-makers' requirements, bridging the gap between technical understanding and real-world policy execution.
- **Public Engagement:** AI-powered geospatial technologies may be used to involve the public in matters such as disaster preparedness, environmental protection, and community development. Collaboration with communication specialists, educators, and public engagement professionals ensures that these tools are accessible, intelligible, and successful for informing and empowering communities.

11.7 Long-term Sustainability and Scalability

- **The Sustainable Development Goals (SDGs):** AI-powered geospatial technologies are critical for monitoring and promoting the SDGs, which include poverty reduction, access to clean water, and environmental protection. Interdisciplinary teams may use AI, geographical data, and policy experience to track progress, assess effect, and scale sustainable solutions internationally.

- **Long-term Impact:** Solving global problems need solutions that are not just technically sound but also socially and environmentally responsible. Interdisciplinary collaboration guarantees that AI-powered geospatial solutions are designed and implemented with a long-term perspective, taking into account economic, social, and environmental considerations.

12. ETHICAL CONSIDERATIONS IN THE FUTURE DEVELOPMENT OF AI-POWERED GEOSPATIAL TOOLS

The future development of AI-powered geospatial tools involves a number of ethical concerns that must be addressed to guarantee that these technologies are utilized ethically and equitably. As these technologies become more widely used in crucial sectors such as disaster response, urban planning, environmental monitoring, and public health, it is necessary to address their societal implications. Here are some important ethical issues in the creation of AI-powered geospatial tools (Radočaj et al., 2021).

13. PRIVACY AND DATA SECURITY

- **Sensitive Data Collection:** AI-powered geospatial technologies frequently rely on massive volumes of location-based data, which may contain sensitive information about persons, communities, and private property. Individual and community privacy is critical, especially when tracking movements, assessing infrastructure, or mapping health data.
- **Informed Consent:** The gathering and use of geospatial data should prioritize informed consent. Individuals and communities should understand how AI systems acquire, retain, and use personal data, especially in sensitive situations such as public health or humanitarian relief.
- **Data Anonymization:** It is critical to balance the benefits of geographic AI applications with the security of individual identities. Developers must guarantee that data is anonymized and safeguarded to avoid the exploitation of personal or location-specific information, which might result in privacy violations or monitoring.

13.1 Bias and Fairness ("Ethical considerations in artificial intelligence development," 2024; R & P, 2024)

- **Algorithmic Bias:** AI algorithms can inherit biases from the data on which they are taught, resulting in unfair or discriminating outputs in geographical applications. For example, AI-powered solutions for urban planning or resource allocation may accidentally favor more affluent locations due to skewed statistics. Biases must be identified and mitigated to guarantee fair and equal outcomes for all populations.
- **Data Representation:** Some locations or communities, especially those in poor nations or rural areas, may have limited access to high-quality geospatial data. This can lead to underrepresentation in AI models, as well as policies and choices that fail to address the concerns of underrepresented people. Ethical development must prioritize inclusive data collection and equitable access to AI-powered tools.
- **Impact on Vulnerable Communities**: AI-powered geospatial tools can have unintended negative consequences for marginalized or vulnerable communities, especially in areas such as disaster response or urban development. Developers must ensure that these tools do not exacerbate existing inequalities or create new forms of discrimination.

13.2 Transparency and Accountability

- **Explainability of AI Models:** Many AI-powered products, particularly those based on deep learning, behave like "black boxes," making it difficult to grasp how choices are made. Transparency in how AI models analyze geographic data is critical for accountability, particularly when these tools are employed for high-risk decision-making, such as disaster response or environmental control.
- **Accountability in Decision Making:** Clear lines of responsibility must be created when AI technologies are employed in domains such as urban planning, public safety, or humanitarian assistance. If an AI model makes a poor forecast, such as underestimating flood threats or misallocating resources during a crisis, it must be clear who is accountable for dealing with the consequences—whether it is the developers, data suppliers, or decision-makers who

13.3 Informed Decision-Making and Public Participation.

- **Public Engagement:** AI-powered geospatial solutions should not only benefit specialists and decision-makers, but also engage the general public in meaningful ways. For example, people impacted by urban expansion or environmental changes should be engaged in talks about how AI-driven geospatial data influences decision-making. Public feedback can help AI systems better connect with the needs and values of local communities.
- **Data Interpretation:** It is vital not to rely too much on AI models, especially when the stakes are high, as in disaster assistance or environmental management. While AI-powered geospatial tools might provide useful insights, they should only supplement, not replace, human judgment and knowledge. Decision-makers must stay critical and knowledgeable about these technologies, including their limitations and potential for mistake.

14. ENVIRONMENTAL AND SOCIETAL IMPACT

- **Sustainability:** The usage of geospatial AI technologies should be consistent with long-term development goals while reducing environmental impact. For example, processing massive datasets or running complicated AI models can leave a considerable environmental imprint. Developers should emphasize energy-efficient algorithms and evaluate their tools' environmental effect, particularly when working on climate-related topics.
- **Social Consequences of Geospatial Monitoring** AI-powered geospatial technologies have the potential to transform society, especially in areas such as land use planning, resource management, and surveillance. Ethical foresight is required to predict and minimize any negative social consequences such as community relocation, gentrification, and loss of privacy as a result of widespread surveillance.
- **Balancing Commercial and Public Interests:** There is an expanding commercial market for geospatial data, with AI businesses providing location-based services to a variety of sectors. However, the public interest must be protected, ensuring that the creation of these instruments emphasizes social benefit over profitability. Ethical norms must be developed to prevent commercial exploitation of geospatial data at the price of privacy, public safety, or environmental protection.

14.1 Global Inequities and Data Sovereignty

- **Digital colonialism** occurs when businesses or governments with considerable technological and financial resources build and manage AI-powered geospatial technologies. This can lead to digital colonialism, in which poor countries rely on other companies for data and AI technologies, giving up sovereignty over how their land, resources, and inhabitants are monitored and governed. Ensuring data sovereignty—the freedom of nations and communities to manage their own geospatial data—is vital to sustaining equity and justice.
- **Global inequity in access to AI tools:** Not every place has equal access to the technology infrastructure required to use AI-powered geospatial applications. Bridging this digital divide is an ethical duty to guarantee that the advantages of modern technologies are worldwide spread and that underdeveloped nations can use

14.2 Long-term Consequences and the Precautionary Principle

- **Unintended consequences:** The long-term social impact of AI-powered geospatial technologies is yet unknown. Ethical development must adopt a precautionary approach, foresee potential negative consequences—such as over-reliance on AI in crucial industries like agriculture or disaster management—and planning for scenarios in which these technologies fail or have unexpected results.
- **Sustainability of AI Tools:** Long-term sustainability must be considered while developing AI-powered geospatial solutions. This involves thinking about how tools will be maintained, updated, and changed over time, as well as how to keep them useful as technology and social demands change.

15. CONCLUSION

The conclusion of this chapter is that the use of artificial intelligence into geospatial analysis marks a significant step forward in geographical data collection, processing, and utilization. Machine learning is transforming geospatial research by automating data-driven processes, revealing new insights from complicated geographical data, and making predictions more accurate. As geospatial data grows in size and complexity, the role of machine learning (ML) will become increasingly important, providing new answers to crucial difficulties in urban planning, environmental management, disaster preparedness, and other areas. Geospatial analysis is a critical tool for comprehending and controlling the physical world, allowing stakeholders to make informed decisions on anything from urban growth to environmental preservation. Deep learning also improves our capacity to monitor, evaluate, and make choices about our surroundings by integrating the power of deep neural networks with geographic data, particularly in sectors like environmental management, urban planning, disaster response, and remote sensing. Remote sensing is an important aspect of geospatial analysis because it provides sophisticated methods for gathering, processing, and evaluating spatial data. Researchers and decision-makers can gain valuable insights into various geographic phenomena by combining remote sensing with GIS and other analytical techniques, allowing them to make more informed decisions in fields such as environmental management, urban planning, agriculture, and disaster response. Predictive analytics is an effective technique for geospatial analysis, offering useful insights into spatial trends and future occurrences. Big data greatly improves geospatial analysis, allowing firms to get important insights from large and diverse datasets. Decision-makers may handle complex issues across several sectors, improve resource management, and, ultimately, improve community quality of life by combining big data technology with geospatial approaches. Interdisciplinary collaboration bridges the gap between technological innovation and practical application, ensuring that AI solutions meet the world's most urgent concerns in an accurate, ethical, sustainable, and egalitarian manner. The ethical issues surrounding the creation of AI-powered geospatial technologies are complex and demand continual study.

REFERENCES

Ahnaf, M. M., Rafizul, I. M., & Shuvo, M. B. (2023). Development of water quality indices for the assessment of groundwater quality: A case study of tubewells adjacent to the waste landfill in Khulna city. *AIP Conference Proceedings*, 2713, 060002. DOI: 10.1063/5.0129962

Anand, A., Batra, G., & Uitto, J. I. (2024). Harnessing Geospatial approaches to strengthen evaluative evidence. *Artificial Intelligence and Evaluation*, 196-218. DOI: 10.4324/9781003512493-10

Arellano, L., Alcubilla, P., & Leguízamo, L. (2023). *Ethical considerations in informed consent*. Ethics - Scientific Research, Ethical Issues, Artificial Intelligence and Education. [Working Title], DOI: 10.5772/intechopen.1001319

Bhambri, P., & Bajdor, P. (Eds.). (2024a). *Handbook of Technological Sustainability: Innovation and Environmental Awareness* (1st ed., p. 412). CRC Press., DOI: 10.1201/9781003475989

Bhambri, P., & Bajdor, P. (2024b). Technological Sustainability Unveiled: A Comprehensive Examination of Economic, Social, and Environmental Dimensions. In Bhambri, P., & Bajdor, P. (Eds.), *Handbook of Technological Sustainability: Innovation and Environmental Awareness* (pp. 80–98). CRC Press., DOI: 10.1201/9781003475989-8

Chithra, N., & Bhambri, P. (2024). Ethics in Sustainable Technology. In Bhambri, P., & Bajdor, P. (Eds.), *Handbook of Technological Sustainability: Innovation and Environmental Awareness* (pp. 245–256). CRC Press., DOI: 10.1201/9781003475989-21

El Baba, M., Kayastha, P., Huysmans, M., & De Smedt, F. (2020). Groundwater vulnerability and nitrate contamination assessment and mapping using drastic and Geostatistical analysis. *Water (Basel)*, 12(7), 2022. DOI: 10.3390/w12072022

Eseosa Halima, I., & Hiroaki, S. (2022). Assessing the disparities of the population exposed to flood hazards in Nigeria. *IOP Conference Series. Earth and Environmental Science*, 1016(1), 012007. DOI: 10.1088/1755-1315/1016/1/012007

Ethical considerations in artificial intelligence development. (2024). Filosofiya Referativnyi Zhurnal, (1). DOI: 10.31249/rphil/2024.01.03

Flood prediction in Nigeria using ensemble machine learning techniques. (2023). *Ilorin Journal of Science, 10*(1). DOI: 10.54908/iljs.2023.10.01.004

Geetha, K., Vigneshwari, J., Bhambri, P., & Thangam, A. (2024). Sustainable Solutions for Global Waste Challenges: Integrating Technology in Disposal and Treatment Methods. In Bhambri, P., & Bajdor, P. (Eds.), *Handbook of Technological Sustainability: Innovation and Environmental Awareness* (pp. 46–56). CRC Press., DOI: 10.1201/9781003475989-5

Gupta, M., & Pandya, S. D. (2024). Predictive modeling of recruitment and selection in campus placement using machine learning algorithms. DOI: 10.2139/ssrn.4862743

Ighile, E. H., Shirakawa, H., & Tanikawa, H. (2022). Application of GIS and machine learning to predict flood areas in Nigeria. *Sustainability (Basel)*, 14(9), 5039. DOI: 10.3390/su14095039

Jurišić, M., Radočaj, D., Plaščak, I., & Rapčan, I. (2022). A UAS and machine learning classification approach to suitability prediction of expanding natural habitats for endangered flora species. *Remote Sensing (Basel)*, 14(13), 3054. DOI: 10.3390/rs14133054

Kibona, I., Mkoma, S., & Mjemah, I. (2011). Nitrate pollution of Neogene alluvium aquifer in Morogoro municipality, Tanzania. *International Journal of Biological and Chemical Sciences*, 5(1). Advance online publication. DOI: 10.4314/ijbcs.v5i1.68095

Li, H., & Bao, J. (2021). Uncertainties in the surface layer physics parameterizations. *Uncertainties in Numerical Weather Prediction*, 229-236. DOI: 10.1016/B978-0-12-815491-5.00008-2

Mkumbo, N. J., Mussa, K. R., Mariki, E. E., & Mjemah, I. C. (2022). The use of the DRASTIC-LU/LC model for assessing groundwater vulnerability to nitrate contamination in Morogoro municipality, Tanzania. *Earth (Basel, Switzerland)*, 3(4), 1161–1184. DOI: 10.3390/earth3040067

Mohammadi, B. (2022). Application of machine learning and remote sensing in hydrology. *Sustainability (Basel)*, 14(13), 7586. DOI: 10.3390/su14137586

Mohana Sundari, V., Ganeshkumar, M., & Bhambri, P. (2024). Environmental Stewardship in the Digital Age: A Technological Blueprint. In Bhambri, P., & Bajdor, P. (Eds.), *Handbook of Technological Sustainability: Innovation and Environmental Awareness* (pp. 57–67). CRC Press., DOI: 10.1201/9781003475989-6

N, D., & J, N. (2023). Review on malware classification with a hybrid deep learning. *Journal of IoT and Machine Learning*, 18-21. DOI: 10.48001/joitml.2023.1118-21

N, D., & J, N. (2023). Review on malware classification with a hybrid deep learning. *Journal of IoT and Machine Learning*, 18-21. DOI: 10.48001/joitml.2023.1118-21

Nayak, D., Surve, N., & Shrivastava, P. (2022). Assessing land use and land cover changes in south Gujarat. *Ecology. Environmental Conservation*, 28(04), 2110–2115. DOI: 10.53550/EEC.2022.v28i04.070

Petrocchi, E., Tiribelli, S., Paolanti, M., Giovanola, B., Frontoni, E., & Pierdicca, R. (2024). GeomEthics: Ethical considerations about using artificial intelligence in Geomatics. *Lecture Notes in Computer Science*, 14366, 282–293. DOI: 10.1007/978-3-031-51026-7_25

R, G., & P, D. L. (1693-1696). R, G. R., & P, L. (2024). Ethical considerations in artificial intelligence development. *International Journal of Research Publication and Reviews*, 5(6), 1693–1696. Advance online publication. DOI: 10.55248/gengpi.5.0624.1453

Radočaj, D., & Jurišić, M. (2022). GIS-based cropland suitability prediction using machine learning: A novel approach to sustainable agricultural production. *Agronomy (Basel)*, 12(9), 2210. DOI: 10.3390/agronomy12092210

Radočaj, D., Jurišić, M., Gašparović, M., Plaščak, I., & Antonić, O. (2021). Cropland suitability assessment using satellite-based biophysical vegetation properties and machine learning. *Agronomy (Basel)*, 11(8), 1620. DOI: 10.3390/agronomy11081620

Rana, A. (2022). Land use and land cover change mapping: A Spatio temporal and correlational analysis of Ramganjmandi Tehsil, Kota, Rajasthan, India. *Ecology. Environmental Conservation*, •••, 1384–1389. DOI: 10.53550/EEC.2022.v28i03.040

Rana, A., & Sharma, R. (2020). Drinking water quality assessment and predictive mapping: Impact of Kota stone mining in Ramganjmandi Tehsil, Rajasthan, India. *Nature Environment and Pollution Technology*, 19(3), 1219–1225. DOI: 10.46488/NEPT.2020.v19i03.036

Rana, R., & Bhambri, P. (2024a). Environmental Challenges and Technological Solutions. In Bhambri, P., & Bajdor, P. (Eds.), *Handbook of Technological Sustainability: Innovation and Environmental Awareness* (pp. 187–200). CRC Press., DOI: 10.1201/9781003475989-17

Rana, R., & Bhambri, P. (2024b). Ethical Considerations in Artificial Intelligence for Environmental Solutions: Striking a Balance for Sustainable Innovation. In Bhambri, P., & Bajdor, P. (Eds.), *Handbook of Technological Sustainability: Innovation and Environmental Awareness* (pp. 389–396). CRC Press., DOI: 10.1201/9781003475989-31

Rinaldi, F. M., & Nielsen, S. B. (2024). Artificial intelligence. *Artificial Intelligence and Evaluation*, 287-308. DOI: 10.4324/9781003512493-14

Ruby, S., Biju, T., & Bhambri, P. (2024). Catalysing Sustainable Progress: Empowering MSMEs through Tech Innovation for a Bright Future. In Bhambri, P., & Bajdor, P. (Eds.), *Handbook of Technological Sustainability: Innovation and Environmental Awareness* (pp. 374–388). CRC Press., DOI: 10.1201/9781003475989-30

Rustamov, J., Rustamov, Z., & Zaki, N. (2023). Green space quality analysis using machine learning approaches. *Sustainability (Basel)*, 15(10), 7782. DOI: 10.3390/su15107782

Rustamov, Z., Rustamov, J., Zaki, N., Turaev, S., Sultana, M. S., Tan, J. Y., & Balakrishnan, V. (2023). Enhancing cardiovascular disease prediction: A domain knowledge-based feature selection and stacked ensemble machine learning approach. DOI: 10.21203/rs.3.rs-3068941/v1

Sharma, R., & Bhambri, P. (2024). Digital Duplicity and the Disintegration of Trust: A Quantitative Inquiry into the Impact of Deep Fakes on Media Sustainability and Societal Equilibrium. In Bhambri, P., & Bajdor, P. (Eds.), *Handbook of Technological Sustainability: Innovation and Environmental Awareness* (pp. 273–291). CRC Press., DOI: 10.1201/9781003475989-24

Stuss, M., & Fularski, A. (2024). Ethical considerations of using artificial intelligence (AI) in recruitment processes. *Edukacja Ekonomistów i Menedżerów*, 71(1). Advance online publication. DOI: 10.33119/EEIM.2024.71.4

Sudarmadji, S., & Santoso, S. (2023, April). The role of nanofluid and ultrasonic vibration in coolant radiator. In *AIP Conference Proceedings* (Vol. 2531, No. 1). AIP Publishing.

Taromideh, F., Fazloula, R., Choubin, B., Emadi, A., & Berndtsson, R. (2022). Urban flood-risk assessment: Integration of decision-making and machine learning. *Sustainability (Basel)*, 14(8), 4483. DOI: 10.3390/su14084483

Thirumalaiyammal, B., Steffi, P. F., & Bhambri, P. (2024). Green Horizons: Navigating Environmental Challenges through Technological Innovation. In Bhambri, P., & Bajdor, P. (Eds.), *Handbook of Technological Sustainability: Innovation and Environmental Awareness* (pp. 292–304). CRC Press., DOI: 10.1201/9781003475989-25

Tohidi, N., & Rustamov, R. B. (2020). A review of the machine learning in gis for megacities application. *Geographic Information Systems in Geospatial Intelligence*, 29-53.

Vigneshwari, J., Senthamizh Pavai, P., Maria Suganthi, L., & Bhambri, P. (2024). Eco-ethics in the Digital Age: Tackling Environmental Challenges through Technology. In Bhambri, P., & Bajdor, P. (Eds.), *Handbook of Technological Sustainability: Innovation and Environmental Awareness* (pp. 201–213). CRC Press., DOI: 10.1201/9781003475989-18

Wei, A., Bi, P., Guo, J., Lu, S., & Li, D. (2021). Modified drastic model for groundwater vulnerability to nitrate contamination in the Dagujia river basin, China. *Water Science and Technology: Water Supply*, 21(4), 1793–1805. DOI: 10.2166/ws.2021.018

Yang, C. (2024). Application and assessment of GIS technology in flash flood risk management. *Sustainable Environment*, 9(1), 26. DOI: 10.22158/se.v9n1p26

Yohana, A. R., Makoba, E. E., Mussa, K. R., & Mjemah, I. C. (2023). Evaluation of groundwater potential using aquifer characteristics in Urambo district, Tabora region, Tanzania. *Earth (Basel, Switzerland)*, 4(4), 776–805. DOI: 10.3390/earth4040042

York, P., & Bamberger, M. (2024). The applications of big data to strengthen evaluation. *Artificial Intelligence and Evaluation*, 37-55. DOI: 10.4324/9781003512493-3

KEY TERMS AND DEFINITIONS

Artificial Intelligence: The use of artificial intelligence into geospatial analysis marks a significant step forward in geographical data collection, processing, and utilization.

Big Data: In geospatial analysis, there is required large data sets from satellites, remote sensors, and social media. It will get through Big Data with the help of AI and ML.

Deep Learning: It is subset of ML which uses neural networks with multiple layers. It is effective for classification of image, characteristics extraction from remote sensing, and segmentation in geospatial analysis.

Geospatial Data: For the geospatial analysis, data includes geographical components like latitude and longitude. It is used in tracking, monitoring, and mapping the environment.

Machine Learning: Machine learning is transforming geospatial research by automating data-driven processes, revealing new insights from complicated geographical data, and making predictions more accurate.

Predictive Analytics: It is used for future prediction of pattern changes in geospatial analysis like weather forecasting, expansion of urban area, and natural disasters.

Remote Sensing: With the use of remote sensing, data is collected about the Earth's surface, which is often required for geospatial analysis. For this optical, radar, and infrared images can included in remote sensing.

Chapter 3
Harnessing AI in Geospatial Technology for Environmental Monitoring and Management:
Applications of AI for Geospatial Data Processing

Monica Gupta
https://orcid.org/0000-0001-6756-4191
Bharati Vidyapeeth's College of Engineering, India

Rupanshi Bhatnagar
https://orcid.org/0009-0005-1790-9442
Ernst & Young, India

ABSTRACT

The chapter explores the role of Geospatial Technology in processing, and analyzing geographically-referenced data from various sources, such as satellite imagery, sensor networks, and climate models. Geospatial Technology tools like GPS, remote sensing, GIS, and LiDAR have proven invaluable in areas like urban planning, disaster management, and environmental conservation. These technologies provide real-time, accurate geographic data, enabling organizations to make informed decisions. Geospatial Technology is widely used in urban development for optimizing infrastructure, tracking deforestation, and monitoring biodiversity. In disaster management, it supports early warning systems and enhances coordi-

DOI: 10.4018/979-8-3693-8054-3.ch003

nation during crisis response. Additionally, it helps manage natural resources and monitor agricultural productivity. The chapter highlights the evolution of Geospatial Technology, emphasizing its growing importance in environmental monitoring and resource management, while showcasing how its applications continue to expand across sectors such as defense, urban planning, and agriculture.

INTRODUCTION

Geospatial data refers to information linked to specific geographic locations and is primarily categorized into vector and raster data. Vector data represents geographic features using points, lines, and polygons, each defined by coordinates, enabling precise representation of elements like roads, boundaries, and landmarks. Raster data, by contrast, represents information as digital images, such as scanned maps, aerial photos, or satellite imagery. Converting vector data to raster data is a crucial process in geospatial analysis. First, an original image is captured, with points marked to identify distinct regions or features. A grid is overlaid, dividing the image into cells, and each cell is assigned a value based on the vector data it covers, transforming points, lines, and polygons into a pixel-based raster format. This creates a grid where each cell corresponds to a specific portion of the vector data. For more clarity, the rasterization process is illustrated in Figure 1. It begins with an input vector image (Figure 1(a)), then mapping a vector data onto a grid (Figure 1(b)) by assigning each cell a value based on corresponding vector attributes, and ultimately producing a raster image (Figure 1(c)).This conversion allows for the integration of various data sources and supports analyses like spatial modeling and remote sensing. Additionally, raster data is vital for visualization in Geographic Information Systems (GIS), as it presents complex phenomena in a user-friendly format for applications such as environmental monitoring, urban planning etc.

Figure 1. (a) Rasterization: Conversion of a Vector image into Raster image (a) Vector Image (b) Rasterization: Conversion of a Vector image into Raster image (b) Vector Image with a grid (c) Rasterization: Conversion of a Vector image into Raster image (c) Raster Image

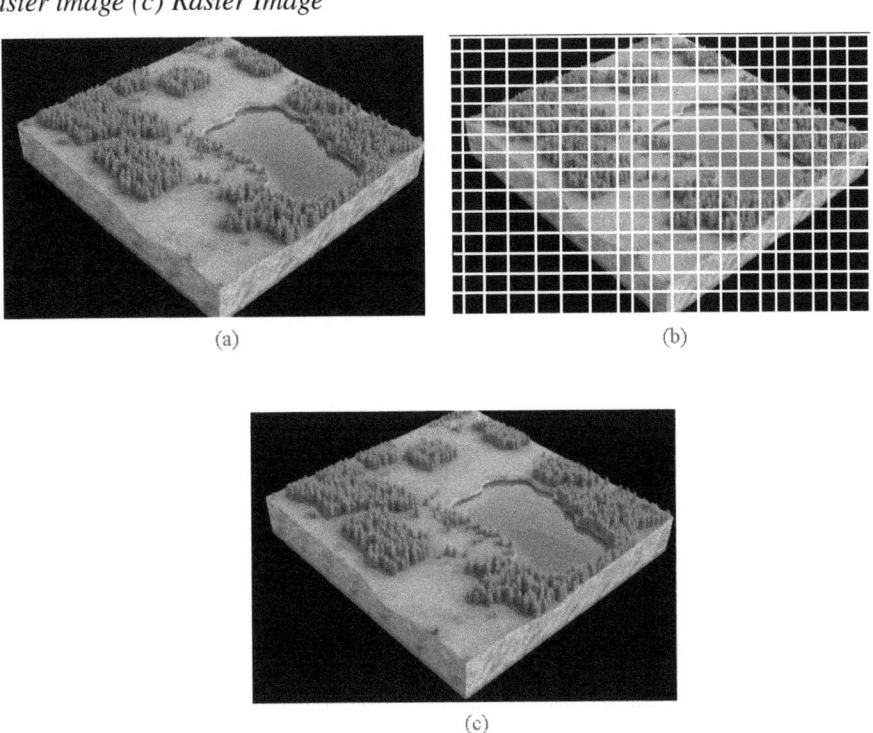

Geospatial Technology (GT) is an interdisciplinary field that integrates various tools and techniques to collect, process, and analyze geographically referenced information. Data is gathered from diverse sources such as satellite imagery, sensor networks, aerial photography, weather and climate models, and even social media platforms. These data streams are then analyzed using advanced technologies, including the Global Positioning System (GPS), remote sensing, Geographic Information Systems, ground control systems, and LiDAR (Light Detection and Ranging). Each of these technologies plays a crucial role in providing accurate, real-time geographic data that can be used across a wide range of applications. It has proven invaluable in sectors such as defense, disaster management, urban planning, and environmental conservation. In defense, for example, it aids in strategic planning and surveillance, while in disaster management, it supports early warning systems and crisis response. Urban planners utilize geospatial tools to map city growth, optimize infrastructure, and manage resources. Environmental scientists depend on this technology for monitoring biodiversity, tracking deforestation, and managing natural resources. Additionally, it is increasingly being applied to agricultural management, where farmers can use precision farming techniques to monitor soil health, optimize water use, and increase crop yields.

With the integration of Artificial Intelligence (AI), the capabilities of geospatial technology are expanding rapidly. AI algorithms enhance the speed, accuracy, and scalability of data analysis, enabling the prediction of trends and the generation of actionable insights. For instance, machine learning models can predict urban expansion patterns, while AI-driven remote sensing can improve environmental monitoring by detecting changes in land use, forest cover, and water bodies. Moreover, the combination of AI and GT is playing a critical role in climate change studies, helping to model the impacts of rising temperatures and extreme weather events.

In a world that increasingly relies on data-driven decisions, GT provides businesses with tools to enhance efficiency, streamline operations, and drive innovation. From optimizing logistics and supply chains to enabling smarter urban development and improving agricultural productivity, GT is essential in shaping the future of both industry and society. This chapter delves into the transformative role of AI-driven geospatial technologies in monitoring, managing, and predicting geographic phenomena.

THE EVOLUTION OF GEOSPATIAL TECHNOLOGY

Geospatial technology has undergone significant evolution, transforming how we collect, analyze, and utilize data about locations on Earth. This technology deals with geospatial data, which encompasses information such as satellite imagery, coordinates, terrain data, and more. Applications of geospatial technology range from urban planning and agriculture to disaster response, navigation, and environmental monitoring. Its continual advancement is opening new possibilities in various sectors.

The origins of geospatial technology can be traced back to the development of tools like Global Positioning Systems and Geographic Information Systems. These innovations revolutionized the way geospatial data is utilized. What was once simple, static mapmaking has evolved into dynamic, interactive, and complex visualizations that offer valuable insights into geographical phenomena.

GPS, originally developed for military navigation in the 1970s, has since become ubiquitous in everyday life. This technology allows for precise location tracking, making it indispensable for a wide range of applications, from navigation systems in vehicles and smartphones to logistics and emergency response. In the transportation industry, GPS has streamlined operations by enabling real-time vehicle tracking and optimizing delivery routes. During natural disasters or emergencies, GPS assists in tracking rescue teams, planning evacuation routes, and delivering aid to affected areas swiftly and efficiently.

While GPS focuses on precise location data, Geographic Information Systems integrate spatial data with various types of information, enabling users to analyze, visualize, and interpret patterns, relationships, and trends in new ways. GIS provides a framework for gathering, managing, and analyzing geospatial data from diverse sources, such as satellite imagery, drone footage, and sensor networks. It enables urban planners to design smarter cities, identify infrastructure needs, and monitor changes in land use (Wang & Liu, 2019).

One powerful example of GIS in action is its use in tracking deforestation. In regions like the Amazon rainforest, satellite imagery combined with GIS data helps scientists and conservationists monitor deforestation rates. Similarly, coastal areas are continuously monitored for erosion, flooding risks, and the effects of rising sea levels, providing crucial information for developing mitigation strategies.

Geospatial technology has been especially transformative in environmental management. It plays a vital role in tracking changes in natural landscapes, such as urban expansion, deforestation, and agricultural development. Satellite imagery, for instance, provides real-time data that helps governments and organizations make informed decisions about resource allocation, conservation, and sustainable development.

In the realm of disaster management, geospatial tools are invaluable. They help predict natural disasters like hurricanes, earthquakes, and floods by providing models based on historical and real-time data. When disaster strikes, geospatial technology supports coordination of relief efforts, enabling efficient deployment of resources and rescuers. It also assists in post-disaster recovery by identifying areas in need of urgent attention, whether it be damaged infrastructure or vulnerable populations.

As geospatial technology continues to evolve, its applications are expected to expand further. With the integration of machine learning and artificial intelligence, geospatial data analysis is becoming faster, more accurate, and increasingly predictive. The potential applications in areas such as precision agriculture, climate modeling, smart city development, and autonomous vehicles are vast, offering even greater societal benefits (Zhu & Li, 2020).

The evolution of geospatial technology marks a critical advancement in how we interact with our environment and manage resources. It has fundamentally reshaped industries, improved disaster preparedness, and enhanced our ability to make data-driven decisions for the future. As this technology continues to evolve, its role in shaping a sustainable and resilient world will only grow.

INTRODUCTION TO ARTIFICIAL INTELLIGENCE

Artificial Intelligence is a branch of computer science focused on developing systems that mimic human intelligence by recognizing patterns, processing large amounts of data, and making informed decisions. Unlike traditional computing methods, AI excels in identifying subtle nuances in data that are often overlooked by conventional approaches. These intelligent systems rely on advanced algorithms, neural networks, and cutting-edge hardware like modern Graphics Processing Units (GPUs) to analyze data from various sources swiftly and accurately, enabling them to deliver real-time predictions and solutions to complex problems.

AI operates through the use of algorithms and machine learning models, which allow systems to learn from historical data without being explicitly programmed for specific tasks (Lee & Tsou, 2020). This process involves neural networks that adjust and optimize themselves based on the data they process, improving accuracy over time. Deep learning, a more advanced form of machine learning, incorporates multi-layered neural networks that can perform highly complex tasks such as image recognition and natural language processing. The role of GPUs is critical in this context, as they accelerate data processing, enabling AI models to analyze vast datasets efficiently and perform real-time decision-making in a wide range of applications (Chen & Yang, 2019).

AI's applications span various industries, making it a transformative force across multiple sectors. In healthcare, for instance, AI has revolutionized diagnostics by analyzing medical images—like X-rays and MRIs—to detect diseases at earlier stages, such as cancer or neurological disorders. AI systems are also being used to streamline drug discovery processes and create personalized treatment plans, improving both patient outcomes and the efficiency of healthcare delivery. In finance, AI plays a pivotal role in fraud detection and risk management by analyzing transaction data in real-time to identify suspicious activities. Financial institutions are also leveraging AI to provide personalized investment strategies by tracking market trends and user behavior. Figure 2 shows the varied applications where AI could be used for performance enhancement such as in automating the process, for increasing the accuracy of future predictions, for reducing the associated risks, for real time monitoring etc.

Figure 2. Benefits of using AI in Geospatial Technology

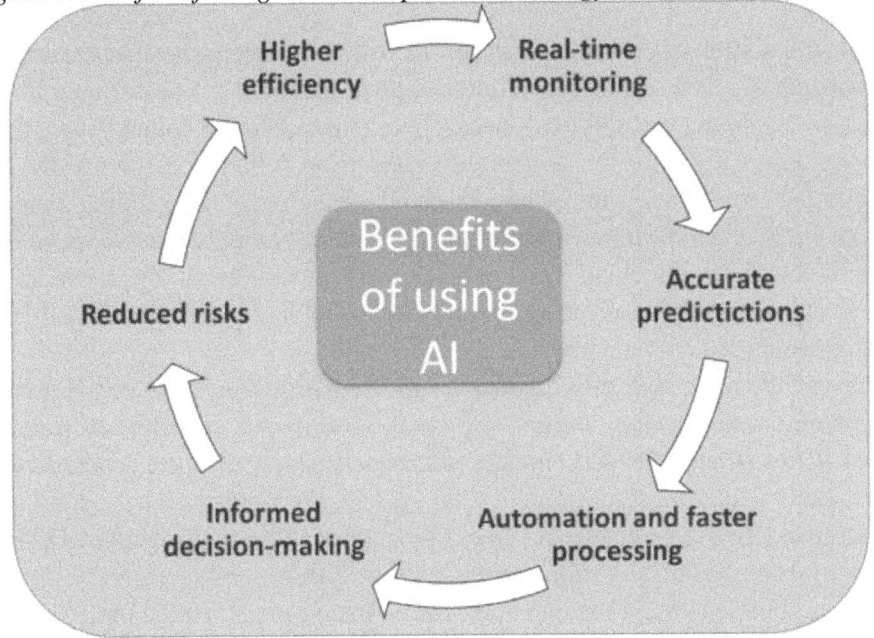

[Diagram: Benefits of using AI — Higher efficiency, Real-time monitoring, Accurate predictictions, Automation and faster processing, Informed decision-making, Reduced risks]

The transportation industry has also been significantly impacted by AI, particularly in the development of self-driving cars. Autonomous vehicles use AI systems to process data from cameras, sensors, and LIDAR to navigate complex environments, interpret road signs, and avoid potential hazards. AI's ability to predict and react to real-time changes in the environment makes self-driving technology a groundbreaking innovation in automotive technology. In entertainment, AI personalizes user experiences by curating recommendations on platforms like Netflix, YouTube, and Spotify. These AI-driven recommendation engines analyze user preferences and habits to suggest content that aligns with individual tastes, enhancing user engagement.

As AI continues to evolve, its influence is expected to expand even further. The integration of AI into fields like robotics, smart cities, and climate change modeling holds immense potential for solving global challenges. By processing larger volumes of data and making more accurate predictions, AI is set to revolutionize decision-making processes across industries, paving the way for innovative solutions and unprecedented advancements in technology. The future of AI promises to transform how we live, work, and interact with the world, opening up new possibilities for efficiency, innovation, and growth across society.

INTEGRATION OF AI AND GEOSPATIAL TECHNOLOGY

The integration of Artificial Intelligence with advanced geospatial technology has ushered in a new era of innovation and problem-solving, transforming industries and expanding our ability to manage and understand the complexities of the environment. AI, known for its ability to learn from data, process it rapidly, and identify patterns, when paired with geospatial technologies like satellite imagery, remote sensing, and Geographic Information Systems, amplifies the scope of data analysis and decision-making. This combination is unlocking solutions to challenges previously considered insurmountable and advancing our understanding of critical environmental and societal trends.

One of the most compelling outcomes of this integration is its transformative impact on urban planning and infrastructure development (Bhaduri & Waddell, 2018). Cities around the world are becoming increasingly complex, with growing populations and infrastructure demands. AI-enhanced geospatial technology is helping to address these challenges by enabling the creation of smart cities. Through the analysis of spatial data from various sources such as satellite imagery, drones, and on-ground sensors, AI can optimize traffic management, public transportation routes, and energy consumption. These technologies predict urban growth patterns, providing decision-makers with the information needed to develop sustainable infrastructure. This results in more efficient use of land, improved public services, and the development of more resilient urban environments that are better equipped to handle population increases, environmental stressors, and resource constraints.

Moreover, the combination of AI and geospatial technology is driving advancements in natural resource management, where the need for accurate and timely data is paramount. The ability to monitor and manage resources such as water, forests, and wildlife has become increasingly important as human activity continues to place pressure on ecosystems. AI-driven analysis of satellite imagery and sensor data is providing real-time insights into the state of natural resources. For example, AI can detect subtle changes in forest cover, allowing conservationists to monitor deforestation in vulnerable areas such as the Amazon rainforest. By analyzing spatial data over time, these systems can track the health of ecosystems, assess the impact of human activities, and propose interventions that help preserve biodiversity and manage natural resources more effectively.

In agriculture, the integration of AI with geospatial technology is creating a revolution in the way farming is managed. Precision agriculture, which uses AI to analyze data on soil health, weather patterns, crop health, and more, is enabling farmers to optimize their use of resources like water and fertilizers. By applying AI models to geospatial data, farmers can develop customized planting schedules, predict pest outbreaks, and implement targeted irrigation strategies (Mollick, Azam,

& Karim, 2023). This not only increases crop yields but also reduces the environmental footprint of farming by minimizing resource waste. For example, satellite data combined with AI analysis can track the moisture content in fields, enabling farmers to apply water only where and when it is needed, leading to water savings and more efficient farming practices. As the global population grows, the demand for food will continue to rise, making AI-driven precision agriculture a critical component of future food security strategies.

Disaster management is another area where the integration of AI and geospatial technology has had a profound impact. Natural disasters such as floods, hurricanes, and earthquakes pose significant risks to lives and property, and the ability to predict, prepare for, and respond to these events is crucial. AI-powered geospatial tools enable real-time monitoring of environmental conditions and provide predictive models that can forecast the likelihood of natural disasters. For example, AI can analyze weather patterns, historical disaster data, and current environmental conditions to predict the path of a hurricane or the likelihood of a flood in a particular area. This information allows authorities to develop more accurate evacuation plans, allocate resources more effectively, and reduce the overall impact of disasters. Furthermore, in the aftermath of a disaster, AI and geospatial technology can be used to assess the extent of the damage, guide rescue operations, and support recovery efforts.

The environmental sector is also benefiting immensely from this technological fusion. Climate change, biodiversity loss, and habitat destruction are pressing global issues that require precise monitoring and management. AI-enhanced geospatial technology enables researchers and policymakers to model and predict the effects of climate change on different ecosystems and regions. For instance, AI can process massive amounts of spatial data to simulate the impact of rising sea levels on coastal areas, helping governments and organizations prepare for future challenges. These models can also assess how climate change might affect agriculture, water resources, and urban environments, allowing for proactive adaptation strategies. By providing accurate, data-driven insights, AI and geospatial technology are critical tools in the fight against climate change and environmental degradation.

Beyond environmental applications, AI and geospatial technology are also revolutionizing industries such as transportation and logistics. In logistics, the ability to analyze spatial data in real-time allows companies to optimize shipping routes, reduce fuel consumption, and improve delivery times. AI can analyze traffic patterns, weather conditions, and road infrastructure to recommend the most efficient routes for vehicles. This not only improves the efficiency of supply chains but also reduces carbon emissions, contributing to more sustainable business practices. In the transportation sector, AI-enhanced geospatial systems are integral to the development of autonomous vehicles. These systems rely on spatial data to navigate, detect obstacles, and make real-time decisions, paving the way for safer and more efficient transportation solutions.

In defense and national security, the integration of AI and geospatial technology has enhanced surveillance, reconnaissance, and threat detection capabilities. By analyzing satellite imagery and other geospatial data, AI can identify potential security threats, monitor border areas, and track movements in conflict zones. The ability to process and analyze large datasets quickly enables defense agencies to make informed decisions, respond rapidly to emerging threats, and protect national interests. Furthermore, AI-powered geospatial tools are being used to track illicit activities such as illegal logging, fishing, and mining, helping to preserve natural resources and combat environmental crime.

The integration of AI and geospatial technology is also making strides in public health. Spatial data analysis can be used to track the spread of diseases, identify hotspots, and model the effects of public health interventions. During the COVID-19 pandemic, for example, AI-driven geospatial models helped public health authorities monitor infection rates, predict outbreaks, and allocate resources to high-risk areas. By combining geospatial data with epidemiological data, AI can provide insights into how diseases spread across regions and populations, enabling more targeted and effective public health responses. This application is not limited to pandemics; it can also be used to combat other infectious diseases, improve vaccination campaigns, and address health disparities in underserved areas.

As AI and geospatial technology continue to evolve, their integration is expected to create even more opportunities for innovation across various sectors. One promising area of development is the use of AI and geospatial data in smart cities. Smart cities leverage technology to improve urban living by enhancing the efficiency of services, reducing energy consumption, and improving the quality of life for residents. By integrating AI and geospatial technology, cities can optimize everything from energy grids and waste management systems to transportation networks and emergency services. For example, AI can analyze real-time data from sensors placed throughout a city to predict electricity demand, helping utility companies manage resources more effectively and reduce energy waste. Similarly, geospatial data can be used to monitor air quality, enabling cities to implement pollution control measures in real-time.

In the field of autonomous systems, the combination of AI and geospatial technology is critical for the development of drones and other unmanned vehicles. Drones equipped with AI-powered geospatial systems can be used for tasks such as aerial surveying, wildlife monitoring, and disaster response. These systems can navigate complex environments, avoid obstacles, and collect valuable spatial data, making them indispensable tools in fields ranging from agriculture to environmental conservation. As drone technology continues to advance, the integration of AI and geospatial data will play a key role in expanding their capabilities and applications.

In conclusion, the integration of AI with advanced geospatial technology is transforming the way we analyze and interact with the world. From urban planning and agriculture to disaster management and environmental conservation, this powerful combination is enabling us to address complex challenges with unprecedented accuracy and efficiency. As AI and geospatial technology continue to evolve, their potential applications will expand even further, offering new solutions to global problems and enhancing our ability to manage and protect the planet's resources. The future of AI and geospatial technology integration is bright, and its impact on society, the environment, and the economy will only continue to grow.

Figure 3 illustrates the operational process of AI-based geospatial technology. It begins with the collection of geospatial data from various sources. This raw data is then pre-processed using application-specific techniques to extract relevant and meaningful information. Once pre-processed, advanced AI algorithms are applied to analyze the data. The analyzed data is then modeled and visualized using sophisticated techniques, enabling clear interpretation. Finally, decisions are made based on the insights derived, and feedback is provided to refine and improve subsequent cycles of operation, ensuring continuous enhancement.

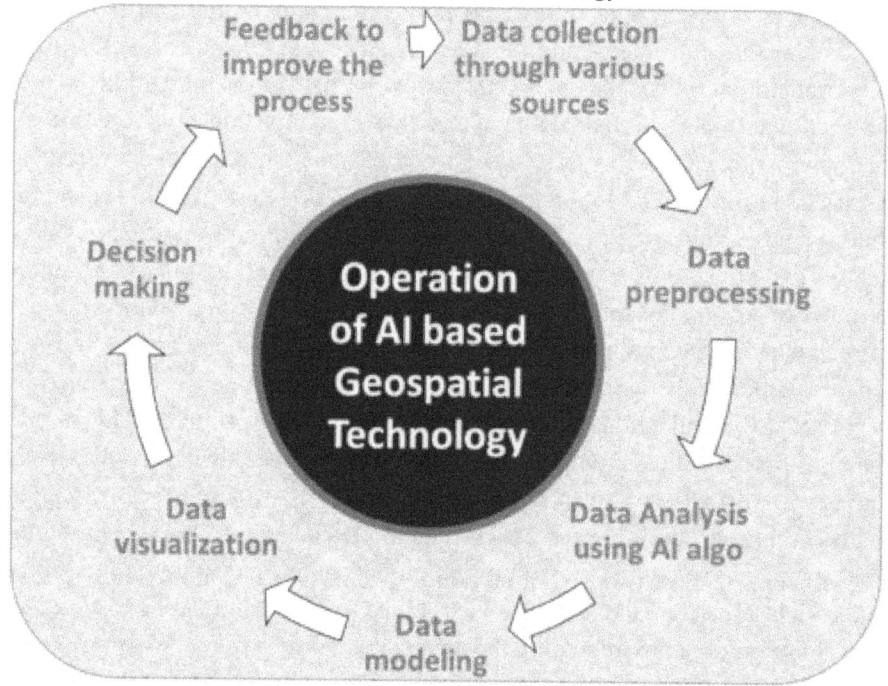

Figure 3. Operation of AI based Geospatial Technology

- Data collection through various sources
- Data preprocessing
- Data Analysis using AI algo
- Data modeling
- Data visualization
- Decision making
- Feedback to improve the process

Operation of AI based Geospatial Technology

APPLICATION OF GT IN ENVIRONMENTAL MONITORING

For many parts of environmental management and monitoring, the amalgamation of geospatial technology and AI is important. Precise forecast of weather patterns, faster detection of changes in the dynamics of land use and prediction of emergency situations such as floods and droughts before they happen can be done by combining these two technologies. This precautionary strategy helps in preventing natural disasters from destructing building and infrastructure and also helps in saving lives. The projection of the consequences of the change in the climate enhanced by AI, additionally play a crucial role in safeguarding the ecosystems and habitats. Artificial Intelligence systems are also capable in predicting the effects of the climate change on biodiversity, vegetation patterns, and sea level rise by accessing large volumes of environmental data. For the manufacturing of flexible plans and regulations, the understanding of these is essential as it lessens the environmental degradation and enhances ecosystem flexibility. The union of GT and AI not only upgrades our capacity to observe and control the environment, but also encourages a higher quality comprehension of consequences of climate change and sustainable resource management techniques. The development of these technologies have extensive potential to solve environmental issues on a global scale and for the future generations it can cultivate a more flexible and sustainable future.

One of the important applications of the merger of GT and AI is anticipating and decreasing the effects of climate change. For the protection of ecosystems and habitats, the forecasting of the consequences of climate change uplifted by AI play an essential role. Evaluating vast amount of environmental makes the AI systems capable to model and foresee the effects of climate change on biodiversity, vegetation patterns, and sea level rise. For avoiding the degradation of environment and to increase the flexibility in the ecosystem, these understanding are important

Soil erosion, water quality, and deforestation are other areas where AI and GT lay a key role along with climate change. Discovering of illegal logging activities, assessing soil health and monitoring pollution levels in water bodies can be done by AI by treating the images captured by the satellites and the data collected by the sensor. Time interventions are enabled by this real-time monitoring and it also helps in the implementation of environmental regulations.

In addition, our capability to control the natural resources sustainability is highly enhanced by the amalgamation of Artificial Intelligence and Geospatial Technology. For precision agriculture, artificial intelligence is important because it can examine the large volumes of data on crop health, patterns of the weather, and soil conditions. With the availability of this information, the farmers are better equipped to decide on the best planting dates, irrigation techniques, and the plans for fertilization. Due to this, farmers are able to boost the agricultural yields while using comparatively less water, fertilizer, and other required resources. This strategy increases more ecologically friendly and sustainable farming methods along with the increase in agricultural output.

AI-powered geographic tools are also essential for disaster management. It can forecast natural disasters like hurricanes, earthquakes, and floods, allowing for the allocation of resource and prompt evaluation strategies. The effects of such calamities are decreased by the systems which offer real-time data and sophisticated prediction models, further saving lives.

All-inclusive, the collaboration of Geospatial Technology and Artificial Intelligence not only increases our ability to monitor and control the environment effectively but also stimulates a higher understanding of the impacts of the climate and the practices for sustainable resource management. As the advancement of these technologies take place, they hold a key assurance in addressing the challenges of global environment and fosters a more adaptable and sustainable future for the generations to come.

Figure 4 highlights the collaboration of GT and AI and how it contributes to various applications such as weather forecasting, land use monitoring, disaster prediction, ecosystem protection, soil and water quality monitoring, precision agriculture, and disaster management. Through enhanced understanding of the effects of climate change and the promotion of sustainable resource management, this union increases our ability to monitor and manage the environment.

Figure 4. Diagram illustrating the integration of Geospatial Technology and Artificial Intelligence in environmental management

APPLICATION OF GT IN ENVIRONMENTAL MANAGEMENT

The evaluation of enormous amounts of data from several sources, which includes the images captured by the satellites, the weather patterns, and the historical records, done by the Artificial intelligence, helps in anticipating and responding to the natural disasters that can take place. Eventually, the skill of evaluating the large volumes of data can save lives and cut down the damage by sanctioning early warning systems and more efficient and effective evacuation plans. Artificial Intelligence is a major constituent of the management of water resource, which further helps in optimizing distribution of water and utilization through demand forecasting, groundwater level forecasting, and rainfall pattern prediction. Furthermore, to recognizing leaks in the water infrastructure, the recommendations made by the AI systems can be effective in making repair plans, which increases the efforts to conserve and save water. Pollution control is another area where AI has proved to shine, the quality of the air water in real time is tracked by the utilization of data analysis. Identification of contaminants and locating their sources with highest precision is one of the capabilities of AI-powered sensors, eventually making it possible for swift regulatory responses and measures, which are further crucial for decreasing the pessimistic effects of the environment. By offering practical insights extracted from the complexed and vast datasets, Artificial Intelligence highly improves our ability to observe and protect the environment. The fusion of artificial intelligence with creative tools including geospatial technology increases our ability to address environmental problems as long as breakthroughs in technology continues. Credits to this fusion, the future generations will benefit the comfort of healthier plant, which will play a key role in motivating sustainable habits globally.

The large amount of data from the images and photographs captured by the satellites, the changes in the patterns of the weather, and the records of the history, can by all analysed by the artificial intelligence, which will highly enhance our ability to understand and respond to the nature calamities. Lastly, the adeptness can save lives and reduce the destruction by permitting the warning systems and increasing efficiency and effectiveness in the evacuation plans. One such example can be that AI systems are able to forecast the duration and strength of storms, which makes it feasible to assign resources and issue evacuation orders in a timely and a precise manner.

For the management of the resources of water, AI is crucial because by the utilization of predictive modelling, AI can examine patterns of the rainfalls, groundwater levels, and demand forecasts, and therefore upgrade appropriate distribution and consumption. Artificial Intelligence algorithms are skilful in identifying leaks in water infrastructure, easing prompt interventions and the making of effective and efficient maintenance plans. This makes a big difference in the brawl of saving water. The utilization of water is effective and sustainable is ensured by AI by taking a proactive approach, which decreases waste and improves overall management of resource. Correspondingly, these technologies are crucial for continuing sustainable practices along with its maintenance, protecting the water resources, and guaranteeing a stable supply of water for all purposes.

At decreases the levels of pollution, AI is outstanding as it utilizes the analysis of data to consistently analyse the condition of the air and the water. Artificial Intelligence - enabled sensors are able to identify pollutants and track their sources down, allowing for timely regulatory responses and mitigation of the effects of environment. For example, AI, can locate the sources of industry of air pollution and predict the areas that are prone to have an air quality which is bad. Revising the policies timely and reacting faster are made possible by this capacity. Artificial Intelligence enhances our capability to handle pollution effectively, eventually resulting in better environmental health and more accurately informed decision-making by offering precise and useful insights (VoPham, Hart, Laden, & Chiang, 2018).

The assessment of extensive amount of environmental data can be done for the prediction of the effects of global warming on ecosystem as AI plays an essential role in reducing climate change. For generating more resilient plans and regulations for guarding biodiversity and further advancing higher environmentally friendly resource management, these realizations are important. The crop yields that might have been affect by the increase in temperatures can be modelled by AI, proposing further changes to the techniques used in agriculture to sustain the result production. The relocation and survival of animal species in changing environment, can be also predicted by artificial intelligence, which eventually helps in the creation of higher effective and successful habitat safeguarding techniques by the conservationists. Artificial intelligence enhances our ability to predict and mark the problems introduced by climate change, nurturing durability and sustainability in a variety of industries, by offering accurate and useful projections.

Normally, artificial intelligence enhances our ability to observe and preserve the environment by extracting useful comprehensions from the complex data sets of the environment (Li et al., 2016). As the advancement of technology takes place, the conjunction of AI with various other tools, such as geospatial technologies, can assist us to find a solution to environmental problems and promote sustainable habits worldwide. Artificial Intelligence combined with the management of environment not only supports long-term planning and sustention initiatives, but also lends a hand with swift crisis response, promising a more resilient and sustainable future for the upcoming generations.

The flow of data from different sources is demonstrated in Figure 5 through AI processing and then it delves into specific application in environmental management. The data sources include the images captured by satellite, patterns of the weather, historical records, and sensor data, which provide important inputs for the systems of the AI. The analysis and interpretation of the data is done in the AI processing system which is useful in extracting beneficial information from the complicated data set. Each application enhances our ability to manage environmental challenges and promotes sustainability.

Figure 5. The comprehensive flow from data sources to AI processing to the applications

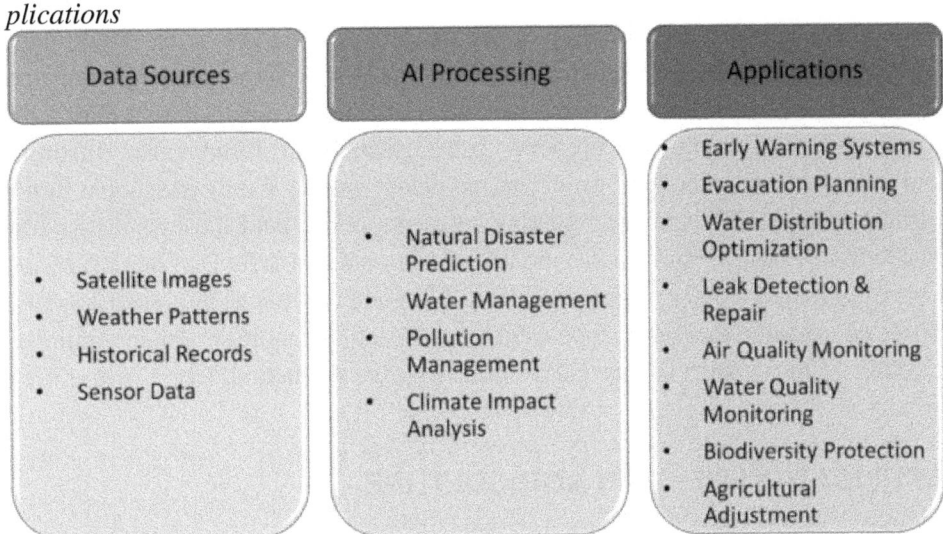

Data Sources	AI Processing	Applications
• Satellite Images • Weather Patterns • Historical Records • Sensor Data	• Natural Disaster Prediction • Water Management • Pollution Management • Climate Impact Analysis	• Early Warning Systems • Evacuation Planning • Water Distribution Optimization • Leak Detection & Repair • Air Quality Monitoring • Water Quality Monitoring • Biodiversity Protection • Agricultural Adjustment

APPLICATION OF GT IN DISASTER MANAGEMENT

Geospatial technology is essential in disaster management, enhancing preparedness, response, and recovery. Tools like GIS, satellite imagery, and remote sensing enable agencies to analyze geographical data and identify areas prone to disasters such as floods, earthquakes, and hurricanes. This technology supports hazard mapping, helping authorities create risk assessments that guide response plans. During emergencies, real-time data collection provides quick situational awareness, allowing responders to allocate resources effectively. For instance, GIS platforms visualize impacted areas, aiding in the coordination of rescue operations. Post-disaster, geospatial technology assists in damage assessment and recovery planning, as agencies use aerial imagery to prioritize efforts and allocate resources. Integrating geospatial data with social media analytics improves public communication, ensuring timely updates for affected communities. Overall, geospatial technology strengthens disaster response and fosters resilience through informed decision-making and stakeholder collaboration.

Recently, geospatial technology has been instrumental in tracking Cyclone Dana, a severe storm from the Bay of Bengal in October 2024. Meteorologists and agencies gain insights into Dana's path, wind speeds, and potential impact areas, particularly along the Odisha and West Bengal coasts. Using satellite imagery, radar, and computer modeling, experts monitor Dana's real-time intensity and trajectory. Satellites like INSAT-3D, along with Doppler radar systems on India's eastern coast, continuously update data, helping the Indian Meteorological Department (IMD) provide accurate forecasts. Currently, Cyclone Dana is moving northwest at 12 km/h and is projected to make landfall near Bhitarkanika National Park with winds up to 120 km/h, affecting areas like Jagatsinghpur, Kendrapara, Bhadrak, and Balasore in Odisha. Real-time tracking platforms like Windy and Zoom Earth further support IMD's efforts, helping residents understand the cyclone's path. The Odisha state government has mobilized National and State Disaster Response Forces, prepared 7,000 cyclone shelters, and begun evacuations in high-risk areas. These coordinated measures, supported by geospatial technology, aim to minimize Cyclone Dana's impact and save lives through proactive actions.

APPLICATION OF GT IN AGRICULTURE

In agriculture, geospatial technology has emerged as a transformative force, optimizing farming practices and enhancing productivity. Precision farming, a key application, utilizes satellite imagery and GIS to gather real-time data on soil conditions, moisture levels, and crop health. This information empowers farmers to make data-driven decisions regarding irrigation, fertilization, and pest management, ultimately leading to increased yields and resource efficiency. For example, farmers can utilize satellite data to monitor crop growth patterns, identify areas needing attention, and implement targeted interventions. Geospatial technology also aids in crop monitoring and management by providing detailed insights into agricultural practices across large areas. AI algorithms can analyze this data to detect early signs of diseases and nutrient deficiencies, enabling timely interventions that mitigate risks and optimize yield. Additionally, predictive analytics based on historical data allows farmers to forecast crop yields and plan planting and harvesting schedules more effectively. By facilitating sustainable land management practices, such as monitoring land use changes and promoting efficient resource allocation, geospatial technology supports the long-term viability of agriculture. Overall, its integration into modern farming is revolutionizing agricultural practices, ensuring food security while minimizing environmental impact.

APPLICATION OF GT IN HEALTHCARE

Geospatial technology has made significant contributions to the healthcare sector by enhancing disease surveillance, resource allocation, and public health interventions. Through GIS and spatial analysis, health organizations can visualize and analyze the distribution of diseases across geographic regions, allowing for more targeted interventions. For instance, during disease outbreaks, GIS mapping can identify hotspots and track the spread of infections, helping public health officials deploy resources where they are most needed. Additionally, geospatial technology aids in health facility planning and management by analyzing population density, accessibility, and healthcare needs, ensuring that medical services are effectively distributed. It enables health authorities to optimize the placement of clinics and hospitals, improving access to care for underserved populations. Moreover, geospatial tools can assist in environmental health assessments by mapping environmental hazards and their impact on communities, facilitating timely interventions to protect public health (Huang & He, 2021). The integration of geospatial data with mobile health (mHealth) applications empowers individuals to access health information and services easily. By providing insights into health trends and disparities, geospatial technology supports informed decision-making in healthcare policies, ultimately leading to improved health outcomes and enhanced community well-being. Overall, its applications in healthcare are critical for advancing public health initiatives and fostering healthier populations.

CHALLENGES AND LIMITATIONS FOR AI

Other than its many advantages, the utilization of artificial intelligence to management of the environment raises ethical questions that need to be answered to ensure secure and unbiased usage of the data. Algorithmic bias is a critical issue whereby AI systems may unintentionally prejudice against particular populations, which requires close monitoring and remediation. Also, the protection of data security and privacy is important, especially when working with the sensitive data of the environment. Powerful data protection procedures and open business processes are demanded to put in place as to maintain integrity and strengthen confidence. To realise the AI's promise completely, the balance of innovation and moral principles is important. To address these problems, we must utilise AI ethically to enhance the environmental management while safeguarding equality and privacy.

Along with the several advantages to integrating artificial intelligence with environmental management, there are also drawbacks. One of the major issues is the unethical use of AI; in particular, making sure that the utilization of data is secure and without bias or prejudice. Due this large complicated and incomplete data, AI systems may unintentionally show algorithmic bias, which may unfairly disadvantage some groups. To resolve this issue, strict and precise regulation and development of transparent, equitable algorithms that prioritize fairness are required.

Additionally, data security and privacy must be ensured in AI applications particularly when handling sensitive environmental data. Robust data protection measures must be implemented to avoid information being misused and accessed without authorization. The risk that AI will be used in ways that compromise privacy or lead to data breaches emphasizes on how essential it is to establish stringent security protocols. Moreover, creating and maintaining AI systems may be extremely expensive and challenging. Access to cutting-edge AI technologies and the knowledge required to use them may be restricted especially in underdeveloped nations potentially exacerbating already-existing disparities. Realizing the full potential of artificial intelligence in environmental management requires a necessary balance between innovation and moral principles. We can ethically use AI to enhance our capability to sustainably manage the environment while preserving justice and privacy, by addressing these issues.

Privacy and data security are vital in AI applications, especially when managing sensitive environmental data. Ensuring that the sensitive data is protected against misuse and unauthorized access is of utmost importance. Stringent privacy rules and strict security measures must be ensured in place in order to guard data, build public confidence, and adhere to legal requirements. Encryption restricted access and regular security audits can all help to protect the integrity of the data. Higher public trust can be achieved through transparent in the collection, storage, and utilization of data. Respecting legal mandates, like GDPR, ensures compliance and raises the credibility of AI projects. Finding a balance between innovation and stringent security and privacy protocols is necessary for the appropriate and effective application of AI in environmental management.

Furthermore, the availability, quantity and quality of data have a major impact on how well AI performs in environmental management. The erosion of prospective advantages of AI can be done by inadequate, expired, or erroneous data, eventually resulting in inaccurate forecasting and conclusions. Continuous efforts to improve data collection validation and updating processes are necessary to establish the accuracy of AI-driven insights. To exchange and preserve correct datasets, this demands in making investments in state-of-the-art monitoring systems, setting up defined data standards, and promoting the participation of stakeholders. By giving data integrity high priority, we can enhance the precision and effectiveness of AI applications leading to more environmentally friendly environmental management strategies and better-informed decision-making.

Another problem is the expensive cost of processing AI systems. Advanced AI algorithms can be highly energy- and processing-intensive which can be a major barrier to their widespread adoption especially in resource-constrained environments. Availability to these technologies may be constrained by the cost of acquiring and maintaining the necessary hardware. The key to resolving this is developing more efficient algorithms that require less processing power. Moreover, by leveraging cloud computing resources, more businesses may use AI without having to make significant investments in local infrastructure. We can make AI technologies more accessible and feasible for a larger range of applications by optimizing algorithm efficiency and leveraging cloud-based solutions.

For the successful integration of AI in environmental management interdisciplinary work is required. To take care of these complex environmental concerns, the experts from AI environmental research policy-making and other relevant fields must work together. By collaborating, we can close knowledge gaps foster innovation and make sure that AI applications are practical and useful in everyday settings. We can produce more comprehensive and effective solutions by the collaboration of knowledge from several fields. These collaborations also facilitate the alignment of AI technologies with regulatory frameworks and environmental goals leading to more sustainable practices and informed decision-making. For environmental concerns to be addressed holistically and to completely benefit from AI, robust stakeholder cooperation is required.

Lastly, the rapid advancement of technology necessitates ongoing education and training. Environmental managers, lawmakers and the general public are among the stakeholders who need to stay informed about AI developments and their implications. Motivating the literacy of the digital and providing ongoing training opportunities can assist artificial intelligence's benefits for environmental management.

To sum up, artificial intelligence has a lot of potential to help with environmental management but there are also challenges that must be taken care of and must be overcome. First steps that are essential are ensuring strong and robust data security promoting interdisciplinary cooperation and finding a balance between innovation and ethical norms. The risks like algorithmic bias and data privacy issues can be decreased by enhancing the dependability of AI systems by following these guidelines. Bridging disciplinary gaps and guaranteeing the effectiveness and utility of AI solutions can be achieved through collaborative efforts. We can effectively leverage artificial intelligence to advance eco-friendly practices and improve the management of environmental resources by listing these issues eventually contributing to a more sustainable, health and well-being-oriented future globally.

Figure 6. Visual representation of key ethical concerns associated with the use of AI in environmental management

```
Core Ethical Concerns ──┬──► Algorithmic Bias
                        ├──► Data Security
                        ├──► Privacy
                        └──► Access inequality
```

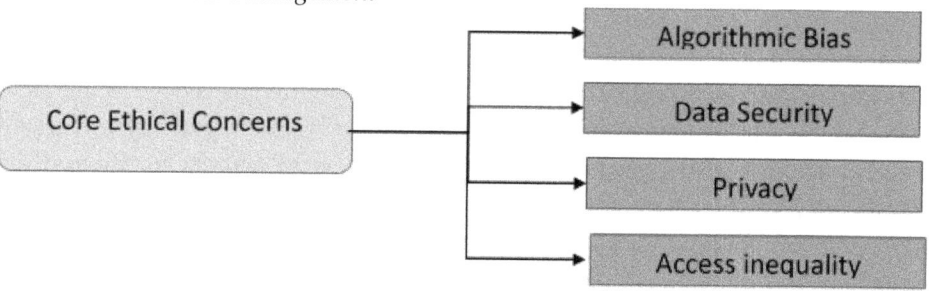

Figure 6 illustrates that although AI can provide significant advantages, it also has some issues that needs to be addressed, such as core ethical concerns, which include algorithmic biases, data security, privacy, and access inequality. Addressing these problems further ensure ethical AI deployment which requires careful and comprehensive consideration of these factors to maintain fairness, security and equality.

CASE STUDIES IN GEOSPATIAL TECHNOLOGY

Since its start in the 1960s, GIS has grown to support areas like urban planning, disaster management, and agriculture. Countries adapt GIS to meet unique challenges: India uses it to map its vast area of 1.3 million square miles and assist its 1.2 billion citizens, and a national GIS platform launched in the 2000s supports decision-making across sectors by providing data accessible to government, businesses, and citizens. Singapore's Urban Redevelopment Authority (URA) uses GIS to manage its limited 279 square miles, ensuring efficient land use and high-quality urban planning reviewed every five years. In the UAE, cities like Dubai and Abu Dhabi use GIS to improve resource management and public services, illustrated by Abu Dhabi's GIS Center for Security, which supports emergency management and inter-agency coordination. These examples show how geospatial technology is reshaping urban planning and improving quality of life.

In Dubai, GIS enhances disaster management and preparedness as the city's rapid growth necessitates proactive planning. Dubai Police, known for advanced systems, use a state-of-the-art headquarters with command and control technologies, including a Barco video wall to visualize real-time threats. Integrating aerial and satellite imagery, they build 3D city models to guide emergency responders effectively. Established in 2007, the Crisis and Disasters Management Department focuses on hazard mapping, risk assessment, and response coordination, aiming to reduce disaster impacts through public awareness and training. Dubai's advanced ambulance services include motorbikes for navigating traffic and helicopters for hard-to-reach areas, supporting both local and regional disaster management.

In agriculture, India is using GIS and precision farming to boost productivity and sustainability. Satellite imagery provides real-time data on soil moisture, crop health, and weather, helping farmers optimize irrigation, fertilization, and pesticide use. This approach has improved yields and efficiency; for example, irrigation advisories based on weather and soil data enhance resource management. Crop monitoring also benefits from satellite data, allowing early detection of diseases or nutrient issues. In the 2017-2018 season, pest forecasting saved significant cotton acreage from pink bollworm, and in 2018, a sowing advisory improved groundnut yields by 10-15% in Anantapur. GIS also supports sustainable land use planning and weather prediction, reducing risks from extreme weather. Market access and supply chain optimization further reduce post-harvest losses, while robotics and drones boost farm efficiency. Altogether, geospatial technology is modernizing Indian agriculture, making it more efficient and sustainable.

CURRENT AND FUTURE CHALLENGES IN IMPLEMENTING GEOSPATIAL TECHNOLOGY

The future of geospatial technology holds transformative potential, with expanded applications in real-time spatial analysis and data integration across various sectors. However, significant challenges accompany this progress, particularly concerning data privacy, data accuracy, and complex integration needs.

Data privacy is a growing concern as GT increasingly incorporates personal data in applications like urban planning, public health, and consumer analysis. With location data being highly sensitive, stringent regulations are necessary to prevent misuse and ensure user consent. Balancing the accessibility of geospatial data with robust privacy protections is a pressing challenge for both public and private sectors (Nguyen & Kim, 2022).

Ensuring accuracy in data collection across diverse environments poses another critical challenge. Differences in terrain, environmental conditions, and sensor reliability can result in inconsistent data, which directly impacts decision-making in areas such as agriculture, infrastructure, and disaster response. Developing new standards and calibration methods is essential to enhance data reliability and ensure that GT applications are effective and dependable.

Furthermore, integrating GT with other digital technologies presents technical and ethical complexities. Coordinating multiple data sources and ensuring compatibility requires standardized frameworks and quality assurance practices. Addressing these challenges will be essential for harnessing the full potential of geospatial technology in the years to come.

FUTURE DIRECTIONS

The anticipated future developments of increasingly sophisticated artificial intelligence algorithms advancements in geospatial technologies and enhanced guidelines for the use of AI in environmental management are expected to bring about significant improvements. The tracking of environmental changes in real time and it can produce more accurate predictions of natural disasters and climate trends can be made possible with the help of these advancements. Brushing up the forecasting models enhance the capacity to recognize and address environmental issues, resulting in responses that are more effective and timelier. Better instruments for waste reduction efficient use of natural resources and sustainable resource management will also become available as AI advances.

As AI technology advances, opportunities to address moral concerns such as making sure that AI applications are transparent accountable and equitable will present themselves as AI technology advances. Holding on to public confidence and ensuring that these technologies be used in charge of the need of the establishment of robust frameworks and explicit rules for the moral and ethical use of AI. The merger of ethical problems in AI enhancement and implementation can promise the realization of benefits without threatening privacy or equity.

The technology evolutions have the prospective to fully alter our understanding of and ability to protect our natural environments. Encouragement for a more resilient and sustainable future can be enabled by facilitating more precise environmental monitoring, higher catastrophe preparedness, and better management of natural resources. As we continue to innovate and address these ethical issues AI and GT will be essential in playing a critical role in creating a healthy planet for future generations.

Upcoming advancements in GT and the development of increasing complex AI algorithms can completely transform the environmental management. Improved real-time monitoring capabilities, more accurate natural disaster prediction, and more efficient decision-making processes for sustainable resource management are all promised by these developments. As AI advances, it will be important to establish stronger standards for its ethical application. It is essential to address concerns such as transparency equity and accountability in order to ensure that these technologies are used responsibly. When collectively considered, these developments will lead to significant discoveries that will strengthen environmental conservation and build a robust and sustainable future.

Initially improved real-time environmental change monitoring functionalities will be accessible. With the company of AI-powered GT, more precise and extensive monitoring of ecosystems, biodiversity, and natural habitats will become enabled. Improved systems for early warning and detection of natural disasters like hurricanes, floods, and wildfires are a component of this. The generation of timely alerts and valuable insights by the prediction can be done by AI systems by analysing vast amounts of data from sensors satellites and historical records. As a result, this allows for proactive reaction measures to be taken, reducing the impact on infrastructure and human life.

Secondly, more precise predictions of climatic patterns and environmental trends are anticipated as AI is expected to become more predictive. In the domains including urban planning, agricultural practices, and sustainable resource management, these advancements will support more informed decisions. For instance, AI, may utilize the analysis of climate data to optimize water management, lessening the effects of drought and boosting agricultural flexibility to changing weather patterns. AI is aimed to enable more efficient resource planning and management by providing exact understanding into the shifting of the environment. This will further help in resolving problems related to limited resources and unpredictable climate. Ultimately, these developments will support more profitable and sustainable business models in a range of sectors.

Furthermore, major developments in the domain will promise to resolve ethical concerns about the use of AI in environmental management. It will make sure that AI systems are transparent, equitable and accountable, which is essential. This demands in encouraging transparency in algorithmic processes, creating strict guidelines for ethical use of data, and working diligently to reduce biases which are all necessary to achieve. By guaranteeing clear guidelines and frameworks we can ensure that AI applications are fair and serve the interests of all stakeholders. These initiatives will help build confidence in AI by promising that the benefits of this technology are taken care of without compromising moral principles. For the complete utilization of AI for sustainable and moral environmental management, solving these ethical problems will be crucial and essential.

Tackling the environmental issues will require interdisciplinary collaboration, as AI advances. This integration of knowledge from environmental science, policymaking, community involvement, and AI research will promote novel approaches and all-encompassing solutions will be produced. By collaborating, we can guarantee that solutions made by AI are cutting edge technologically, socially and ecologically responsible. The combination of this range of perspectives can develop state-of-the-art AI applications that are also in line with more general societal and environmental goals. This type of collaboration will support the creation of solutions that points complex problems holistically balancing and finding technological advancement with moral and practical concerns for a healthy and sustainable future.

In summary, artificial intelligence is very promising for the bright future in environmental management. Our understanding and observation along with conservation of the natural world will be fundamentally altered by the advancements in Geospatial Technology and Artificial Intelligence as well as by interdisciplinary cooperation and higher ethical standards. These developments will enhance our capability to predict events more precisely, and make the best use of the resources at hand and adjust to changing conditions. Integration of information from environmental science policy-making community involvement and AI research will result in cutting-edge and socially responsible AI solutions. This approach will move us nearer to a self-reliant future in which technology is crucial in preserving the environment and maintaining a more robust and sustainable society. Successful solutions to environmental issues and improving and sustaining the Earth for future generations may come from a combination of ethical behaviour and innovation.

CONCLUSION

The significant boosting of efficacy and efficiency of artificial intelligence is changing and revolutionizing monitoring and management of the environment. Leading this change are innovations like automated data analysis which enable prompt detection of environmental changes and quicker response to hazards like natural calamities and problems related to pollution. Higher improvements in accuracy and scalability may be anticipated AI and GIS technologies continue to develop further. These advancements will enable and make it possible to monitor ecosystems more comprehensively which will lead to better decision-making for the management and conservation of the environment.

Utilizing AI with real-time data from the satellite sensors and other resources is particularly intriguing. Through the combination of these, dynamic environmental models are resulted that have the ability to foresee and proactively mitigate environmental issues. For example, AI can forecast air pollution levels, track deforestation, and monitor water quality by the analysis of massive amounts of data collected from the sensor and the satellite. It can also provide us with useful information resulting in timely responses.

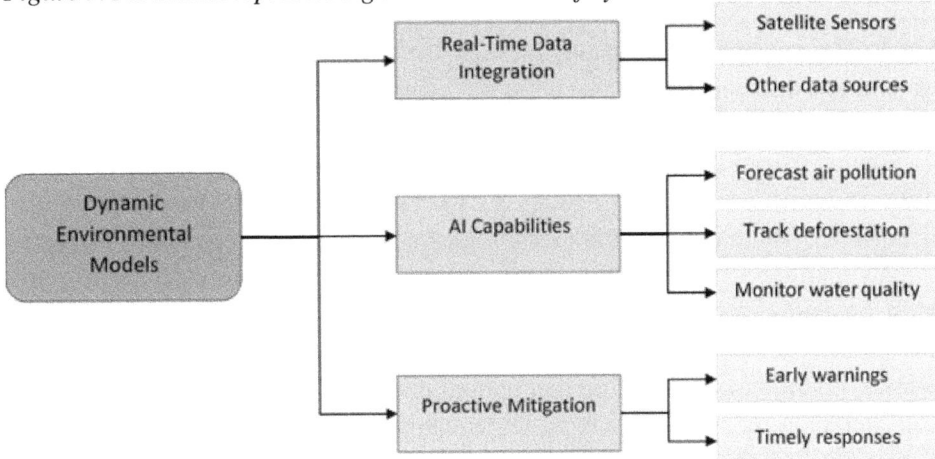

Figure 7. Flowchart representing the constituents of dynamic environmental models

Figure 7 illustrates that models use real-time data and AI capabilities to understand and forecast environmental changes, enabling prompt and more effective responses to mitigate their impacts proactively.

Moreover, a lot of capability are there resolve issues raised by the sustainability due to the ever-evolving potential of AI. It can enhance the efficiency of resources and promote behaviour which are sustainable by increasing allocation of resources and creating and advancing polies based on data-driven insights. To support environmental sustainability, AI algorithms for instance may improve the energy efficiency of buildings or optimize irrigation systems to conserve water.

To summarize, AI is transforming environmental management by creating tools for faster responses, decision-making driven by the data and more precise control and monitoring. As technology is collaborated and integrated with real-time data, it possesses sophisticated analytical skills it can address environmental concerns and further promote sustainability as it continues to develop.

Artificial Intelligence is transforming the monitoring and management of environment by increasing the effectiveness and efficiency across different fields. Novel innovations like automated analysis of data ensure and enable faster detection of changes in the environment along with prompt responses to the warnings and threats such as natural calamities and incidents related to pollution.

The developments in artificial intelligence and geographic technology will lead to more precise and scalable environmental management. These improvements will make it even possible to monitor ecosystems with a higher range of comprehensiveness, which will eventually facilitate higher and better informed decisions related to strategies of management and conservation. Combining artificial intelligence with geographic data will allow us to make more precise evaluations and projections improving our capability to respond to environmental issues. It is anticipated that this development will significantly increase the ability to preserve and protect ecosystems and manage and control natural resources further leading to more details and informed environmental stewardship which will also be sustainable.

When artificial intelligence is integrated with real-time data from sensors, satellites and other sources, the creating of dynamic environmental models have a lot of potential. These cutting-edfe technologies when collaborated allow us to build models that expect environmental dangers and take proactive steps to decrease them. Sustaining global environmental initiatives and increasing resilience require this capacity. AI-driven models, for instance, can locate changes in ecosystems, monitor pollution levels, and predict natural disasters by analysing real-time data. This facilitates more effective management strategies along with faster and timely responses. With this combination, we can be much better able to address problems related to environment and move towards a more flexible and sustainable future.

With the advancement of various capabilities of artificial intelligence, numerous opportunities arise to address issues related to sustainability. By applying and utilizing data-driven insights we can improve the creation of policies and the allocation of resources. Natural resource management and ecosystem conservation are enhanced by this strategy which and enhances stewardship of the environment. Decision-making which is accurate and that supports long-term ecological sustainability and health can be enables by the analysis which is AI-driven. We can solve pressing problems and contribute to the development of a more flexible and sustainable future by integrating this cutting-edge technology into environmental plans.

To conclude, technology has the potential to significantly reshape how humans perceive and protect the natural world, as demonstrated by its impact on environmental management. Together, we can strive for a more sustainable future where innovation plays a vital role in safeguarding our planet through thoughtful and wise application of technological advancements.

REFERENCES

Bhaduri, B., & Waddell, P. (2018). Advances in spatiotemporal big data analytics for urban planning. *Computers, Environment and Urban Systems*, 69, 1–11. DOI: 10.1016/j.compenvurbsys.2018.03.003

Chen, L., & Yang, L. (2019). Deep learning for geospatial data analysis: A review. *Remote Sensing*, 11(12), 1432. DOI: 10.3390/rs11121432

Huang, J., & He, Y. (2021). Spatial-temporal modeling for environmental hazard prediction using deep learning. In *Proceedings of the IEEE Conference on Computer Vision and Pattern Recognition* (pp. 6342-6351). IEEE. DOI: 10.1109/CVPR46437.2021.00634

Lee, K., & Tsou, M.-H. (2020). Machine learning techniques for geospatial data classification: A comparative study. In *Proceedings of the ACM SIGSPATIAL International Conference on Advances in Geographic Information Systems* (pp. 82-91). ACM. https://doi.org/DOI: 10.1145/3382324.3382364

Li, S., Dragicevic, S., Castro, F. A., Sester, M., Winter, S., Coltekin, A., Pettit, C., Jiang, B., Haworth, J., Stein, A., & Cheng, T. (2016). Geospatial big data handling theory and methods: A review and research challenges. *ISPRS Journal of Photogrammetry and Remote Sensing*, 115, 119–133. DOI: 10.1016/j.isprsjprs.2015.10.012

Mollick, T., Azam, M. G., & Karim, S. (2023). Geospatial-based machine learning techniques for land use and land cover mapping using a high-resolution unmanned aerial vehicle image. *Remote Sensing Applications: Society and Environment*, 29, 100859. DOI: 10.1016/j.rsase.2022.100859

Nguyen, T., & Kim, J. (2022). Advances in spatial data integration and analytics: A survey of current research. In *Proceedings of the International Conference on Geographic Information Science* (pp. 1-10). https://doi.org/DOI: 10.1109/GIScience54041.2022.00001

VoPham, T., Hart, J. E., Laden, F., & Chiang, Y.-Y. VoPham. (2018). Emerging trends in geospatial artificial intelligence (geoAI). *Environmental Health*, 17(1), 40. Advance online publication. DOI: 10.1186/s12940-018-0386-x PMID: 29665858

Wang, Y., & Liu, W. (2019). Real-time geospatial data processing using edge computing. In *Proceedings of the IEEE International Conference on Big Data* (pp. 1025-1034). IEEE. DOI: 10.1109/BigData47090.2019.9006068

Zhu, X., & Li, S. (2020). Integrating geospatial big data and artificial intelligence for smart city applications: Challenges and opportunities. *International Journal of Applied Earth Observation and Geoinformation*, 88, 102035. DOI: 10.1016/j.jag.2019.102035

Chapter 4
Geospatial Technologies for Smart Cities

Amit Sai Jitta
https://orcid.org/0009-0000-1076-7713
Indiana University, Bloomington, USA

Vijaya Kittu Manda
https://orcid.org/0000-0002-1680-8210
PBMEIT, India

Theodore Tarnanidis
https://orcid.org/0000-0002-4836-3906
International Hellenic University, Greece

ABSTRACT

Geospatial technologies, both traditional and modern technologies, have changed the way spatial data is collected, stored, processed, analyzed, and visualized for decision-making in Smart cities. Popular geospatial technologies are geospatial information systems (GIS), remote sensing, and global positioning systems (GPS). Computing technologies have undergone rapid development in recent times. Artificial Intelligence (AI), the Internet of Things (IoT), Big Data, and others are used alongside geospatial technologies for improved decision-making by city planners and administrators. These technologies help Smart cities offer various services to the citizens, promote a circular economy, and be sustainable. Real-time data processing and predictive analytics help proactively manage infrastructure, optimize resource allocation, and enhance overall urban resilience. The chapter is novel in its comprehensive overview of geospatial technologies and data in smart cities, integrating both traditional and modern technologies and emphasizing the significance of geospatial data visualization.

DOI: 10.4018/979-8-3693-8054-3.ch004

1. INTRODUCTION TO GEOSPATIAL TECHNOLOGIES IN SMART CITIES

1.1 Defining Smart Cities

Smart cities (SCs) reflect an evolutionary approach to urban growth and are considered a new form of traditional cities (Marzouk & Othman, 2020; Zhou et al., 2021). By using advanced technologies, SCs are undergoing digital transformations. They can provide personalized and accessible services to citizens, thereby improving the satisfaction and well-being of their citizens. Because the definition of Smart cities changes based on social, technical, and cultural perspectives, there is no universal definition. Figure 1 shows the logical progression of a city (Petrov et al., 2024). Smart Cities optimize urban infrastructure efficiency, promote sustainable economic growth, and build a circular economy. These cities integrate knowledge from multiple domains to create an interconnected and intelligent urban environment (Rani et al., 2021). This information remarkably improves decision-making in transportation, energy usage, waste management, health, public safety, and quality of education and governance. These cities have to move from being digital to sustainable (Zheng et al., 2020) and consequently need to zero net carbon emissions, reduce energy consumption, improve operational efficiency, address the pressing challenges of rapid urbanization and climate change, and enhance the quality of life (Sánchez et al., 2024).

Figure 1. Logical Progression of a City into a Smart City

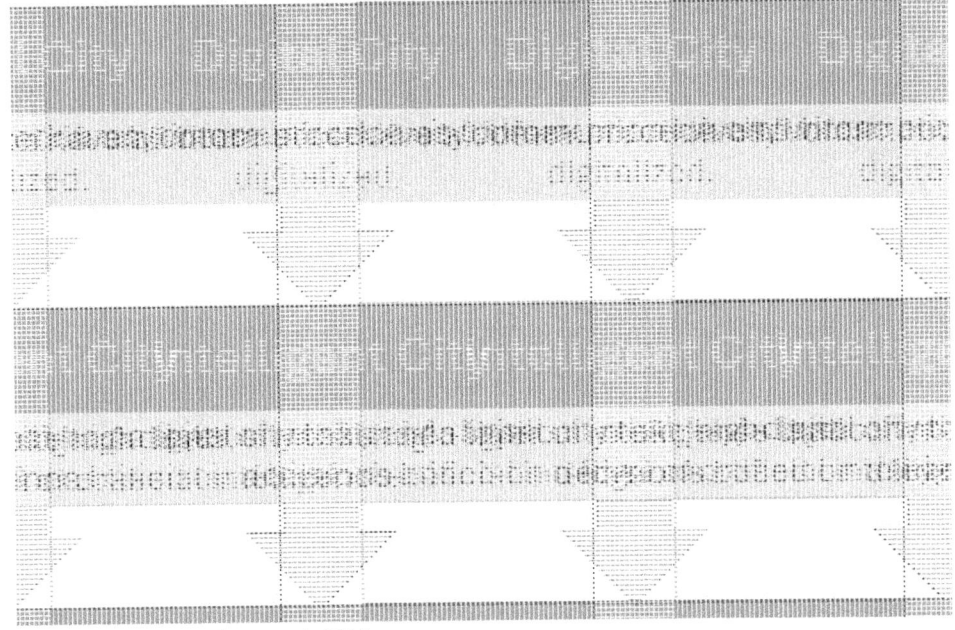

Source: Based on (Petrov et al., 2024)

Geospatial, geographic, or spatial data is information associated with explicit geographical or location coordinates (Kraak & Ormeling, 2020). Objects specific to Earth can be referred to it. Spatial data can be vector data (points, lines, polygons) or raster data (a grid of cells or pixels corresponding to a geographic attribute). Satellites, ground surveys, and other capturing devices are used to acquire this data (Nagavi et al., 2024). Smart cities can use information and communication technologies (ICT) for urban planning and manage and use resources effectively to run city operations and services (Zheng et al., 2020). Infrastructure improves transportation and logistic networks and creates sustainable environments with livable communities. These reach the United Nations Sustainable Development Goal (SDG) 11 (Costa et al., 2024). SCs are recognized as processes for social innovation and promote inclusive and sustainable urban development. Integrating advanced technologies and following data-driven solutions becomes essential for creating more resilient and sustainable communities as cities grow and face increasingly complex challenges.

1.2 Geospatial Technology Applications in Smart Cities

Geospatial Technologies are critical in urban infrastructure, such as gas, water supply, wastewater, and new road layouts (Marzouk & Othman, 2020). Geospatial data is essential and will be integral to various systems used to run Smart city infrastructure. Figure 2 summarizes the various applications of geospatial technologies in Smart cities.

Figure 2. Applications of Geospatial technologies in Smart cities

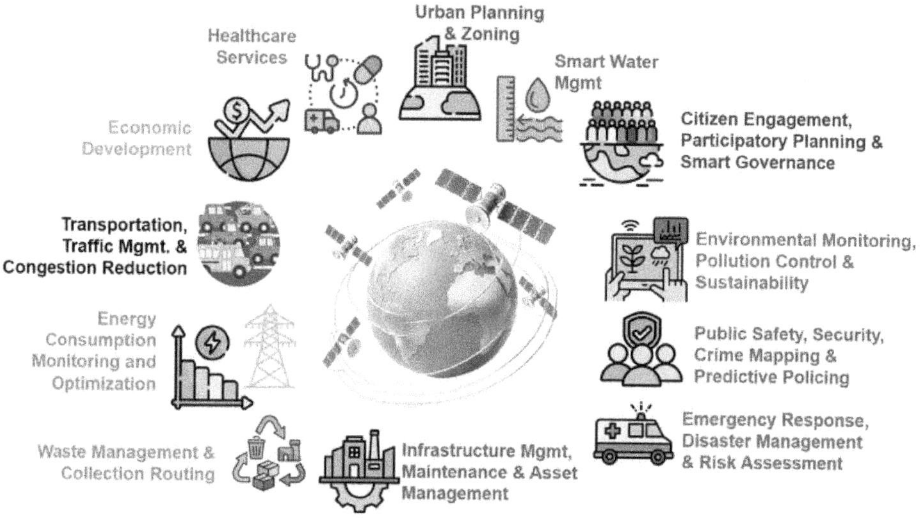

The following heads help understand their role:

1. **Planning**: Geospatial data can identify suitable locations for new infrastructure, such as roads, bridges, and utilities, and put land to proper use. It helps design infrastructure resilient to natural disasters and other hazards (Zhou et al., 2021). It helps disaster preparedness, response, and recovery efforts (Dhananjaya et al., 2021). It contributes to the United Nations SDG goals (Nagavi et al., 2024). For example, Bangladesh built and used disaster risk mitigation to predict floods and give early warning from recurring natural disasters (Uddin & Matin, 2021). Geospatial technology helped map hard zonation and flood shelter (Balasubramanian, 2024). Urban landscapes can be prioritized to develop green urban infrastructure (Borisova et al., 2024). Green infrastructure can help protect urban biodiversity (Ribeiro et al., 2024). African countries benefited from AI-driven wildlife conservation monitoring. The City of San Francisco and other

Bay Area cities used geospatial data to design a new transportation system that reduces traffic congestion and minimizes Traffic-Related Air Pollution (TRAP) (Chiang, 2021). Table 1 illustrates how Smart cities and geospatial technologies contribute to specific United Nations Sustainable Development Goals (SDGs).

Table 1. United Nations Sustainable Development Goals (SDGs) Goal Contributions of Smart Cities and Geospatial Technologies

SDG	Goal	Contribution of Smart Cities	Contribution of Geospatial Technologies
SDG 1	No Poverty	Promotes equitable access to city services, improving livelihoods and quality of life through digital infrastructure.	Maps vulnerable populations to optimize resource allocation and monitor poverty-related trends.
SDG 2	Zero Hunger	Enhances urban agriculture through Smart technologies, optimizing food supply chains.	Facilitates precision agriculture and monitors food production areas for improved yields.
SDG 3	Good Health and Well-being	Implements Smart healthcare systems and telemedicine and monitors public health trends.	Tracks health infrastructure, disease outbreaks, and environmental impacts on health.
SDG 4	Quality Education	It provides digital platforms for accessible education and Smart infrastructure for schools.	Maps access to educational resources and identifies underserved areas.
SDG 5	Gender Equality	Utilizes data to identify and close gender gaps in urban planning and services.	Provides location-based analysis to identify gender-based inequalities in service access.
SDG 6	Clean Water and Sanitation	Optimizes water management through Smart systems for efficient distribution and waste management.	Maps water resources and helps monitor water quality and access in urban and rural areas.
SDG 7	Affordable and Clean Energy	Promotes energy-efficient buildings, smart grids, and renewable energy use in urban areas.	Supports location analysis for renewable energy sources like solar and wind.
SDG 8	Decent Work and Economic Growth	Fosters innovation, Smart economic zones, and digitally inclusive workplaces.	Identifies economic disparities, opportunities, and industrial development zones.
SDG 9	Industry, Innovation, and Infrastructure	Integrates IoT, Smart transport, and urban infrastructure management systems.	Maps and monitors infrastructure development and promotes innovation clusters.
SDG 10	Reduced Inequalities	Enhances accessibility to city services for all, promoting inclusivity through data-driven policy.	Maps inequality across regions, identifying marginalized groups and ensuring inclusive growth.
SDG 11	Sustainable Cities and Communities	Builds resilient urban environments through smart mobility, waste management, and housing.	Enables urban planning, disaster management, and monitoring of housing and infrastructure.

continued on following page

Table 1. Continued

SDG	Goal	Contribution of Smart Cities	Contribution of Geospatial Technologies
SDG 12	Responsible Consumption and Production	Monitors and reduces waste through Smart systems, promoting sustainable consumption in cities.	Tracks resources and waste management to optimize production and consumption patterns.
SDG 13	Climate Action	Reduces urban carbon footprint through Smart energy, transportation, and emission management.	Maps climate risks and monitors carbon emissions and natural disaster vulnerability.
SDG 14	Life Below Water	Manages marine pollution and coastal protection through Smart city systems.	Maps and monitors marine ecosystems, pollution sources, and coastal habitats.
SDG 15	Life on Land	Protects urban green spaces and biodiversity with Smart environmental monitoring.	Monitors deforestation, biodiversity loss, and land-use changes to support conservation.
SDG 16	Peace, Justice, and Strong Institutions	Enhances urban safety and governance through data-driven policies, Smart surveillance, and transparent systems.	Maps crime hotspots, tracks governance efficiency and supports transparent decision-making processes.
SDG 17	Partnerships for the Goals	Promotes collaboration through data-sharing, Smart city networks, and global partnerships.	Provides geospatial data-sharing platforms for global partnerships and policy alignment.

2. **Development**: Geospatial data can help in tracking the condition of existing infrastructure. City areas that require repairs on a priority basis, such as road potholes, can be identified, and maintenance or repair works can be undertaken (Gavali et al., 2024). Similarly, infrastructure systems can be optimized to run at the best performance levels. The cities of London and Newcastle use the geospatial layout of utility infrastructure networks to manage and maintain their aging water infrastructure. Water leak detection and other problems in the water distribution system can be detected using sensors and geospatial data (Ji, 2020).
3. **Management**: Geospatial data can be used to manage infrastructure smoothly and economically (Krishna et al., 2024). Smart Transportation Systems use the data for traffic management, mobility, and routing of autonomous vehicles. It helps respond to emergencies like road accidents, natural disasters, or public health crises. It can locate and track affected areas and help in relief efforts. For example, the City of New York City, USA, uses geospatial data to respond to medical emergencies. The city has developed a system that creates geospatial "hot spots" that track disease outbreaks and their spread. The system also expedites treatment and organizes emergency response efforts (Kost, 2020).

4. **Optimization**: By properly using geospatial data, city planners can allow for the intelligent distribution, utilization, and management of city resources. For example, it can help Smart cities adjust energy distribution based on the power consumption patterns of the city residents, manage water resources at the micro level, and monitor air quality. The City of Chicago uses geospatial data to optimize its energy distribution based on the power consumption patterns of its residents. The city has installed smart meters in homes and businesses throughout the city. These meters collect real-time energy consumption data and send it for analysis using geospatial techniques. The analysis provides details of high energy consumption areas and the reasons for such high/abnormal consumption and allows the city administration to devise appropriate strategies.

Geospatial data enables data integration from multiple sources to perform spatial, temporal, and thematic analytics captured using multiple technologies. Doing so gives insights into efficient planning and management processes. The efficiency of logistics, transportation networks, energy distribution systems, water supply, and waste management is improved (Shahat Osman & Elragal, 2021). Urban planners can use geospatial data for three core reasons:

1. Model infrastructure scenarios
2. Predict growth patterns
3. Identify optimal locations for new developments.

Key geospatial technologies involved are:

1. **Geographic Information Systems (GIS)** use hardware, software, and data to handle all geographically referenced information. It helps perform spatial analysis (such as proximity analysis, overlay analysis, and network analysis), data visualization, and data management. GIS professionals use GIS services to determine property boundaries and land subdivisions and guide the construction sites. Real-time GIS provides high throughput, and high-speed processing of GIS data streams are topics of research interest (Li et al., 2020).
2. **Remote Sensing (RS)** captures geographic information without having physical contact within Earth observation (EO) bounds. Sensors installed on top of drones, aircraft, and satellites collect data from the surface of Earth. It uses various forms of RS, such as Optical RS, Microwave RS, Hyperspectral RS, and others. Visible, infrared, and microwave sources capture multispectral data. Changes in the Earth's surface because of natural disasters, land use changes, and environmental degradation can detected. Drone surveys can provide high-resolution aerial imagery and cover large areas quickly and efficiently.

3. **Global Positioning Systems (GPS)** uses satellite-based navigation systems to provide location and time information anywhere on or near the Earth's surface. It has applications in navigation, surveying, mapping, geocaching, and disaster management (Yabe et al., 2022). Precision GPS technology provides highly accurate location data. The precision levels are down to centimeters now, making it suitable for various industries and applications.

Using these, urban planners can capture and analyze data related to land use, transportation networks, environmental features, and infrastructure systems and make informed decisions. Maps and 3D models aid in the visual representation of spatial data. The models present a shared understanding of urban dynamics and support participatory planning processes for stakeholders. High-resolution terrain models and hydrological data help cities develop robust mitigation strategies and optimize emergency response protocols.

1.3 Objectives and Scope of the Chapter

This work first explains what smart cities are all about and highlights the need, importance, and various applications of geospatial technology in smart cities. Section 2 outlines various geospatial technologies emphasizing the significance of geospatial data visualization. It covers key aspects such as data collection, processing, and visualization techniques, along with an introduction to widely used geospatial data management tools and effective data storage strategies. Section 3 focuses on integrating geospatial data with modern disruption technologies such as IoT, Big Data, AI, and Cloud Storage. Figure 3 shows traditional and modern technologies used in geospatial analysis.

Figure 3. Traditional & Modern Technologies in Geospatial Analysis

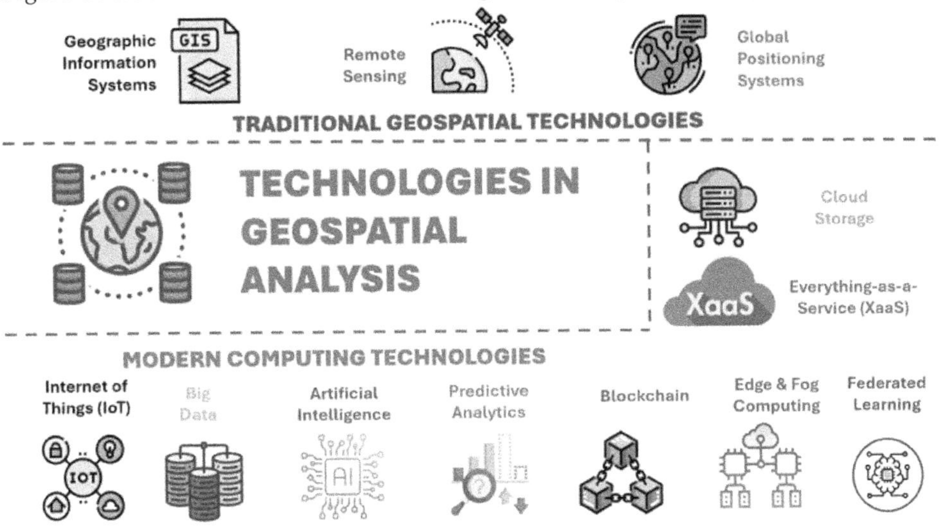

2. GEOSPATIAL TECHNOLOGIES AND DATA SOURCES FOR SMART CITIES

This section covers the fundamentals of geospatial technologies and data used in Smart Cities, laying the technological foundations of the concepts.

2.1 Types of Geospatial Data

Geospatial data involves information from a diverse range of sources. Each source contributes to the spatial intelligence essential for developing and managing Smart cities. Geospatial data sources can be broadly classified into:

1. Traditional geospatial data sources
2. Modern geospatial data sources

Combining traditional and new geospatial data sources provides a wealth of information for Smart cities. Table 2 shows the types of vital geospatial data and their use cases in Smart Cities.

Table 2. Types of Geospatial Data and Their Use Cases in Smart Cities

Data Source	Description	Few Key Applications
Satellite Imagery	High-resolution data from orbiting satellites	Urban planning, disaster management
Geographic Information System (GIS)	Joins together spatial and non-spatial data	Land use mapping, environmental monitoring
Radio Detection and Ranging (RADAR)	Data is collected from airborne and satellite sources.	Traffic management and monitoring, Smart parking systems, Pedestrian and Cyclist Safety, environment monitoring
Light Detection and Ranging (LIDAR)	3D mapping, topographical assessments	Infrastructure development, flood prediction
Uncrewed Aerial Vehicles (UAVs)	Real-time high-resolution imagery	Traffic monitoring, structural inspections

Popular traditional geospatial data sources are:

1. **Satellite imagery**: Several countries launched artificial satellites that orbit the Earth and send in images. These images are a valuable source of geospatial data for Smart cities because they provide low to high-resolution real-time data. Such data is comprehensive and gives coverage and temporal analysis capabilities. It is more suitable for large-scale monitoring.

Various sources for collecting satellite imagery are:

a. Landsat (of NASA/USGS; such as Landsat 8)
b. Copernicus and Sentinel (of European Space Agency); Sentinel-2 Satellite images are famous amongst Land Use Land Cover (LULC)
c. Moderate Resolution Imaging Spectroradiometer (MODIS) from NASA satellites (like Terra and Aqua)
d. WorldView satellites (e.g., WorldView-1, WorldView-2, and WorldView-3) are operated by Maxar Technologies

USGS Earth Explorer provides data for STRM DEM 1-ARC. It provides details such as slope, elevation, and drainage density. Precipitation (rainfall) data can be collected from Climate Hazards Group InfraRed Precipitation with Station (CHIRPS) (Tetteh et al., 2024). Google Earth Engine (GEE) processes Sentinel-1 Synthetic Aperture Radar (SAR) images available in the public domain (Uddin & Matin, 2021). Satellite imagery can be used in:

a. Urban planning (to map land use)

b. Identification of environmental hazards (environmental condition monitoring)
c. Monitor changes in the built environment (disaster management), such as studying turbid waters like rivers (Wu et al., 2024).
d. Other related decision-making (Qudus et al., 2024)

2. **Geographic Information System (GIS)**: GIS systems can integrate spatial and non-spatial data in vector and raster formats and allow complex spatial analyses and visualization. It can represent various features, such as roads, buildings, and utilities. GIS data can help infrastructure development and resource management by creating maps, analyzing spatial relationships, and developing models. Various GIS systems called Participatory Geographic Information Systems (PGIS) are popular. It refers to using GIS in a participatory manner, where local communities, stakeholders, or the public are involved in collecting, managing, and analyzing spatial data. PGIS promotes inclusion in GIS and makes data available for marginalized and underserved populations (Malakar & Roy, 2024). 3D seismic imaging involves sending sound waves into the ground and analyzing the reflections. Geologists use these to create detailed subsurface maps by visualizing geological formations with high precision. Reflection seismology and tomography techniques are often used.

3. **Light Detection and Ranging (LIDAR)**: LIDAR is a remote sensing technology wherein a laser beam measures the distance between the sensor and the ground. It is used for 3D mapping, spectral analysis capabilities, and terrain modeling that helps in accurate topographical assessments and urban design. Building highly accurate 3D models (such as urban landscapes) of the built environment can be done with LIDAR. LIDAR data can help in urban development, building monitoring, infrastructure assessment, and disaster management.

4. **Uncrewed Aerial Vehicles (UAVs)**: UAVs or drones generate ultra-high-resolution imagery and 3D models from difficult or dangerous areas. They provide real-time data acquisition. UAVs generally have various sensors, such as cameras, thermal, and multispectral sensors. They offer high flexibility in data collection. Detailed aerial imagery and hyperspectral remote sensing can be used for structural inspections, urban construction, and planning (B. Yang et al., 2022). They also provide real-time updates, such as traffic monitoring.

Smart cities also collect public and open data from web portals, census and demographic data, meteorological and climate data, hydrological data, and transportation and infrastructure data. Some research studies pondered upon the idea of digital twins of urban areas and suggested the use of sensors and instruments such as aerial photography and orthophotomosaic creation (raster image mosaic made by merging or stitching orthophotos), Aerial LIDAR scanning, Simultaneous Localization and Mapping (SLAM) technology for handheld LIDAR scanning, and Global Navigation Satellite Systems (GNSS) measurements to determine control points (Petrov et al., 2024). Apart from these traditional geospatial data sources, Smart cities are also generating new geospatial data from several modern sources, including:

1. **IoT Devices and Sensors**: Smart cities install and maintain a network of Internet of Things (IoT) devices, wireless sensor devices (WSD), and other ground-based sensors (Krishna et al., 2024). These devices capture real-time spatial data on traffic flow, air quality, noise levels, and energy consumption. The captured data helps identify areas that require improvements.
2. **Social media**: Citizen participation forms the human and social capital of a Smart city (Kaluarachchi, 2022). Increased social media penetration means increased generation of rich geospatial data collected. Human activity and movement are also usually captured along with citizen perceptions, comments, and preferences. This data helps understand how people use the city and identify areas where there is a need for new or improved infrastructure. Social media data provide rich sources of geo-referenced user-generated content along with insights into urban dynamics and citizen sentiments.
3. **Mobile devices**: Mobile devices provide precise location information on the movement patterns of people and vehicles. GPS and cellular network towers capture this data almost automatically (Yabe et al., 2022). The data complements geospatial data and helps understand service utilization. It helps understand how people move around the city and identify areas where there is a need for new or improved infrastructure, such as transportation paths and places to sit and relax.
4. **Metaverse, Augmented Reality (AR), and Virtual Reality (VR)**: Interactive metaverse platforms like Mona and Voxels create virtual space using digital photogrammetry with images captured from drones and 3D Gaussian splatting (3DGS). The immersive visualization tools offered by AR and VR and other Web3 technologies like Blockchain (Yadav & Sagi, 2024) can enhance the geospatial data applications for urban planning and public engagement (Abramov et al., 2024). These diverse geospatial data sources collectively support the development of Smart cities by enabling data-driven strategies for sustainable and resilient urban environments.
5. **Crowd-sourced Data:** Crowd-sourcing platforms like OpenStreetMap enrich geospatial databases with up-to-date information on urban features and infrastructure. Fast infrastructure-building countries like Qatar benefited from such systems where commuters' mobile data helped other commuters understand

road diversions. Mobile devices and social media platforms form crowd-sourced geospatial information. They give insights into human behavior, mobility patterns, behavioral trends, public sentiment, and event detection.

6. **Other Sources:** Terrestrial laser scanning adds to street-level details, while thermal imagery facilitates urban heat island analysis. Hyperspectral sensors provide hyperspectral imaging by capturing and processing images across a wide range of electromagnetic spectrum bands far beyond visible light. This data can provide rich spectral details, material composition, and their physical and chemical properties that can be used in urban surface studies (B. Yang et al., 2022) and in seed and food storage (Liao et al., 2023). Some critical foundational layers are:
   a. Administrative boundaries
   b. Cadastral data
   c. Transportation network information form.

Integrating these diverse data types can give a comprehensive geospatial framework for informed urban planning and management in Smart city contexts.

2.2 Data Collection and Processing Techniques

Geospatial data collection for Smart cities uses multiple devices, processes, and methods. These include remote sensing devices, IoT sensors, fog nodes, GPS-enabled devices, and mobile devices (Badidi et al., 2020). These methods allow for acquiring high-resolution geospatial data essential for urban planning and management. Data is cleaned and preprocessed to ensure accuracy and reliability. Data preprocessing techniques address data heterogeneity. The techniques include data normalization and standardization (Saha et al., 2021). Removing duplicates and correcting errors are part of the process. Standardized data is necessary because it is often collected from different sources and at different times.

Popular geospatial data collection methods include aerial surveys, satellite imagery, and ground-based surveys. Aerial surveys and satellite imagery provide extensive coverage and are particularly useful for monitoring large urban areas. The satellites orbiting the Earth provide satellite imagery. Ground-based surveys offer high precision and are ideal for detailed mapping tasks. Data management in Smart cities requires a robust digital infrastructure and storage strategies. Cloud-based solutions come with scalability and flexibility. These platforms provide real-time data processing and integration from diverse sources. Advanced data management techniques such as data warehousing and big data analytics will handle the enormous volume of data generated.

Some things to consider in data collection and processing of geospatial data are:

1. **Data quality and accuracy**: Geospatial data quality and accuracy are critical because of the heavy dependency on them by Smart city applications. Data quality can vary depending on the data source and data collection methods. Data accuracy might depend on positional, temporal, and attribute accuracy.
2. **Data cleaning:** Automated anomaly detection algorithms and machine learning models can identify and rectify inconsistencies in data. Spatial interpolation methods such as kriging and inverse distance weighting address gaps in geospatial datasets. Dimensionality reduction techniques, including principal component analysis and t-SNE, optimize high-dimensional geospatial data for efficient processing.
3. **Data transformation**: This process involves converting the data into an appropriate format for analysis. It can involve various techniques, such as rescaling, normalization, and aggregation. For example, data collected from GPS trackers on vehicles, traffic cameras, and road sensors would be of different formats and scales requiring data transformation.
4. **Data integration**: SCs collect data from different integrated sources to get a unified city picture. Data integration can be challenging but is inevitable in developing practical Smart city applications.
5. **Data management infrastructures**: Data infrastructure includes the hardware and software used to store, process, and analyze data. Smart cities should use distributed storage systems and cloud-based platforms to handle the volume and velocity of data. Geospatial databases, such as PostGIS and MongoDB, offer specialized indexing and querying capabilities for spatial data. Data lakes and data warehouses facilitate the integration of heterogeneous data sources while ensuring scalability and fault tolerance. Data storage strategies such as Cloud, On-premises, and Hybrid storage are popular.
6. **Data privacy and security**: Geospatial data can contain sensitive information about people and places. Hence, data managers/service providers should take measures to protect data from unauthorized access and use.

2.3 Geospatial Data Visualization

Maps, charts, and graphs help represent geospatial data visually in Smart cities. Cartography is the study and practice of making and using maps (Kraak & Ormeling, 2020). It involves interpreting complex spatial datasets and supporting decision-making processes in Smart cities (Costa et al., 2024). Various techniques and methods help in the dynamic exploration of data. They help effectively communicate spatial information and allow interpretation and analysis of spatial data.

These include cartographic principles, including symbology, color theory, and visual hierarchy, amongst others.

More advanced techniques have been developed and used in recent times. These include choropleth mapping, heat maps, and 3D visualization. These represent various data types, including point, line, and polygon. The techniques enhance the interpretability of complex urban datasets. Interactive visualization methods use dynamic filtering and temporal visualization (animations) and update the dashboards. All these tools' fundamental principles are to provide clarity, accuracy, scalability, and the intuitive representation of spatial relationships. The goal is that the visual representations are both comprehensible and precise.

2.4 Considerations for Visualizing Geospatial Data

Geospatial data visualization helps explore and communicate with data and to solve problems. A Geospatial analyst or a GIS specialist typically does geospatial data visualization. Geospatial analysis conducts several types of geospatial analysis, but two statistical techniques require special mention:

1. Spectral Analysis
2. Time Series Analysis

These are closely related statistical techniques used to analyze data collected over time. Spectral analysis identifies the periodic components of a time series. Time Series Analysis is the study of how a variable changes over time. It is used to identify trends, patterns, and seasonality in time series data and can even help forecast future values. Map representations show the rate of change per time unit in a geospatial context. Both these analyses use grid cells for geospatial visualization. A grid cell is the smallest regional unit when performing trend rate analysis. Each grid cell is assigned a unique identifier containing a data value time series. Trend rates are calculated for each grid cell. Trend lines are used to mark the individual grid cells. The grid cells are then decomposed with the seasonal periods specified through spectral analysis. Effective geospatial data visualization requires careful consideration of the following three principles:

1. **Choosing the correct visualization technique**: Two critical factors that determine the selection of correct visualization are:
   a. The data type of the data being used to create the visualization
   b. The purpose of the visualization.

For example, maps are a good choice for visualizing the location of features. Charts and graphs are a good choice for visualizing trends and patterns.

2. **Using the correct visual variables**: Visual variables, such as color, shape, and size, encode information in a visualization. For example:
   a. Color can represent different categories of data.
   b. Shapes can represent different types of features.
3. **Creating clear and concise visualizations**: Visualizations should be clear and concise so that end users can easily understand and grasp the information. Hence, simple and uncluttered designs are to be used. Unnecessary visual elements are to be avoided.

2.5 Best Practices for Smart City Data Visualization

Geospatial data visualization developers use User-Centered Design (UCD) principles for data visualization in Smart city applications, as explained in Figure 4. This will make the data accessible, actionable, and meaningful for the stakeholders. The primary goal is to customize the visualizations according to needs and usage.

Figure 4. User-Centered Design (UCD) principles in Geospatial Data Visualization

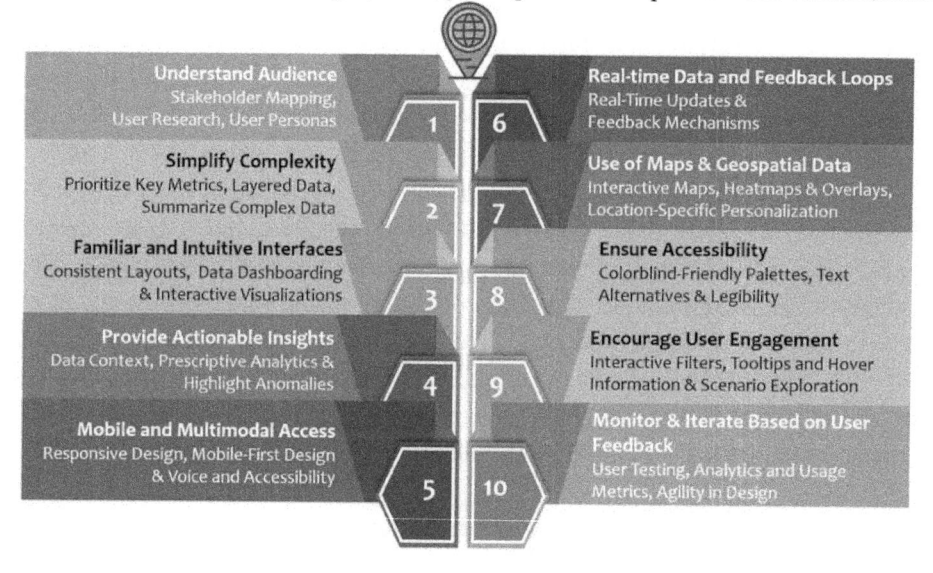

#	Principle	Details
1	Understand Audience	Stakeholder Mapping, User Research, User Personas
2	Simplify Complexity	Prioritize Key Metrics, Layered Data, Summarize Complex Data
3	Familiar and Intuitive Interfaces	Consistent Layouts, Data Dashboarding & Interactive Visualizations
4	Provide Actionable Insights	Data Context, Prescriptive Analytics & Highlight Anomalies
5	Mobile and Multimodal Access	Responsive Design, Mobile-First Design & Voice and Accessibility
6	Real-time Data and Feedback Loops	Real-Time Updates & Feedback Mechanisms
7	Use of Maps & Geospatial Data	Interactive Maps, Heatmaps & Overlays, Location-Specific Personalization
8	Ensure Accessibility	Colorblind-Friendly Palettes, Text Alternatives & Legibility
9	Encourage User Engagement	Interactive Filters, Tooltips and Hover Information & Scenario Exploration
10	Monitor & Iterate Based on User Feedback	User Testing, Analytics and Usage Metrics, Agility in Design

Some best practices for effective data visualization include:

1. Visual variables, such as color, shape, and size, can encode different information types in a visualization.
2. Interactive visualizations allow users to explore the data in more detail. Users can easily zoom in and out of a map or filter the data by different criteria.
3. Visualizations should help explain and provide content for the data being rendered. Adding a legend, a scale bar, or extra text to clarify the data often helps.
4. The integration of multi-source data can provide a comprehensive view of urban phenomena.
5. The tools should allow for scalability and cross-platform accessibility.
6. The tools should be customized with the end-user in mind. The end-users could be policymakers, urban planners, or citizens, who require a different level of usability and comprehension. Ensuring user-centric designs enhances the utility of visualizations.
7. Employing standardized data formats and adhering to visualization guidelines, such as those proposed by ISO/IEC, ensures consistency and interoperability.
8. Big Data analytics and machine learning techniques can uncover hidden patterns and trends, driving informed decision-making.

2.6 Geospatial Data Management Tools

Prominent tools, platforms, and databases for geospatial data management are:

1. **Industry-standard GIS platforms and databases** like ESRI's ArcGIS Pro, Quantum GIS (QGIS; open-source GIS software), Carto (geospatial data analysis and visualization that integrates data from different sources like IoT and sensors), Feature Manipulation Engine (FME; a platform for integrating, transforming, and automating geospatial data workflows) and Tableau (commercial visualization software) offer excellent visualization capabilities. They are helpful, especially for spatial analysis and mapping.
2. **GIS Databases** are ESRI ArcGIS Online, OpenStreetMap (OSM; crowd-sourced map data), Natural Earth (public domain map data), Mapbox (open-source platform for custom maps, location search, and navigation data), HERE Technologies (mapping, location, and navigation services, along with APIs for real-time traffic, route optimization, and geocoding), Mapillary (tool for street-level imagery collection and mapping).
3. **Geospatial Knowledge Graphs (GKG)** such as WorldKG, Yago2Geo, and FineGeoKG (Wei et al., 2023)
4. **Database Management Tools** include the traditional RDBMS approach (PgRouting within Postgres), spatial database (PostGIS), and a NoSQL graph database (Neo4j) (Dia et al., 2024; Ji, 2020).

5. **Web-based tools** such as Mapbox and Carto are now available to enable the creation of responsive, shareable visualizations.
6. **Open-source formats, standards, libraries** like D3.js, CesiumJS (JavaScript library for 3D geospatial visualization in web browsers), and Deck.gl (WebGL-powered framework for large-scale data visualization, often used in mapping and geospatial analytics), Turf.js (JavaScript library for advanced geospatial analysis), Leaflet (open-source JavaScript library for mobile-friendly interactive maps), Geospatial Data Abstraction Library (an open-source library for reading and writing raster and vector geospatial data formats), GeoJSON (format for encoding geographic data structures using JSON), SensorThings API (open standard for managing IoT data with geospatial features), CityGML (data model and XML-based format for representing 3D city models). Software developers enjoy flexibility for custom visualization solutions with these libraries.
7. **Cloud-based platforms** such as Google Maps, Google Earth Engine (GEE), and Microsoft's Azure Maps have become famous. They have features that can visualize large-scale urban datasets, such as:
   a. Scalability
   b. Ability to handle large datasets
   c. Help create interactive maps.

Web mapping tools are increasingly accessible and user-friendly, allowing non-experts to navigate maps and query data easily. They now offer advanced map creation and online geoprocessing features beyond default functions. These capabilities allow users to generate new data and insights directly within the platform. (González et al., 2020).

2.7 Geospatial Data Storage Strategies

Proper data storage strategies are required to ensure data accessibility and security. Distributed storage systems and data lakes are commonly used to store large datasets. These are efficient for retrieval and analysis. Data storage solutions have evolved and are strong enough to meet the unique requirements of geospatial Big Data. Specialized spatial databases and data lakes enable storing and retrieving diverse geospatial data types at unprecedented scales, including vector and raster data. These systems support spatial indexing and querying for efficient data access and analysis.

3. INTEGRATING GEOSPATIAL DATA WITH MODERN TECHNOLOGIES IN SMART CITIES

Modern technologies can use geospatial data to help build the sub-systems and domains of Smart cities (Rani et al., 2021). AI, Big Data, and IoT are already proven to be good tools for wetland management (L. K. Sharma & Naik, 2024). These technologies allowed for sophisticated data collection and analysis and improved decision-making. Using real-time geospatial information will allow us to respond to infrastructure demands instantly. One example of this is to reroute road traffic during peak congestion.

Consequently, geospatial data is instrumental in advancing urban infrastructure towards greater sustainability, resilience, and adaptability in rapidly evolving urban environments. The level of urbanization is measured using Multi-Criteria based Decision Making (MCDM) such as Entropy-Weight and Technique for Order of Preference by Similarity to an Ideal Solution (TOPSIS) (Roy & Ray, 2024). Analytic Hierarchy Process (AHP) can help address problems related to assigning weights to each criterion. For example, the criterion that influences flooding are precipitation, drainage density, LULC, elevation, and slope (Tetteh et al., 2024). Multi-criteria analysis and assessments with GIS data have applications in green infrastructure and others (Ribeiro et al., 2024). Inspired by MCDM, a whole branch called multi-criteria spatial analysis methods has evolved (González et al., 2020). Table 3 briefly describes modern technologies used in geospatial applications.

Table 3. Modern Technologies for Integrating Geospatial Data

Technology	Brief Description
Internet of Things (IoT)	Network of connected devices to collect real-time urban data, such as in city service management
Big Data	Analyzing large and complex dataset processing to identify patterns and trends
Artificial Intelligence (AI)	Algorithms and models learn from data and make predictions. Machine Learning (ML) will train computers to perform analysis and get insights for decision-making.
Blockchain	Providing secure and transparent data sharing and has applications in secure land records, transparent urban governance
Edge Computing	Processing data close to its source to reduce latency
Federated Learning	Collaborative model training across multiple stakeholders while preserving data privacy
Parallel Computing	Speeds up geospatial data analysis by processing large datasets simultaneously across multiple processors, enabling faster and more complex simulations, modeling, and spatial analysis.
Cyber-infrastructure	Provides necessary networked computing resources, data storage, and tools for collaborative geospatial research, enabling real-time data sharing and large-scale geospatial computations across distributed systems.

3.1 Internet of Things (IoT) for Smart Cities

IoT involves a collection of objects that collect data and transmit it to users using the internet without human intervention. They can capture large quantities of geospatial data (Silva & Holanda, 2022). So, IoT is a perfect tool for data-driven decision-making for Smart cities.

Industry forecasts say that by 2025, 75 billion IoT-connected devices will be in use worldwide, reflecting a nearly threefold increase in the IoT installation base since 2019. In 2020, the total number of IoT-connected devices was 26.66 billion, with 5.4 billion installed in China alone. While IoT devices generated 13.6 zettabytes of data in 2019 alone, they would generate 79.4 zettabytes in 2025 (FinancesOnline, 2024; Statista, 2024). They use sensors and devices that collect real-time data on urban conditions, such as air quality, traffic flow, and energy consumption.

Geospatial data will call for deploying IoT devices across urban landscapes and accessing their data. They help monitor and manage city services in real-time. The synergy between geospatial data and advanced computational methods enables the extraction of actionable insights from complex spatial datasets. This integration supports proactive maintenance strategies, predictive analytics for infrastructure resilience, and the development of Smart grid technologies.

3.2 Big Data in Geospatial Analysis

Big Data revolutionized the field of geospatial analysis, which improved our understanding of and management of urban environments (Shahat Osman & El-ragal, 2021). The integration of Big Data with data from geospatial analysis will bring immense value to Smart cities. Big Data refers to datasets that are too large and complex to be processed using traditional methods. Such datasets typically can contain billions or even trillions of records. Big Data tools and techniques help analyze large and complex datasets to identify patterns and trends. India designated 100 Smart cities and began sharing 3,571 catalogs and 5,363 resources under its Datasmart Cities Strategy (MHUA, 2024). Efficient geospatial analytics on time series Big Data are now available, which allows running a hybrid geospatial proximity query on time series data (Al Jawarneh et al., 2022).

Large-scale data processing plays a crucial role in handling the immense volume, velocity, veracity, value, and variety of geospatial data generated by Smart cities. Advanced processing techniques are to be used to extract meaningful insights. Similarly, advanced distributed computing frameworks, such as MapReduce (best suited for partitioning data), Apache Hadoop (popular for storing big data in large-distributed files), and Apache Spark (famous for its in-memory computing), are used to process geospatial Big Data efficiently (Li et al., 2020). Such frameworks are used in real-time analysis and decision-making. Smart cities can process and interpret vast amounts of location-based information using Big Data. Urban planners benefit by gaining insights into patterns, trends, and correlations across vast geographic regions as part of their decision-making processes. Big Data brings two specific advantages to geospatial analysis:

1. New insights from the spatial distribution of phenomena are drawn. For example, researchers can identify human movement and interaction patterns by analyzing extensive social media data.
2. Help build new geospatial analysis methods. For example, ML algorithms can be trained on Big Data to identify objects in satellite imagery or to predict future events.

Though there are challenges in Big Data implementations, they can potentially revolutionize the field of geospatial analysis. By providing new insights and enabling new methods, Big Data can help us better understand phenomena' spatial distribution and solve many problems.

3.3 AI, ML, and DL in Geospatial Data Management for Smart Cities

Urban planners can use geospatial data to design layouts that optimize resource use. Artificial Intelligence (AI) uses algorithms and models to learn from data and make predictions or decisions. Smart cities underscore the need for 'hard' physical infrastructure and thus require a case for potential investment of resources (Masik et al., 2021). AI-based simulation models can predict long-term benefits that can accrue from the investments. AI and geospatial technologies can transform supply chain transparency and visibility by enhancing goods tracking, equipment maintenance forecasting, and route optimization. Studies confirmed that doing so promotes a circular economy and makes the city self-sustainable to the extent possible. (Ioanid & Andrei, 2024).

Machine Learning (ML) techniques are applied to multi-temporal satellite imagery and LiDAR data to detect and predict urban sprawl patterns. Some popular ML techniques are Convolutional Neural Networks (CNNs) and random forest classifiers. These models incorporate spatial predictor variables, including proximity to existing urban centers, transportation networks, and topographic features, to generate probabilistic maps of future urban expansion. Geospatial data with ML algorithms such as Gradient Boosting Machine (GBM) enhances bathymetric mapping precision. They can handle high-dimensional heterogeneous data integration and depth prediction. Consequently, better hydrological dynamics and more accurate predictions, such as turbid waters of the channel, can be obtained (Wu et al., 2024). Predictive infrastructure components or equipment maintenance will likely fail, allowing cities to schedule maintenance and proactively avoid costly breakdowns.

Deep learning architectures strengthen advanced spatial regression models to predict population density changes across urban landscapes. Socio-economic indicators are layered on top of high-resolution remote sensing data. With this, the models can provide granular insights into future demographic distributions, informing decisions on public service allocation and urban zoning policies. Linear 4.0 projects are a collection of thirty-seven Industry 4.0 technologies, including AI, Big Data management, and Intelligent systems that extend the concept of sustainable development to Smart cities (Sánchez et al., 2024). AI systems can be input with free or low-cost Earth Observation (EO) and geospatial data. The AI system can help in Sentinel-2 and morphometric analysis to map urban areas and extract information supporting the Sustainable Development Goals (SDG) indicator 11.1.1 (Tareke et al., 2024). Extraction capabilities of computer vision and prediction capabilities of machine learning and deep learning allow for decision support systems (DSS) capable of analyzing, interpreting, and predicting complex spatial phenomena (Balasubramanian, 2024).

ML and DL techniques can be used with object-oriented feature extraction (OOFE) algorithms to improve feature extraction accuracy. Similarly, hybrid approaches combining convolutional neural networks (CNNs) with traditional object-based methods have significantly improved land cover classification accuracy. Adaptive multi-resolution segmentation (MRS) techniques can improve handling complex landscapes. Spectral indices improve land cover types' differentiation (Qudus et al., 2024). Two popular ones are:

1. Normalized Difference Vegetation Index (NDVI)
2. Texture features derived from gray-level co-occurrence matrices (GLCM)

NLP and Deep Transfer Learning (DTL) are used to build AI-based Chatbots. These chatbots are integrated into geo portals to improve geospatial services and promote human-computer interactions (HCI). Chatbots can significantly reduce the average time required to access geospatial data and processing services by more than 50% (Vahidnia, 2024). AI-based techniques such as 3D point clouds and geo-spatial digital twins are generic components of geospatial AI. These techniques help efficiently build and maintain base data for geospatial digital twins like virtual 3D cities, indoor, or building information models (Döllner, 2020).

3.4 Predictive Analytics for Smart Cities

Predictive analytics tools improve urban development decision-making (Huang et al., 2021), leading to better outcomes for residents and businesses. These tools analyze historical data and identify patterns to forecast future trends and plan accordingly. Geospatial data contains information about the location of people, objects, and events. A bundle of such data is organized logically and can form a dataset collected from satellite imagery, sensors, and historical records. AI algorithms can analyze complex datasets and help identify patterns and trends that inform urban planners about future growth areas. The result of such analysis is proactive decision-making. Some applications of predictive analytics in Smart cities are:

1. Cities can use geospatial data to track the growth of its population over time. Historical data can be analyzed based on birth, death, and migration patterns. The data can help forecast future population growth and plan for constructing new housing, schools, and other infrastructure.
2. Cities could use geospatial data to track the movement of traffic based on volume and congestion in real-time. The data helps forecast future traffic patterns and plans for constructing new roads and highways.
3. Predictive analytics can improve urban development by planning new parks and green spaces. By simulating various scenarios, AI can assess the impact of different development strategies, helping planners to optimize resources and mitigate risks. It helps identify areas at risk of flooding or other natural disasters.

Location-based Big Data (LocBigData) combines location/activity sensing technologies and location-based services (LBS) (Huang et al., 2021). So, with AI-driven predictive analytics, cities can anticipate growth and create resilient, adaptive infrastructures that are well-equipped to handle future challenges.

3.5 Real-time Data Processing and Decision Support Systems (DSS)

City planners and administrators depend on real-time data processing and DSS to make timely decisions (Kaluarachchi, 2022). Their activities include monitoring urban dynamics, detecting anomalies, and responding to events as they unfold. They need systems that help collect, process, and analyze data in real-time and help them make informed decisions. The systems enable continuous feed, analysis, and utilization of spatiotemporal information from various data streams. Advanced geospatial analytics techniques such as spatial clustering, hot spot analysis, and predictive modeling can help extract and analyze the data. A DSS typically does three things (Gupta et al., 2022):

1. Use urban data, identify spatial relationships and dependencies, and process it in real time.
2. Analyze the data using various computing technologies.
3. Offer location-aware recommendations by considering.

These systems can interpret the processed geospatial information to suggest optimal courses of action for city managers and policymakers. These systems often incorporate spatial decision trees, MCDA, and location-allocation models to evaluate complex urban scenarios. Integrating real-time geospatial data processing with intelligent decision support mechanisms enables Smart cities to respond proactively to urban challenges, optimize resource allocation, and enhance overall urban livability. Integrating geospatial data into DSS allows for visualizing spatial patterns and trends. The systems help dynamically create and update maps and models reflecting the current state of the urban environment. Graphical dashboards keep refreshing with city statistical data and trends, allowing the city authorities to make informed, data-driven decisions (Rojas et al., 2020). Technologies such as IoT devices (Han et al., 2020), Fog computing (Patel et al., 2021), and Blockchain (Marsal-Llacuna, 2020) help with the SC dashboards. The decisions are vital to enhance sustainability, safety, and citizens' overall quality of life. Thus, real-time data processing is the backbone of Smart city operations, transforming geospatial data into actionable intelligence.

3.6 Blockchain for Smart Cities

Blockchain is a distributed ledger technology that provides secure and transparent data storage and sharing abilities. It provides enhanced security, transparency, decentralization, and interoperability when working with geospatial data (Chafiq et al., 2024). Blockchain helps store geographical location data such as WGS84 coordinates, Geohash, and Mapcode related to land administration as a part of e-governance (Tahar et al., 2024). Non-fungible tokens (NFT) can represent ownership of physical and digital assets and can be used in land administration and other Smart City applications (Musamih et al., 2024). Blockchains such as Solana and CORE are suitable for implementing a land administration system (Ilesanmi & Timothy, 2024).

3.7 Edge and Fog Computing for Smart Cities

Edge Computing (EC) allows faster on-site data preprocessing because it is close to data. Doing so reduces latency and bandwidth requirements and aids in faster decision-making. Fog computing promotes the Internet of Vehicles (IoV) concept by communicating traffic congestion status to drivers, informing them of road accidents and evacuation routes, and finding parking spots, among others (Badidi et al., 2020). Similarly, smart grid solutions such as power distribution, advanced metering, and micro-grid systems are also popular (Minh et al., 2022). Edge, Fog, and Cloud Computing form an excellent technology stack. This stack tends to be closer to the user, reducing computing costs and improving server response time. They also can address the limitations of Big Data in building Smart city applications (Badidi et al., 2020; Minh et al., 2022).

3.8 Federated Learning for Smart Cities

Federated Learning (FL) analyzes data across multiple locations (and shared collaborative learning) without transferring the actual data to a central server (or between satellites) (Khasgiwala et al., 2022). FL uses decentralized ML models. It enables collaborative model training across multiple smart city stakeholders while preserving data privacy and sovereignty (Ramu et al., 2022). Its applications in Smart cities come due to privacy concerns, data heterogeneity, and the need for real-time insights from various sources such as IoT sensors, traffic systems, and environmental monitors. FL brings a privacy-preserving solution, scalability, and classification, making it ideal for easy sharing of RS and crowd-sourced data (Moreno-Álvarez et al., 2024).

4. CONCLUSION

The growing urban population requires creating sustainable, resilient, and efficient ecosystems called Smart cities. As Smart cities develop, geospatial technologies and data will be increasingly crucial. City planners must use geospatial data to enhance urban planning and optimize resource allocation. The quality of life of citizens will eventually improve by offering them quality services. They provide critical insights for city planners and administrators to make key city infrastructure development and management decisions. Geospatial Analysts integrate geospatial data with both traditional and advanced technologies to draw predictive analytics and insights. Smart cities can achieve sustainable development and resilience using advanced geospatial data visualization, spatial analysis techniques, and real-time data processing. Data scientists try to create synergy between geospatial data sources and emerging computing technologies. The synergy opens doors for opportunities for data-driven decision-making in urban environments. The ongoing evolution of geospatial AI and machine learning algorithms promises to enhance human abilities to use spatial analysis and urban optimization. As urban areas expand, the strategic deployment of geospatial data will become more necessary. The technologies help cities address the complexities of modern urbanization, ensuring efficient, equitable, and sustainable urban growth.

REFERENCES

Abramov, N., Lankegowda, H., Liu, S., Barazzetti, L., Beltracchi, C., & Ruttico, P. (2024). Implementing Immersive Worlds for Metaverse-Based Participatory Design through Photogrammetry and Blockchain. *ISPRS International Journal of Geo-Information*, 13(6), 211. DOI: 10.3390/ijgi13060211

Al Jawarneh, I. M., Bellavista, P., Corradi, A., Foschini, L., & Montanari, R. (2022). Efficient Geospatial Analytics on Time Series Big Data. *ICC 2022 - IEEE International Conference on Communications*, 3002–3008. DOI: 10.1109/ICC45855.2022.9839005

Badidi, E., Mahrez, Z., & Sabir, E. (2020). Fog Computing for Smart Cities' Big Data Management and Analytics: A Review. *Future Internet*, 12(11), 190. DOI: 10.3390/fi12110190

Balasubramanian, S. (2024). AI-driven Geospatial Decision Support Systems for Sustainable Development and Natural Resource Management. *International Journal of Artificial Intelligence In Geosciences*, 2(1), 1–11.

Borisova, B., Semerdzhieva, L., Dimitrov, S., Valchev, S., Iliev, M., & Georgiev, K. (2024). Geospatial Prioritization of Terrains for "Greening" Urban Infrastructure. *Land (Basel)*, 13(9), 1487. DOI: 10.3390/land13091487

Chafiq, T., Azmi, R., Fadil, A., & Mohammed, O. (2024). Investigating the Potential of Blockchain Technology for Geospatial Data Sharing: Opportunities, Challenges, and Solutions. *Geomatica*, 100026(2), 100026. Advance online publication. DOI: 10.1016/j.geomat.2024.100026

Chiang, K. (2021, November 24). *Combating Traffic-Related Pollution in the Bay Area*. https://storymaps.arcgis.com/stories/8c1fb9facc774ada87d92e900e421b45

Costa, D. G., Bittencourt, J. C. N., Oliveira, F., Peixoto, J. P. J., & Jesus, T. C. (2024). Achieving Sustainable Smart Cities through Geospatial Data-Driven Approaches. *Sustainability (Basel)*, 16(2), 640. DOI: 10.3390/su16020640

Dhananjaya, A. S., Vinayak, B. K., Mahajan, S., & Agrawal, S. (2021). Geospatial Technology in Disaster Management: Harnessing Spatial Intelligence for Effective Preparedness, Response, and Recovery. *International Journal of Open Publication and Exploration*, 9, 36–41.

Dia, F., Bayar, N., & Abdellatif, T. (2024). From Functional Requirements to NoSQL Database Models: Application to IoT Geospatial Data. In Mosbah, M., Kechadi, T., Bellatreche, L., Gargouri, F., Guegan, C. G., Badir, H., Beheshti, A., & Gammoudi, M. M. (Eds.), *Advances in Model and Data Engineering in the Digitalization Era* (Vol. 2071, pp. 224–236). Springer Nature Switzerland., DOI: 10.1007/978-3-031-55729-3_18

Döllner, J. (2020). Geospatial Artificial Intelligence: Potentials of Machine Learning for 3D Point Clouds and Geospatial Digital Twins. *PFG –. Journal of Photogrammetry, Remote Sensing and Geoinformation Science*, 88(1), 15–24. DOI: 10.1007/s41064-020-00102-3

FinancesOnline. (2024). *35 IoT Device Statistics You Must Read: 2024 Data on Market Size, Adoption & Usage*. https://financesonline.com/iot-device-statistics/

Gavali, V., Jagtap, A., Jagzap, G., Garud, A., & Digraskar, V. (2024). Pothole Detection a Geospatial Approach to Prioritize Road Repairs. In Tripathi, S. L. (Ed.), *Emerging trends in IoT and Computing Technologies*. CRC Press. DOI: 10.1201/9781003535423-41

González, A., Kelly, C., & Rymszewicz, A. (2020). Advancements in web-mapping tools for land use and marine spatial planning. *Transactions in GIS*, 24(2), 253–267. DOI: 10.1111/tgis.12603

Gupta, S., Mittal, S., & Agarwal, V. (2022). Identification and Analysis of Challenges and Their Solution in Implementation of Decision Support System (DSS) in Smart Cities. In Gaur, L., Agarwal, V., & Chatterjee, P. (Eds.), *Decision Support Systems for Smart City Applications* (1st ed., pp. 99–118). Wiley., DOI: 10.1002/9781119896951.ch6

Han, Q., Nesi, P., Pantaleo, G., & Paoli, I. (2020). Smart City Dashboards: Design, Development, and Evaluation. *2020 IEEE International Conference on Human-Machine Systems (ICHMS)*, 1–4. DOI: 10.1109/ICHMS49158.2020.9209493

Huang, H., Yao, X. A., Krisp, J. M., & Jiang, B. (2021). Analytics of location-based big data for smart cities: Opportunities, challenges, and future directions. *Computers, Environment and Urban Systems*, 90, 101712. DOI: 10.1016/j.compenvurbsys.2021.101712

Ilesanmi, K. S., & Timothy, O. I. (2024). Possibility of Land Ownership Transaction with Non-Fungible Token Technology: Minting Survey Plan. *African Journal on Land Policy and Geospatial Sciences*, 7(2), 488–497. DOI: 10.48346/IMIST.PRSM/AJLP-GS.V7I2.41704

Ioanid, A., & Andrei, N. (2024). *Artificial Intelligence and Geospatial Technologies for Sustainable Maritime Logistics. Case Study: Port Of Constanta, Romania*. 15. https://marlog.aast.edu/files/marlog13/MARLOG13_paper_112.pdf

Ji, Q. (2020). *Geospatial Inference and Management of Utility Infrastructure Networks* [Ph. D. Thesis, School of Engineering, New Castle University]. https://theses.ncl.ac.uk/jspui/handle/10443/4985

Kaluarachchi, Y. (2022). Implementing Data-Driven Smart City Applications for Future Cities. *Smart Cities*, 5(2), 455–474. DOI: 10.3390/smartcities5020025

Khasgiwala, Y., Castellino, D. T., & Deshmukh, S. (2022). A Decentralized Federated Learning Paradigm for Semantic Segmentation of Geospatial Data. In Vasant, P., Zelinka, I., & Weber, G.-W. (Eds.), *Intelligent Computing & Optimization* (Vol. 371, pp. 196–206). Springer International Publishing., DOI: 10.1007/978-3-030-93247-3_20

Kost, G. J. (2020). Geospatial Hotspots Need Point-of-Care Strategies to Stop Highly Infectious Outbreaks. *Archives of Pathology & Laboratory Medicine*, 144(10), 1166–1190. DOI: 10.5858/arpa.2020-0172-RA PMID: 32298139

Kraak, M.-J., & Ormeling, F. (2020). *Cartography: Visualization of Geospatial Data* (4th ed.). CRC Press., DOI: 10.1201/9780429464195

Krishna, E. S. P., Praveena, N., Manju, I., Malathi, N., Giri, R. K., & Preetha, M.E. S. Phalguna Krishna. (2024). IoT-Enabled Wireless Sensor Networks and Geospatial Technology for Urban Infrastructure Management. *Journal of Electrical Systems*, 20(4s), 2248–2256. DOI: 10.52783/jes.2395

Li, W., Batty, M., & Goodchild, M. F. (2020). Real-time GIS for smart cities. *International Journal of Geographical Information Science*, 34(2), 311–324. DOI: 10.1080/13658816.2019.1673397

Liao, X., Liao, G., & Xiao, L. (2023). Rapeseed Storage Quality Detection Using Hyperspectral Image Technology—An Application for Future Smart Cities. *Journal of Testing and Evaluation*, 51(3), 1740–1752. DOI: 10.1520/JTE20220073

Marsal-Llacuna, M.-L. (2020). The people's smart city dashboard (PSCD): Delivering on community-led governance with blockchain. *Technological Forecasting and Social Change*, 158, 120150. DOI: 10.1016/j.techfore.2020.120150

Marzouk, M., & Othman, A. (2020). Planning utility infrastructure requirements for smart cities using the integration between BIM and GIS. *Sustainable Cities and Society*, 57, 102120. DOI: 10.1016/j.scs.2020.102120

Masik, G., Sagan, I., & Scott, J. W. (2021). Smart City strategies and new urban development policies in the Polish context. *Cities (London, England)*, 108, 102970. DOI: 10.1016/j.cities.2020.102970

MHUA. (2024). *Open Data Platform: India Smart Cities*. Smart Cities Mission Data Portal. https://smartcities.data.gov.in/

Minh, Q. N., Nguyen, V.-H., Quy, V. K., Ngoc, L. A., Chehri, A., & Jeon, G. (2022). Edge Computing for IoT-Enabled Smart Grid: The Future of Energy. *Energies*, 15(17), 6140. DOI: 10.3390/en15176140

Moreno-Álvarez, S., Paoletti, M. E., Sanchez-Fernandez, A. J., Rico-Gallego, J. A., Han, L., & Haut, J. M. (2024). Federated learning meets remote sensing. *Expert Systems with Applications*, 255, 124583. DOI: 10.1016/j.eswa.2024.124583

Musamih, A., Dirir, A., Yaqoob, I., Salah, K., Jayaraman, R., & Puthal, D. (2024). NFTs in Smart Cities: Vision, Applications, and Challenges. *IEEE Consumer Electronics Magazine*, 13(2), 9–23. DOI: 10.1109/MCE.2022.3217660

Nagavi, J. C., Shukla, B. K., Bhati, A., Rai, A., & Verma, S. (2024). Harnessing Geospatial Technology for Sustainable Development: A Multifaceted Analysis of Current Practices and Future Prospects. In Sharma, C., Shukla, A. K., Pathak, S., & Singh, V. P. (Eds.), *Sustainable Development and Geospatial Technology* (pp. 147–170). Springer Nature Switzerland., DOI: 10.1007/978-3-031-65683-5_8

Patel, M., Mehta, A., & Chauhan, N. C. (2021). Design of Smart Dashboard based on IoT & Fog Computing for Smart Cities. *2021 5th International Conference on Trends in Electronics and Informatics (ICOEI)*, 458–462. DOI: 10.1109/ICOEI51242.2021.9452744

Petrov, S., Dimitrov, S., & Ihtimanski, I. (2024). Integrated application of geospatial technologies for digital twining of urbanized territories for microscale urban planning. In Michaelides, S. C., Hadjimitsis, D. G., Danezis, C., Kyriakides, N., Christofe, A., Themistocleous, K., & Schreier, G. (Eds.), *Tenth International Conference on Remote Sensing and Geoinformation of the Environment (RSCy2024)* (p. 4). SPIE. DOI: 10.1117/12.3034288

Qudus, T., Ade, S., Modibbo, M. A., Aleem, K. F., Lawal, M. A., & Musa, L. A. (2024). Advancements and Innovations in Object-Oriented Feature Extraction Algorithms for Nigeria-Sat2 Data: A Comprehensive Review. *International Journal of Research Publication and Reviews*, 5, 815–829.

Ramu, S. P., Boopalan, P., Pham, Q.-V., Maddikunta, P. K. R., Huynh-The, T., Alazab, M., Nguyen, T. T., & Gadekallu, T. R. (2022). Federated learning enabled digital twins for smart cities: Concepts, recent advances, and future directions. *Sustainable Cities and Society*, 79, 103663. DOI: 10.1016/j.scs.2021.103663

Rani, S., Mishra, R. K., Usman, M., Kataria, A., Kumar, P., Bhambri, P., & Mishra, A. K. (2021). Amalgamation of Advanced Technologies for Sustainable Development of Smart City Environment: A Review. *IEEE Access : Practical Innovations, Open Solutions*, 9, 150060–150087. DOI: 10.1109/ACCESS.2021.3125527

Ribeiro, M. P., De Melo, K., Chen, D., & Valente, R. A. (2024). Prioritization of New Green Infrastructures Aimed at Protecting Urban Biodiversity. *IGARSS 2024 - 2024 IEEE International Geoscience and Remote Sensing Symposium*, 5447–5452. DOI: 10.1109/IGARSS53475.2024.10640597

Rojas, E., Bastidas, V., & Cabrera, C. (2020). Cities-Board: A Framework to Automate the Development of Smart Cities Dashboards. *IEEE Internet of Things Journal*, 7(10), 10128–10136. DOI: 10.1109/JIOT.2020.3002581

Roy, S. S., & Ray, R. (2024). An Integrated Approach of Multi-Criteria Decision Analysis and Spatial Interaction on Urbanization using Geospatial Techniques—A Case Study of Barasat Sub-division, North 24 Parganas District, West Bengal, India. *Journal of Interdisciplinary and Multidisciplinary Research*, 19(8), 102–116.

Saha, R., Misra, S., & Deb, P. K. (2021). FogFL: Fog-Assisted Federated Learning for Resource-Constrained IoT Devices. *IEEE Internet of Things Journal*, 8(10), 8456–8463. DOI: 10.1109/JIOT.2020.3046509

Sánchez, O., Castañeda, K., Vidal-Méndez, S., Carrasco-Beltrán, D., & Lozano-Ramírez, N. E. (2024). Exploring the influence of linear infrastructure projects 4.0 technologies to promote sustainable development in smart cities. *Results in Engineering*, 23, 102824. DOI: 10.1016/j.rineng.2024.102824

Shahat Osman, A. M., & Elragal, A. (2021). Smart Cities and Big Data Analytics: A Data-Driven Decision-Making Use Case. *Smart Cities*, 4(1), 286–313. DOI: 10.3390/smartcities4010018

Sharma, L. K., & Naik, R. (2024). *Conservation of Saline Wetland Ecosystems: An Initiative towards UN Decade of Ecological Restoration.* Springer Nature Singapore., DOI: 10.1007/978-981-97-5069-6

Silva, D. S., & Holanda, M. (2022). Applications of geospatial big data in the Internet of Things. *Transactions in GIS*, 26(1), 41–71. DOI: 10.1111/tgis.12846

Statista. (2024). *Internet of Things (IoT) connected devices installed base worldwide from 2015 to 2025*. https://www.statista.com/statistics/471264/iot-number-of-connected-devices-worldwide/

Tahar, A., Mendy, G., & Ouya, S. (2024). Efficient and Optimized Geospatial Data Representation in Blockchain-Based Land Administration. In X.-S. Yang, S. Sherratt, N. Dey, & A. Joshi (Eds.), *Proceedings of Ninth International Congress on Information and Communication Technology* (Vol. 1014, pp. 519–535). Springer Nature Singapore. DOI: 10.1007/978-981-97-3562-4_41

Tareke, B., Filho, P. S., Persello, C., Kuffer, M., Maretto, R. V., Wang, J., Abascal, A., Pillai, P., Singh, B., D'Attoli, J. M., Kabaria, C., Pedrassoli, J., Brito, P., Elias, P., Atenógenes, E., & Santiago, A. R. (2024). User and Data-Centric Artificial Intelligence for Mapping Urban Deprivation in Multiple Cities Across the Globe. *IGARSS 2024 - 2024 IEEE International Geoscience and Remote Sensing Symposium*, 1553–1557. DOI: 10.1109/IGARSS53475.2024.10640428

Tetteh, A. T., Moomen, A.-W., Yevugah, L. L., & Tengnibuor, A. (2024). Geospatial approach to pluvial flood-risk and vulnerability assessment in Sunyani Municipality. *Heliyon*, 10(18), e38013. DOI: 10.1016/j.heliyon.2024.e38013 PMID: 39381211

Uddin, K., & Matin, M. A. (2021). Potential flood hazard zonation and flood shelter suitability mapping for disaster risk mitigation in Bangladesh using geospatial technology. *Progress in Disaster Science*, 11, 100185. DOI: 10.1016/j.pdisas.2021.100185

Vahidnia, M. H. (2024). Empowering geoportals HCI with task-oriented chatbots through NLP and deep transfer learning. *Big Earth Data*, •••, 1–41. DOI: 10.1080/20964471.2024.2403166

Wei, B., Guo, X., Wu, Z., Zhao, J., & Zou, Q. (2023). Construct Fine-Grained Geospatial Knowledge Graph. In El Abbadi, A., Dobbie, G., Feng, Z., Chen, L., Tao, X., Shao, Y., & Yin, H. (Eds.), *Database Systems for Advanced Applications. DASFAA 2023 International Workshops* (Vol. 13922, pp. 267–282). Springer Nature Switzerland., DOI: 10.1007/978-3-031-35415-1_19

Wu, Z., Liu, Y., Fang, S., Shen, W., Li, X., Mao, Z., & Wu, S. (2024). Integration of Geographic Features and Bathymetric Inversion in the Yangtze River's Nantong Channel Using Gradient Boosting Machine Algorithm with ZY-1E Satellite and Multibeam Data. *Geomatica*, 100027(2), 100027. Advance online publication. DOI: 10.1016/j.geomat.2024.100027

Yabe, T., Jones, N. K. W., Rao, P. S. C., Gonzalez, M. C., & Ukkusuri, S. V. (2022). Mobile phone location data for disasters: A review from natural hazards and epidemics. *Computers, Environment and Urban Systems*, 94, 101777. DOI: 10.1016/j.compenvurbsys.2022.101777

Yadav, A., & Sagi, S. (2024). Management Cases Studies and Technical Use Cases on Web 3. In Darwish, D. (Ed.), (pp. 267–287). Advances in Web Technologies and Engineering. IGI Global., DOI: 10.4018/979-8-3693-1532-3.ch013

Yang, B., Wang, S., Li, S., Zhou, B., Zhao, F., Ali, F., & He, H. (2022). Research and application of UAV-based hyperspectral remote sensing for smart city construction. *Cognitive Robotics*, 2, 255–266. DOI: 10.1016/j.cogr.2022.12.002

Zheng, C., Yuan, J., Zhu, L., Zhang, Y., & Shao, Q. (2020). From digital to sustainable: A scientometric review of smart city literature between 1990 and 2019. *Journal of Cleaner Production*, 258, 120689. DOI: 10.1016/j.jclepro.2020.120689

Zhou, Q., Zhu, M., Qiao, Y., Zhang, X., & Chen, J. (2021). Achieving resilience through smart cities? Evidence from China. *Habitat International*, 111, 102348. DOI: 10.1016/j.habitatint.2021.102348

KEY TERMS AND DEFINITIONS

Artificial Intelligence (AI): The simulation of human intelligence by machines for decision-making, pattern recognition, and predictive analytics. In the context of geospatial data for smart cities, AI enhances the efficiency of urban infrastructure management by enabling real-time analysis and predictive modeling.

Big Data: Big Data refers to the large and complex data sets generated by smart cities. Big Data Analytics is an advanced analytical method for city management and services to examine large datasets and identify patterns, trends, and associations.

Data Visualization: Data Visualization is the graphical representation of data to enhance understanding and decision-making. In smart cities, visualizing geospatial data helps planners and policymakers easily interpret complex spatial patterns, improving communication and resource management.

Geographic Information System (GIS): GIS is a computer system/technology/software platform designed to collect, capture, store, manipulate, process, analyze, manage, integrate, visualize, and present spatial or geographic data. GIS systems began evolving in the 1980s and are now mainstream technology.

Geospatial AI: Geospatial AI integrates Artificial Intelligence techniques with geospatial data to analyze, model, and interpret spatial patterns and relationships. It enables automated analysis of location-based data for applications such as urban planning, environmental monitoring, and infrastructure management in smart cities.

Geospatial Technologies: A collection of tools and technologies used to collect, store, analyze, and visualize spatial data. These include GIS, remote sensing, and GPS. In the context of smart cities, geospatial technologies help map urban areas, monitor infrastructure, and enable data-driven decision-making.

Global Positioning System (GPS): GPS is a satellite-based navigation system that provides geolocation and time information. In smart cities, GPS helps track transportation networks, enable autonomous vehicles, and improve location-based services for citizens.

Internet of Things (IoT): IoT are interconnected devices containing sensors that collect and exchange data using the internet. These are crucial Smart city infrastructures that help in real-time decision-making. IoT devices monitor real-time conditions (e.g., traffic flow, energy consumption) and contribute to the automation and optimization of urban services.

Remote Sensing (RS): The process of collecting information about an object or area from a distance (such as from Earth's surface) without making physical contact. It is often used in mapping and monitoring the physical characteristics of urban areas from satellites or aerial sensors such as aircraft.

Smart Cities: Urban areas that use modern computing technologies such as IoT, AI, and geospatial systems to enhance the quality of life, optimize infrastructure, and promote sustainability. These cities aim to improve services like transportation, healthcare, energy management, and public safety using real-time data.

Chapter 5
PATCH.AI:
Forest Cover Virtualization on Digital Maps Using Satellite Imagery (Google Maps API)

R. Parvathi
Vellore Institute of Technology, India

V Pattabiraman
https://orcid.org/0000-0001-8734-2203
Vellore Institute of Technology, India

B. Shakthi
Vellore Institute of Technology, India

ABSTRACT

In a situation where the severe threats of down environmental challenges are prevalent, there is no doubt that there has never been a time when innovative tools for the forest ecosystems monitoring and management are relevant. PATCH.AI stands out as a pioneering technology that enables the integration of satellite imagery and AI algorithms to execute a through coverage status survey of forests. PATCH.AI provides analytical solutions based on the state-of-the-art Google Maps API. This technology makes use of high-definition satellite photographs that allow for the precise visualization and analysis of the given area. The core functions of PATCH.AI are in getting accurate details on areas that are forested or not through the artificial intelligence technology. Through the use of the latest in image processing technology, the system automatically recognizes and colors any forest regions in green, and thereafter, the background is plainly white, indicating the absence of forest.

DOI: 10.4018/979-8-3693-8054-3.ch005

INTRODUCTION

In an age when ecological sustainability is the vital ingredient, one can count the time of PATCH.AI as the momentous step ahead in ecological monitoring and management. Mueller, J. P. (2006) This innovative app utilizes the Google Maps API extensively and it enables the collection of satellite images that have enough detail and a broader field of view. Using the advanced field of image processing, PATCH.AI expands and increases its capacity to collect and analyze data from forest coverage across diverse landscapes.

One of the main components of the technology used by PATCH.AI is its highly evolved algorithmic infrastructure. This network of computations closely studies satellite images and separates green forested areas from those that are not. Through the code coloring of the forests in vibrant green and contrasting them against the barren lands in stark white, PATCH.AI provides a clear and immediate visual with the green forests and the white barren land. This is really helpful in the identification of forested area in an ease manner and makes a very strong visual communication tool for the condition of the earth forest cover.

The fact that it is PATCH.AI's system real-time functionality is also extended to the quantification of the green vegetation coverage. It figures out percentage of the screen displayed that is covered with forest and thereby gives users an instant and accurate estimate of forest density. This function can be particularly helpful for people with a vary of interests ranging from environmentalists, land managers, and researchers who need detailed data as their basis to form decisions on time.

PATCH.AI is a shining example of such constantly growing technology in the world of environmental technology. By integrating pivotal satellite imagery with the most advanced image processing technique, it becomes a unique device for the monitoring and management of forest ecosystems. Its user-friendly interface and almost real-time analysis capabilities makes it an indispensable tool for those who care about the well-being of our natural assets and make studies about them as well.

PROBLEM STATEMENT AND MOTIVATION

Global forest cover reduced at the quickest rate is one of the major environmental problems of the modern era. Forests have several functions that are vital for the ecosystem as they help preserve biodiversity and they also act as carbon sinks reducing climate change. Nevertheless, forestry clearing, and forest decline remain to be a problem since urbanization, agriculture, logging, and other anthropogenic activities are still taking place. Sustainability of forests is the most challenging because their efficient monitoring and management for future generations is very difficult.

Conventional systems of forest monitoring are usually costly, take much time and effort, and the resulting information may be too little for effective quick decision making. The development of a computerized system that can be used in a widespread way as an aid in monitoring forest coverage and health is needed urgently. This is where PATCH.AI comes in as we think of electric urban delivery solutions to support the conservation of green cover.

PATCH.AI, the new AI platform is guided by the wish to develop the Future where forest maintenance is positive instead of negative. Using real time FIVA analysis PATCH.AI emboldens stakeholders to discover deforestation actions in time, monitor reforestation operations and plan sustainable land use. Our service PATCH.AI aims at filling a gap that exists today between environmental data collected and reported in real time and what is being addressed by the current tools.

The app implies a Google Maps API that helps to retrieve the satellite images for forests combined with advanced image processing methodologies. This solves the problem of using obsolete and ineffective forest monitoring methods. By automatically colourising tree areas and showing the coverage of green vegetation in live time, PATCH.AI disrupts the conventional approach to environmental data entirely.

In short, PATCH.AI, the idea has an attitude that technology has the potential of doing good rather than harm, with the environment being a major sector that can be touched. It is a case of human intelligence triumphing over the natural elements, and technology that can tackle environmental problems. PATCH.AI is more than just a machine; it is a collection of forces that is heading towards a new direction – a journey towards sustainability and climate resilience.

LITERATURE SURVEY

Technology has been an area of inquiry overlapped by the environmental conservation and the studies got focused on the design of appropriate instrumentation that can be used for the monitoring and managing of forest ecosystems. Throughout the entire section covering different literature reviews shows that there are many different methods that have been proposed to fight deforestation and forest degradation.

The use of remote sensing technology is an outstanding sample of information in this field, and it is applied to forest monitoring. Survey results have shown a strong performance of satellites when it comes to delivering forest cover and health status across vast areas. One illustration is where Tariq et al. (2023) employed Landsat data to map different global forests' reduction, clearly presenting the power of satellite-based monitoring systems. Other than that, the research outcomes of Singh, et al.

(2012) were very much as how accuracy in distinguishing deforestation with the application of high resolution imagery is imperative.

Simultaneously, Li, H et al (2021) the feature image processing algorithms have evolved which allow for much higher accuracy of the monitored area. One of the machine learning techniques used in the research, for instance, is convolutional neural networks (CNNs), which is very efficient in classifying the land cover types on the high level base. In a study, Blaga et al.(2023) presented the use of CNNs classifying forested areas from non-forested regions, this in turn is relevant for designing the automated forest classifiers.

In addition, the merge of GIS with telemetric data is another major area of scientific space investigation. The study by Ganz et al. (2020) is one of the many examples that provide evidence to the fact that GIS systems are powerful combination of the two which are in using the satellite imagery to analyze spatial patterns of deforestation and to inform conservation strategies.

In the domain of the real-time monitoring, the penetration of applications such as patch. ai stands for the further paving the revolutionary path for utilization of technology for saving the environment. Through ongoing advances made by being carried over in the sequence of research, PATCH.AI becomes one part of where it includes data processing and visualization happening in real-time and is meant to be usable with a great deal of ease.

The purpose of the PATCH.AI is an extension of the studies that were done, collectively, showing the importance of the availability of the monitoring systems that are easy to use and reliable. According to literature, integration of satellite imagery with advanced imaging and real-time functionality is defiantly going to be a net positive to the future of environmental conservation and sustainable land management.

Overall, the literature review indicates the changing face of forest monitoring techniques with time. The history of the research goes from using satellites data to the latest development about the real-time analysis. Rai, P. K. (2013), the direction taken by research is the one that involves a very integrated and complex use of technology in the fight against deforestation. PATCH.AI is pioneering the forces of nature in the realm of environmental technology, being the apex of several years of research and innovation.

NOVELTY

PATCH.AI introduces several novel elements to the field of forest cover visualization:

Integration with Google Maps API: Unlike conventional ways that involves pure usage of remote sensing data, Artificial Intelligence-based Patch.AI integrates Google Maps API very well into easy-to-use interface to achieve popularity throughout the world.

Advanced Image Processing: Data analytics comprises the image processing modern methods for an automatic classifier which distinguishes between the visible and non-visible forest, thus increasing the accuracy and precision of the forest cover analysis Knauer et.al(2019)

Real-Time Vegetation Coverage Percentage: PATCH.AI brings the monitor of percent vegetation cover in green color that is the main key factor for the plants and businesses. This real-time capacity facilitates first-hand evaluation and judicious actions.

Colorization for Enhanced Visualization: Manual coloring of forest regions by green and other areas by white illustrating data is very important for individuals who need to respond to emergencies quickly and accurately.

Application in Environmental Conservation: Of value comes the provision of a tool of assessment of the forest cover rapidly. This will to a significant extent foster the initiatives for environmental conservation by aiding in monitoring of deforestation and afforestation activities.

Land Management and Ecological Research: The project's strengths lie in the soil and climate research and the gathering of correct and updated information on the forests in order to generate sustainable development and maintain diversity of flora and fauna.

PATCH.AI marks a revolutionary transition in ramifications of GIS and remote sensing for monitoring the environment. Being the first to employ the combination of Google Maps API with cutting-edge image processing and data visualization technological techniques reinforces the uniqueness of the tool compared to already existing methodologies, which has triggered high interest in the scientific community, policymakers and conservationists alike.

The following is a cohesive exposition emphasizing the current status on forest cover visualisation technology as well as the field specific value of PATCH.AI. The innovatively incorporated aspects of your project will make it potentially useful one of the tools used globally to protect and effectively manage forest ecosystems.

Dataset Acquisition Links

https://console.cloud.google.com/google/maps-apis/api-list

GOOGLE API MAP SCRAPING

It is indisputable that the current world is facing problems that include environmental destruction due to the negative impact of industrial activities and urbanization, hence, there is a need for such an approach. On the other hand, Jahromi et al(2021) there are numerous innovative tools for the forest ecosystem monitoring and management which have never been as relevant. Not only PATCH.AI appears as a unique tech that allows to make satellite imagery and AI algorithms an engine of a detailed status of forests survey, but also it is remarkable for its high accuracy concerning any status of forest. PATCH.AI utilizes its Maps API to the full extent and builds its analytical models based on Google's latest technologies. This system as per Aszkowski et al(2023) is based on an interaction of satellite high-resolution imagery, which may capture images of the whole area with all possible nuances and details. Artificial intelligence is the key technology that PATCH.AI uses to get details on forests which are not in existence or have been cut down in the process of doing their core functions. Using the systems latest in image processing technique, the forests are automatically colorized in green and subsequently marked as to covering the neighborhood, while on the other hand, the background is a plain white, signifying the lack of forest. It is the this meted out quality that makes this statistical more clearly be understood by every individual irrespective of whether he or she is literate of not.

METHODOLOGY

Data Selection and Attributes

The PATCH.AI project leverages a combination of Geographic Information Systems (GIS) tools and advanced image processing techniques to analyze and visualize forest cover from satellite imagery. Workflow and methodology is shown in Figure 1.
key components of the methodology:

1. Data Acquisition and Preprocessing:1. Data Acquisition and Preprocessing:
   - Acquire high-resolution satellite images as data sources. These images are used as foundation so that the data for assessment can be calibrated.

- Install QGIS, an open-source GIS platform for data acquisition, research, and visualization of spatial data.
- Take advantage of the SCP(Seni-Automatic Classification Plugin) tool that are embedded in the QGIS environment in order to speed up the whole process. SCP has a command foe image processing: downloading, preprocessing, and postprocessing of satellite imagery.
- Bring the Aircraft image into phase by taking care of atmospheric interference, sensor noise, and other distortions.

2. Image Segmentation:
   - An image segmentation, which is a very important stage, is namely a stage. Rather, it merges the likelihood of neighboring clusters of pixels that present similar spectral and shape features.
   - A Mean Shift filter is available in SCP (segmentation image) using a moving window for computing average pixel values. They are referred to as "Means Shift". This leads to the creation of neighborhoods at a similar level with average color tone.
   - Tweak settings such as spectral detail (proportion of finding spectral differences), spatial detail (weighting between proximity features), and minimum segment size.

3. Supervised Classification:
   - By using the SCP, the researcher can make the segmented image controllable in a supervised manner. Instead of painting in every pixel, it paints in huge chunks which saves time and readjusting.
   - Such specific object is extricated to use the set of classes by the classifier.
   - And this novelty is in segmenting of area, which makes classification accurate by reducing noise level.

4. Colorization and Visualization:
   - Use green to symbolize the forested areas and white to indicate the non-forested regions; try not to get too technical.
   - Clarify the image visibly using transparency placed over the segment to ensure the objects are aligned.
   - The real time coverage of live green vegetation, automatically calculated and reflected in the screen through a dynamic way.

5. Integration with Google Maps API:5. Integration with Google Maps API:
   - Screen the PATCH.AI app is what makes this app ahead of its competitors for it uses widely- accessed Google Maps API. This user interface is friendlier in the sense that is does usability and accessibility.

6. Applications and Impact:
   - **Environmental Conservation:** PATCH.AI aids in rapid assessment of forest cover, supporting conservation efforts.

- **Land Management:** Provides valuable insights for sustainable development.
- **Ecological Research:** Facilitates biodiversity studies and ecological monitoring.

Figure 1. Workflow Diagram

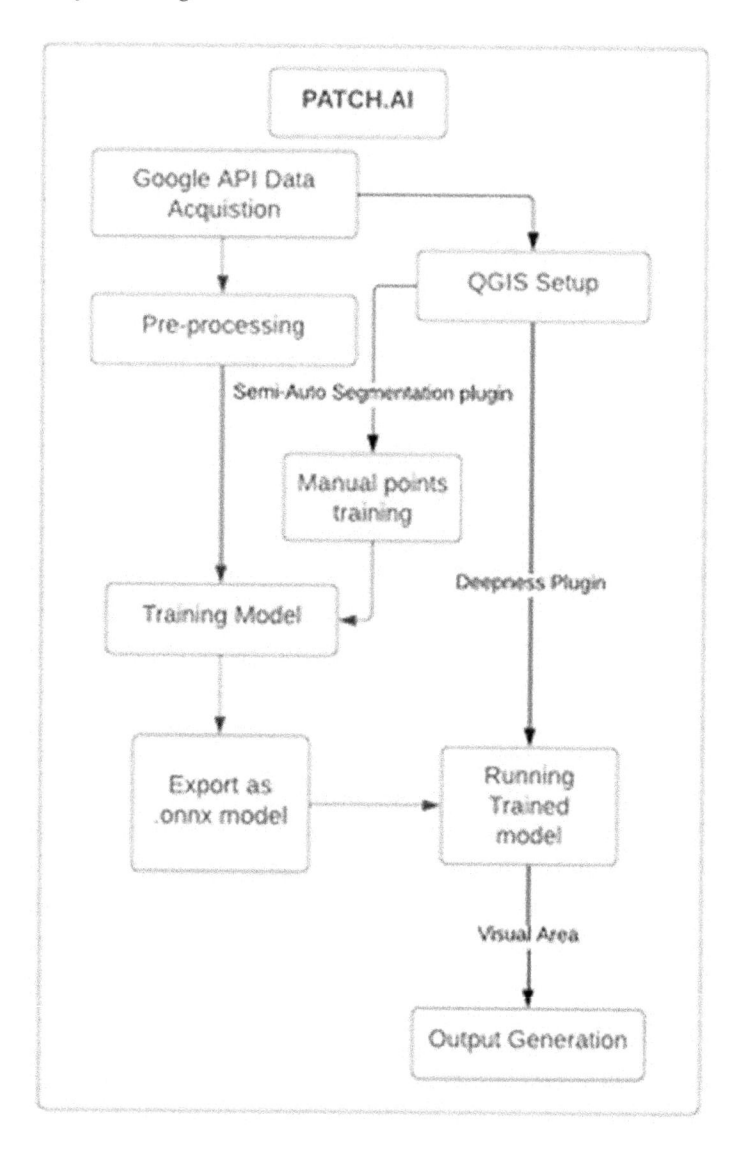

DATA PRE-PROCESSING AND ML MODEL

For our project of PATCH.AI, data preprocessing is a crucial foundation that makes the way for the machine learning models to work with this satellite images. And at that, image, which is resized to a standard size, is used that is important for the consistency to be met, which is the necessary input requirement of the model. Thereafter, normalization process is implemented to adjust the pixel values all over the updated size, converting om to a range which will be helpful with the model identifying patterns including lighting and color variations, which it then can effectively discern.

KNN algorithm alongside with land cover segmentation is a statistical technique used for data analysis, a machine learning approach that categories the data based on the similarity with its neighbors. It's pertinent to your project to differentiate between forested areas from urban zones, it classifies the land cover in every satellite imagery based on the type.

Another important model used for this task is CNN which uses higher features learned from pixels to classify the character with very high accuracy. The trained CNN architecture output is stored in .onnx(Open Neural Network Exchange) format which makes it portable and deploys its network across several different processes.

Moreover, the training and validation data subsets are separated from the testing data portion symbolically. Thereby, this provides an opportunity to evaluate it more correctly and serves as a baseline for its general performance upon being exposed to other data which it has not seen yet. Such a step is to evaluate the model's ability to divide a classified land cover into different types that are present in diverse settings.

MODULES

Data Acquisition Module: The module is considered the opening step for the whole project and strives to procure all high-quality satellite data with the aid of different platforms. There are many such algorithms which guarantee that the images will not only appear on the screen in detail but also at the appropriate resolution for in depth analysis. The module's functionality is equipped with a selection of the named satellites, downloading of the images and preparing them for the next steps.

Data Preprocessing Module: The next crucial step is to process the raw satellite images by way of transforming into proper sizes, normalizing, and also data cleaning. It is assigning photos to a regular scale and correcting pixels to one range which is imperative for the proper processing by the machine learning algorithms.

Machine Learning Module: This module is where the key analytical algorithms of the project, the KNN based in landcover segmentation and CNN for models that have patterns, are housed. It operates on the preclassified images, computes their spectral features, and acquires the understanding of the messes that are characteristic of the spaceborne data.

Image Segmentation Module: The satellite imagery is parsed to generate meaningful patterns here. The module groups areas in the image based on their spectral characters and next spatial distribution. It can create separated segments where each type of land cover is represented. The next step of the workflow is vital to get reliable classification for forest zones.

Classification and Visualization Module: After segmentation, this module is using classification expert to do the land cover analysis and it displays the results in visual mode. As for subject coloring, it specifically depicts the segmented sections in a way that improves the map's clarity for the readers and provides an easy and quick understanding about the forest cover distribution patterns.

Integration Module: Making sure that Google Maps API is error free and user experience is smooth, this module seamlessly ties the project features to the interface that is easy to use. Through satellite images visualization, the tool enables users to easily navigate the images and directly inspect the data displayed.

Real-Time Analysis Module: And what is special for PATCH.AI, this module is a % of green vegetation cover indicator in real-time. Such a product helps in getting on-the-fly aerial views of the forest coverage which is very useful for monitoring and into making particular decision.

Output Generation Module: The last section of the project encompasses modules that transmit results, e.g., comprehensive maps, statistical reports, and other indicators. A significant aspect of AI is that it combines the information obtained after analyzing the data into user-friendly formats for interpretation as well as further use.

VISUALIZATIONS

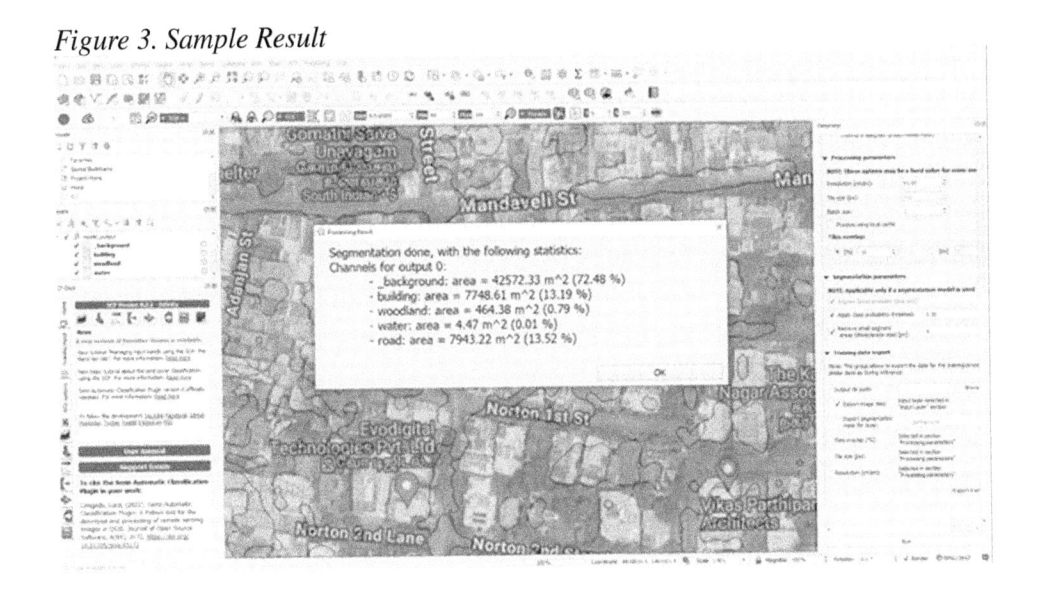

Figure 2. Result of Segmentation

Figure 3. Sample Result

147

RESULTS AND DISCUSSION

The PATCH.AI system successfully demonstrated its capability to visualize forest cover by processing satellite imagery through its various modules. The sample results are shown in Figure 2 and Figure 3. The data acquisition module efficiently gathered high-resolution images from multiple satellite sources. The preprocessing module standardized these images, making them suitable for analysis by the machine learning algorithms. The KNN and CNN models, housed within the machine learning module, effectively segmented and classified the land cover types, distinguishing forested areas from other land covers with high accuracy.

The image segmentation module's performance was crucial, as it grouped pixels into meaningful segments that represented different land cover types. This step was essential for the accurate classification of forested areas. The classification and visualization module then color-coded these segments, providing an intuitive map that clearly displayed the distribution of forest cover.

The integration module allowed for seamless interaction with the Google Maps API, enhancing the user experience by providing a familiar interface to interact with the visualized data. The real-time analysis module's innovative feature of calculating the percentage of green vegetation cover in real-time provided immediate insights into the extent of forest coverage, which is essential for monitoring and decision-making processes.

The output generation module compiled the analyzed data into user-friendly formats, such as detailed maps and statistical reports, summarizing the findings from the forest cover analysis. These outputs are invaluable for stakeholders involved in environmental conservation and land management.

The discussion revolves around the implications of the results and the potential applications of the PATCH.AI system. The high accuracy of the land cover classification suggests that the system can be reliably used for monitoring deforestation and afforestation activities. The real-time analysis feature could be particularly useful for rapid response scenarios, such as detecting illegal logging activities or natural disasters affecting forested areas.

The system's ability to integrate with the Google Maps API means that it has the potential for widespread adoption, given the popularity and accessibility of Google Maps. This could democratize access to forest cover data, allowing not just researchers and policymakers but also the general public to engage with environmental monitoring efforts.

The project also highlights the importance of data preprocessing in GIS projects. Properly resized and normalized images are crucial for the performance of machine learning models, as seen in the effectiveness of the KNN and CNN algorithms used in PATCH.AI. The methodology employed in the project could serve as a blueprint for future GIS projects that aim to integrate machine learning for environmental analysis.

thus, the PATCH.AI project stands as a significant advancement in the application of GIS and remote sensing for environmental monitoring. Its innovative approach to integrating advanced image processing techniques with user-friendly interfaces sets it apart from existing methodologies. The project's success opens up new avenues for research and development in the field of environmental GIS and paves the way for more sophisticated tools for land management and ecological research. The implications of this project extend beyond forest cover analysis, suggesting a promising future for the integration of GIS and machine learning in various domains of environmental conservation.

HARDWARE AND SOFTWARE REQUIREMENT

- Python: Utilize packages such as TensorFlow, Torch, GeoPandas, and Keras for model training. (Basic usage)
- TensorFlow: Implement the Convolutional Neural Network (CNN) algorithm using the Keras.Sequential method to train the model with various features. (minimum of 16GB RAM is required and for faster training times minimum 1660 GPU is recommended)
- GeoPandas: Use to access and manipulate the shape files exported from QGIS. (minimum of 8GB Ram is required)
- Torch/Torchvision: Export the trained model in ONNX format for compatibility with the Deepness plugin. (minimum of 16GB RAM is required and for faster training times minimum 1660 GPU is recommended)
- Deepness Plugin: Upload the trained model into QGIS and visualize the data.

Different kinds of vegetation can be segregated
Different kinds of vegetation can be segregated using Satellite Imagery

1. Data Coverage: Provides broad coverage of large areas and broad coverage of large areas.
2. Spatial Resolution: Typically ranges from meters to tens of meters Offers high-resolution data at ground level
3. Temporal Resolution: Can provide frequent revisits (daily to weekly) Offers continuous real-time data collection
4. Cost: Relatively lower cost for wide-area coverage, Higher initial and maintenance costs per sensor

5. Data Accessibility Generally accessible through satellite providers Requires physical access for maintenance and data retrieval
6. Environmental Conditions Unaffected by weather conditions during acquisition Susceptible to environmental factors (e.g., weather, vegetation growth) affecting data quality.
7. Applications: Suitable for large-scale monitoring and analysis Ideal for precise, localized monitoring and intervention.
8. Data Integration Can be integrated with GIS and remote sensing tools Requires integration with local monitoring systems and networks.
9. Monitoring Flexibility Limited flexibility in monitoring specific locations Provides flexibility in selecting specific monitoring points
10. Accuracy and Precision Moderate to high accuracy depending on resolution and processing techniques Offers high accuracy for specific parameters within monitored areas

CONCLUSION

PATCH.AI project which has evolved from the utilization of Geographic information systems (GIS) and machine learning in environmental monitoring, which is a big step. Given that the project incorporates complex image processing methods together with the powerful Google Map library, a new perspective on mapping the natural cover by satellite has therefore been shown. This system can do an on-the-spot analysis of how much of the landscape is covered with plant life and thus provides a potent new instrument for quick appraisal and decision making in conservation of the environment. The procedure employed to show what land is covered by plants was this: procuring the satellite photographs using software QGIS and the Semi-Automatic Classification Plugin as well as, Machine learning models like KNN Forest cover accuracy is hence critical in observing trends such as may be caused by deforestation, afforestation or by any other natural disturbances on forested areas.

As a result of the continuous experience and the subsequent integration, the classification, and the visualization module with the use of the graphics illustrate again and again the future of widespread adoption of this project. Enabling people to engage in processes of environmental monitoring not only through logging forest cover data would not be a job for researchers, policymakers but for everyone as well. With that, it is seen that the project also stresses the data pre-processing importance for GIS projects. A crucial issue concerning machine learning algorithms implemented at PATCH.AI relates indeed to the necessity to use normalized and re-sized images for processing. With the project approach taken as a model, it is also demonstrated that further GIS projects that seek to integrate ML for environmental analysis can be easily executed.

In conclusion, PATCH.AI stands as a testament to the power of combining GIS with machine learning to create robust tools for environmental conservation. Its success opens new avenues for research and development in the field of environmental GIS and paves the way for more sophisticated tools for land management and ecological research. The implications of this project extend beyond forest cover analysis, suggesting a promising future for the integration of GIS and machine learning in various domains of environmental conservation.

REFERENCES

Blaga, L., Ilie, D. C., Wendt, J. A., Rus, I., Zhu, K., & Dávid, L. D. (2023). Monitoring forest cover dynamics using orthophotos and satellite imagery. *Remote Sensing (Basel)*, 15(12), 3168. DOI: 10.3390/rs15123168

Ganz, S., Adler, P., & Kändler, G. (2020). Forest cover mapping based on a combination of aerial images and Sentinel-2 satellite data compared to National Forest Inventory data. *Forests*, 11(12), 1322. DOI: 10.3390/f11121322

Knauer, U., von Rekowski, C. S., Stecklina, M., Krokotsch, T., Pham Minh, T., Hauffe, V., Kilias, D., Ehrhardt, I., Sagischewski, H., Chmara, S., & Seiffert, U. (2019). Tree species classification based on hybrid ensembles of a convolutional neural network (CNN) and random forest classifiers. *Remote Sensing (Basel)*, 11(23), 2788. DOI: 10.3390/rs11232788

Li, H., Hu, B., Li, Q., & Jing, L. (2021). CNN-based individual tree species classification using high-resolution satellite imagery and airborne LiDAR data. *Forests*, 12(12), 1697. DOI: 10.3390/f12121697

Mueller, J. P. (2006). Mining Google web services: building applications with the Google API. John Wiley & Sons. Jahromi, M. N., Jahromi, M. N., Zolghadr-Asli, B., Pourghasemi, H. R., & Alavipanah, S. K. (2021). Google Earth Engine and its application in forest sciences. Spatial Modeling in Forest Resources Management: Rural Livelihood and Sustainable Development, 629-649. Aszkowski, P., Ptak, B., Kraft, M., Pieczyński, D., & Drapikowski, P. (2023). Deepness: Deep neural remote sensing plugin for QGIS. *SoftwareX*, 23, 101495.

Rai, P. K. (2013). Forest and land use mapping using Remote Sensing and Geographical Information System: A case study on model system. *Environmental Skeptics and Critics*, 2(3), 97.

Singh, V., & Dubey, A. (2012). Land use mapping using remote sensing & GIS techniques in Naina-Gorma Basin, part of Rewa district, MP, India. *International Journal of Emerging Technology and Advanced Engineering*, 2(11), 151–156.

Tariq, A., Jiango, Y., Li, Q., Gao, J., Lu, L., Soufan, W., Almutairi, K. F., & Habib-ur-Rahman, M. (2023). Modelling, mapping and monitoring of forest cover changes, using support vector machine, kernel logistic regression and naive bayes tree models with optical remote sensing data. *Heliyon*, 9(2), e13212. DOI: 10.1016/j.heliyon.2023.e13212 PMID: 36785833

KEY TERMS AND DEFINITIONS

Data Visualization: Different charts used to visualize the Forest land coverage

Ecological Monitoring: Monitoring mechanism for aquatic and terrestrial ecosystems

Ecological Research: Research to analyse the aquatic and terrestrial ecosystems

Forest Coverage: Land area covered by Forest

Google Maps API: to get rich data, coverage and reliability

Image Processing: Processing the various algorithms to find the forest details from Google Map

PATCH.AI: Application to monitor the forest

Satellite Images: Visible imagery is very useful for distinguishing clouds, land, and sea/ocean.

Chapter 6
Optimizing Flood Risk Management Through Geospatial AI and Remote Sensing

G. Prabhanjana
Christ University, India

Yashas Shetty
Christ University, India

Daksh Vats
Christ University, India

Rajesh Kanna Rajendran
 https://orcid.org/0000-0001-7228-5031
Christ University, India

ABSTRACT

Flooding presents an increasing threat to communities globally, intensified by climate change and urban expansion. Effective flood risk management requires precise and timely information to guide decision-making and planning. This chapter explores the use of geospatial artificial intelligence (AI) in combination with remote sensing to enhance flood risk management. Advanced AI techniques are applied to analyze satellite and aerial images, enabling more accurate identification of flood-prone areas and prediction of potential flood events. Machine learning is utilized to integrate historical and real-time data, improving flood prediction models and evaluating the effectiveness of various mitigation strategies. A decision-support system is developed

DOI: 10.4018/979-8-3693-8054-3.ch006

to leverage this technology, providing valuable insights for policymakers, emergency responders, and urban planners. This chapter demonstrate that the integration of geospatial AI and remote sensing can significantly advance flood risk management, offering a more proactive and resilient approach to addressing this critical issue.

INTRODUCTION

Artificial intelligence (AI) involves the development of intelligent machines capable of performing tasks that typically require human intelligence. With the rapid advancements in machine learning (ML), a branch of AI focused on developing models that learn from data, there is growing interest in exploring how these technologies can help address climate change challenges. While recent discussions have emphasized the potential of ML techniques for improving weather and climate modeling, there is also significant value in applying AI/ML to assess the risks and impacts of climate change, particularly in relation to extreme events like floods, droughts, heatwaves, and wildfires. The scope of AI's application extends beyond long-term predictive assessments, which often span decades, to include shorter timescales of days to months. This is crucial for climate resilience planning across various sectors, such as energy distribution, logistics, agriculture, supply chain management, and infrastructure, where timely predictions can inform proactive decision-making and risk mitigation strategies. By leveraging AI, organizations can better anticipate and adapt to the impacts of climate variability and extreme weather events, enhancing their ability to cope with future challenges.

Floods rank among the most common and destructive natural disasters, leading to significant loss of life and widespread damage to infrastructure around the globe. The intensifying impacts of climate change are expected to increase the frequency and severity of extreme weather events, including flooding, thereby raising risk levels in already vulnerable regions. This growing threat highlights the urgent need for precise and reliable flood hazard modeling to aid disaster risk reduction efforts and promote sustainable development. The field of flood hazard modeling has seen remarkable progress due to advancements in geospatial technologies and computational methods. Machine learning, in particular, has emerged as a promising tool for flood modeling because of its capacity to manage the complex, nonlinear relationships between flooding events and their contributing factors. Despite these advancements, there remain substantial research challenges, such as limited model interpretability, the need for more rigorous validation, and insufficient sensitivity analysis concerning the conditioning factors used as model inputs.

Accurate flood hazard mapping is essential for effective disaster risk management and flood prevention. However, traditional flood mapping techniques often fall short in addressing the complexities of flood dynamics. Conventional statistical and analytical methods frequently rely on limited in situ data, fail to account for interactions between various flood-causing factors, and struggle to adequately represent nonlinear processes involved in flooding. Addressing these limitations is critical for improving flood risk assessment and implementing effective mitigation strategies. Floods pose a growing threat to communities worldwide, exacerbated by the dual pressures of climate change and rapid urbanization. As the frequency and intensity of flooding events increase, the need for precise, timely, and actionable information to support flood risk management becomes ever more crucial. Traditional approaches to flood management, which often rely on historical data and simplistic modeling, struggle to keep pace with the evolving nature of flood risks. To address these challenges, the integration of advanced technologies such as geospatial artificial intelligence (AI) and remote sensing is emerging as a transformative solution.

Geospatial AI leverages machine learning algorithms to analyze large-scale spatial data, while remote sensing provides continuous monitoring of Earth's surface through satellite and aerial imagery. When combined, these technologies offer a powerful approach to detecting and predicting flood-prone areas with a high degree of accuracy. By incorporating both historical and real-time data, machine learning models can enhance flood prediction capabilities, supporting a more proactive response to potential flood events. This approach also enables a more comprehensive assessment of mitigation strategies, ensuring that resources are allocated effectively and emergency measures are optimized. The development of decision-support systems that integrate geospatial AI and remote sensing further enhances flood risk management. These systems provide critical insights for decision-makers, including policymakers, emergency responders, and urban planners, allowing for informed planning and timely interventions. Through this integration, the potential to shift from reactive to anticipatory flood management is significantly improved, paving the way for communities to build resilience against future flood threats. This chapter delves into the methodologies and applications of geospatial AI and remote sensing, showcasing how their synergy can drive innovation in flood risk management and contribute to a more resilient society.

RELATED WORK

Floods are recognized as one of the most prevalent and destructive natural hazards, causing significant social, economic, and environmental impacts worldwide. The intensifying effects of climate change are increasing the frequency and magnitude of extreme weather events, including floods, exacerbating vulnerabilities in many regions (Alfieri et al., 2018). Traditional flood risk management strategies, which often rely on historical flood data and empirical models, have been found inadequate to cope with the rapidly changing risk patterns (Schumann, 2011). This has led to an increased focus on advanced technologies such as remote sensing and AI to improve flood modeling and prediction capabilities.

The availability of high-resolution remote sensing data has significantly enhanced flood risk assessment. Modern satellites and aerial platforms provide continuous spatial coverage, enabling the monitoring of hydrological conditions and land surface changes in near-real time (Schumann & Di Baldassarre, 2010). Synthetic Aperture Radar (SAR) and optical imagery are widely used in flood mapping due to their ability to penetrate cloud cover and capture flooded areas with high accuracy (Pulvirenti et al., 2016). Furthermore, digital elevation models (DEMs) derived from remote sensing have become crucial for flood hazard mapping, providing valuable terrain information used in hydrological and hydraulic modeling (Tarboton, 1997).

The integration of geospatial data with remote sensing has advanced flood risk modeling from traditional deterministic approaches to more data-driven methodologies. For instance, Li et al. (2020) demonstrated the use of remote sensing to detect flood extents and estimate flood depth through multispectral and SAR imagery. These geospatial techniques offer a comprehensive view of flood-affected regions and allow for more effective flood management.

Machine learning (ML) and artificial intelligence (AI) have shown great potential for improving flood prediction models by handling the complex and nonlinear relationships between various flood-influencing factors. ML algorithms such as Random Forest, Support Vector Machine, and Neural Networks have been used to predict flood susceptibility and inundation patterns based on hydrological, meteorological, and topographical data (Tehrany et al., 2015).

Research has shown that ML techniques can outperform traditional statistical methods in predicting flood events due to their ability to learn from large datasets and model intricate dependencies (Mosavi et al., 2018). For example, deep learning approaches like convolutional neural networks (CNNs) have been applied to extract features from remote sensing images, improving flood detection accuracy (Zhao et al., 2019). Additionally, ensemble methods have been used to combine multiple models, further enhancing the predictive performance and robustness of flood forecasts (Zhang et al., 2018).

Despite significant progress in using AI and remote sensing for flood management, several challenges persist. One major issue is the interpretability of machine learning models, which often function as "black boxes" with limited insight into the underlying decision-making process (Sameen & Pradhan, 2019). Additionally, there is a lack of standardized validation procedures, making it difficult to assess the generalizability of flood models across different geographical regions (Alfieri et al., 2018).

Another critical limitation is the quality and availability of input data. Flood modeling relies heavily on accurate input data, such as rainfall measurements, soil moisture, and land cover information. However, in many regions, especially in developing countries, data scarcity and poor-quality records can impair the reliability of flood prediction models (Hoch et al., 2019). Furthermore, sensitivity analysis regarding the conditioning factors used in flood models is often inadequate, potentially leading to biased predictions (Pourghasemi et al., 2017).

The integration of geospatial AI with remote sensing provides an advanced framework for flood risk management, offering improvements in both prediction accuracy and response capabilities. This approach involves using AI algorithms to analyze satellite and aerial imagery for detecting flood-prone areas, forecasting flood events, and monitoring changes in water levels (Huang et al., 2020). For example, geospatial AI can be used to map flood susceptibility by combining remote sensing data with geographic information system (GIS) layers, such as land use and elevation data (Rahmati et al., 2019).

Recent studies have demonstrated the effectiveness of geospatial AI in automating the processing of large datasets and integrating multiple sources of information for more comprehensive flood modeling (Sampson et al., 2015). By utilizing both historical and real-time data, AI-driven models can dynamically adjust to evolving flood risks, making them valuable tools for decision-support systems used by policymakers and emergency responders (Huang et al., 2020). Additionally, remote sensing provides the spatial context necessary for validating AI-generated predictions, ensuring that flood management strategies are based on accurate and up-to-date information.

Panahi et al. integrated machine learning algorithms, specifically Random Forest and convolutional neural networks (CNNs), with remote sensing data to improve flood susceptibility mapping. By analyzing topographical, hydrological, and land use data, the hybrid model showed enhanced accuracy in identifying areas prone to flooding, thereby demonstrating the utility of combining different machine learning techniques for complex environmental datasets (Panahi et al., 2022).

Liu et al. applied deep learning techniques to multi-source remote sensing data, including synthetic aperture radar (SAR) and optical imagery, for near-real-time flood detection. The study showed how AI could automate flood extent mapping, providing faster response times for emergency management during flood events (Liu et al., 2021).

Sun et al. focused on the integration of climate data with AI for better flood prediction. Using Long Short-Term Memory (LSTM) networks and geospatial analysis, the study predicted flood events based on dynamic climate patterns, achieving higher prediction accuracy by incorporating climate-informed data (Sun et al., 2022).

Youssef et al. used a hybrid AI approach combining Support Vector Machine (SVM) and Geographical Information Systems (GIS) to map flood susceptibility in urban areas. This integration allowed for more detailed flood risk assessments, particularly in densely populated regions where urban planning requires precise risk mapping (Youssef et al., 2023).

Ensemble learning techniques were employed by Cheng et al., who combined various remote sensing datasets with Gradient Boosting algorithms to improve flood hazard modeling. The study showed that using ensemble learning methods enhanced the model's accuracy across different regions, indicating the benefit of integrating multiple data sources (Cheng et al., 2022).

Graph neural networks were used by Zhang et al. to model the spatial dependencies and connectivity of water flow in flood-prone areas. This approach improved the prediction of flood propagation and its potential impacts on communities, providing insights into the dynamics of flood spread (Zhang et al., 2023).

Huang et al. developed a flood forecasting system that incorporated both hydrological data and satellite-derived precipitation data, demonstrating the benefits of combining traditional and remote sensing data for more accurate and timely predictions, particularly in regions with limited ground-based observations (Huang et al., 2021).

Jafarzadeh et al. proposed a hybrid AI model that integrated fuzzy logic with machine learning for flood risk assessment. By using historical flood data along with satellite-based topographical and meteorological data, the model could adapt to various flood scenarios, enhancing risk management strategies (Jafarzadeh et al., 2023).

Khosravi et al. employed transfer learning techniques to enable AI models trained in one region to be applied effectively in another. This approach addressed the challenges of data scarcity and variations in flood characteristics, thus expanding the applicability of flood susceptibility models (Khosravi et al., 2022).

These recent studies illustrate how different AI and remote sensing approaches contribute to the advancement of flood risk management by improving prediction models, detection methods, and susceptibility mapping through innovative combinations of data and techniques.

EMPOWERING GEOAI(GEOSPATIAL ARTIFICIAL INTELLIGENCE)

Empowering GeoAI involves leveraging various cutting-edge technologies to analyze, interpret, and utilize geospatial data for a wide range of applications. At the heart of this ecosystem is Machine Learning (ML), which plays a crucial role in identifying patterns and relationships within geospatial datasets. ML algorithms enable GeoAI to perform predictive modeling, classification, and clustering tasks across diverse data types, including multispectral and hyperspectral satellite imagery, GPS data, and geotagged social media posts. This ability to make sense of complex geospatial information enhances decision-making processes in areas like urban planning, disaster response, and environmental monitoring.

Deep Learning adds another layer of capability by excelling at handling unstructured and complex geospatial data such as satellite imagery, aerial photos, and multitemporal data cubes. These models can automatically identify features, detect changes, and extract relevant information from vast datasets without requiring manual intervention. As a result, tasks like land cover classification, infrastructure assessment, and change detection are streamlined, enabling more efficient management of resources and timely responses to evolving situations.

Computer Vision enhances the analysis of visual geospatial data by allowing GeoAI to interpret images from satellites, drones, and other sources with remarkable precision. Using advanced algorithms, GeoAI can identify land cover types, map urban areas, and monitor changes in infrastructure over time. This capability is crucial for applications like land use mapping, disaster management, and smart city planning, where visual data provides critical insights. Meanwhile, Natural Language Processing (NLP) techniques extend GeoAI's reach by enabling the extraction of geospatial information and trends from unstructured text data, such as social media posts or news articles, supporting tasks like geospatial trend analysis and sentiment-based urban planning.

The integration of Remote Sensing, IoT, Big Data, and Cloud Computing ensures that GeoAI can efficiently manage the large volumes of geospatial data generated by modern technologies. Remote sensing provides comprehensive earth monitoring through various sensors on satellites and aircraft, while IoT devices continuously collect location-based data on environmental conditions, infrastructure, and human activities. Cloud computing offers the scalability needed to process and store this massive influx of data, making real-time analysis and decision-making feasible. Foundation models further empower GeoAI by providing pre-trained deep learning models that can adapt to different tasks with minimal additional training, boosting efficiency and versatility across a broad range of geospatial applications. Together, these technologies create a robust framework for data-driven decisions and proactive interventions in an increasingly dynamic world.

GEOSPATIAL AI AND REMOTE SENSING: TRANSFORMING SPATIAL ANALYSIS

The fusion of Geospatial Artificial Intelligence (AI) and remote sensing has dramatically reshaped how we collect, analyze, and apply spatial data, offering advanced tools for monitoring environmental changes, predicting natural disasters, and managing urban development. These technologies, when integrated, provide a robust framework for addressing complex problems across diverse fields, from disaster management to climate science.

Geospatial AI

Geospatial AI refers to the use of AI techniques—such as machine learning, deep learning, and computer vision—to process and interpret data that includes location information. This type of data encompasses a range of geographic and spatial attributes linked to natural or human-made features on the Earth's surface. Geospatial AI can identify patterns, cluster similar regions, detect changes over time, and generate predictive models that account for spatial relationships. These capabilities are applied across various domains, including land use classification, risk assessment, and environmental monitoring.

The value of geospatial AI lies in its ability to automate the extraction of information from massive datasets and enhance the accuracy of predictive models. Unlike traditional statistical methods, AI can handle complex, non-linear relationships and vast amounts of data, making it suitable for analyzing intricate environmental patterns and dynamics.

Remote Sensing

Remote sensing is the technology used to collect information about the Earth's surface from a distance, without making direct contact. This is typically achieved using sensors on satellites, aircraft, drones, or other platforms that gather data in various forms—optical images, radar signals, thermal measurements, etc. Remote sensing is essential for observing and measuring environmental conditions such as land cover, vegetation health, water levels, and atmospheric characteristics.

Different remote sensing methods offer unique benefits:

- **Optical Remote Sensing** involves using sensors that detect light in the visible, near-infrared, or shortwave infrared range to capture detailed images of the Earth. It is widely used for applications like monitoring land cover changes, assessing crop health, and mapping urban areas.
- **Synthetic Aperture Radar (SAR)** uses microwave signals that can penetrate clouds, fog, and even the forest canopy, making it useful for observing terrain changes, flood extents, and geological shifts.
- **Thermal Infrared Remote Sensing** measures the heat emitted from objects, allowing for the detection of temperature variations in land and water, forest fire monitoring, and urban heat island analysis.

THE POWER OF COMBINING GEOSPATIAL AI AND REMOTE SENSING

The integration of geospatial AI with remote sensing has unlocked new possibilities for analyzing spatial data and making informed decisions. Here's how the combination of these technologies drives advancements in various areas:

1. **Environmental Monitoring**

Geospatial AI can process satellite imagery to detect changes in land use, deforestation, or desertification in near-real-time. AI-driven models can automatically classify land cover, identify hotspots of environmental change, and generate insights that help in tracking the health of ecosystems. This capability is essential for managing natural resources and protecting biodiversity.

2. **Disaster Risk Management**

During natural disasters such as floods, wildfires, or earthquakes, geospatial AI can rapidly analyze remote sensing data to map affected areas and predict future events. For instance, using SAR and optical imagery, AI models can detect flood extent and track wildfire progression, providing vital information to emergency responders and decision-makers. This improves the speed and efficiency of disaster response and recovery efforts.

3. **Urban Planning and Development**

With up-to-date satellite imagery, geospatial AI enables accurate mapping of urban growth, infrastructure development, and changes in land use. Predictive models can forecast urban expansion and help city planners optimize land use, manage resources more effectively, and address the needs of growing populations.

4. **Agricultural Management**

In agriculture, combining remote sensing data from satellites like Landsat or Sentinel with AI algorithms helps monitor crop conditions, assess soil moisture, and predict yields. This allows farmers to optimize irrigation schedules, detect early signs of pest outbreaks, and maximize productivity, leading to more sustainable farming practices.

5. **Climate Change Research**

The integration of remote sensing and AI is crucial for studying the long-term impacts of climate change. It allows scientists to observe melting glaciers, rising sea levels, and shifts in weather patterns. By processing historical data, AI models can simulate future climate scenarios, providing valuable insights into potential risks and aiding in climate adaptation strategies.

CHALLENGES AND OPPORTUNITIES

Despite significant progress in combining geospatial AI and remote sensing, challenges remain:

- **Data Quality and Accessibility**: High-resolution and consistent data are crucial for accurate modeling, but in some regions, data may be scarce or of low quality. Addressing these gaps is essential for improving model reliability.
- **Complexity of AI Models**: Some AI models, especially deep learning techniques, can be difficult to interpret. This "black box" nature can limit their usefulness in critical applications where understanding the model's decision-making process is necessary.
- **Computational Demands**: Processing large volumes of remote sensing data requires substantial computing resources, which may pose a challenge for organizations with limited access to high-performance computing facilities.

The convergence of geospatial AI and remote sensing has brought about a new era in spatial analysis, offering unprecedented capabilities for understanding and addressing complex environmental and societal challenges. These technologies enable more accurate mapping, monitoring, and prediction of natural and human-induced events, providing critical insights for disaster management, sustainable development, and climate adaptation. As they continue to evolve, geospatial AI and remote sensing will play an increasingly vital role in shaping future decision-making processes and managing the Earth's dynamic landscape.

ADAPTIVE NEURO-FUZZY INFERENCE SYSTEM

ANFIS, or Adaptive Neuro-Fuzzy Inference System, is an advanced version of the Fuzzy Inference System (FIS) that integrates artificial neural networks to enhance its performance. FIS is widely recognized for its capability to address complex real-world problems by utilizing IF-THEN rules, which facilitate decision-making in uncertain environments. However, traditional FIS has limitations when it comes to producing accurate results for unforeseen circumstances. To overcome these challenges, FIS was optimized through the incorporation of neural network principles, resulting in the development of ANFIS.

ANFIS excels at solving linear load problems by leveraging a multilayer network structure. The parameters within the FIS can be adjusted through a learning process that combines least squares methods with backpropagation techniques, allowing the system to learn effectively from the input data. The architecture of the ANFIS consists of five primary layers in addition to the input layer. If each input variable is linked to two fuzzy sets, the system is capable of generating a total of 16 inference rules.

In Layer 1, known as the fuzzification layer, each node functions as a membership function—commonly using types like Gaussian or sigmoid functions. Layer 2 employs fuzzy T-norm operations, where each node executes a product operator to assess the fuzzy rules. In Layer 3, the outputs from the T-norms are normalized to ensure they sum to one. Layer 4 performs a linear combination of the input variables, leading to the final predictions. Finally, in Layer 5, the system computes the predicted output by taking a weighted average of the outputs derived from the various rules (see Figure 1 for a visual representation of the structure).

Figure 1. The structure of ANFIS for flood susceptibility models

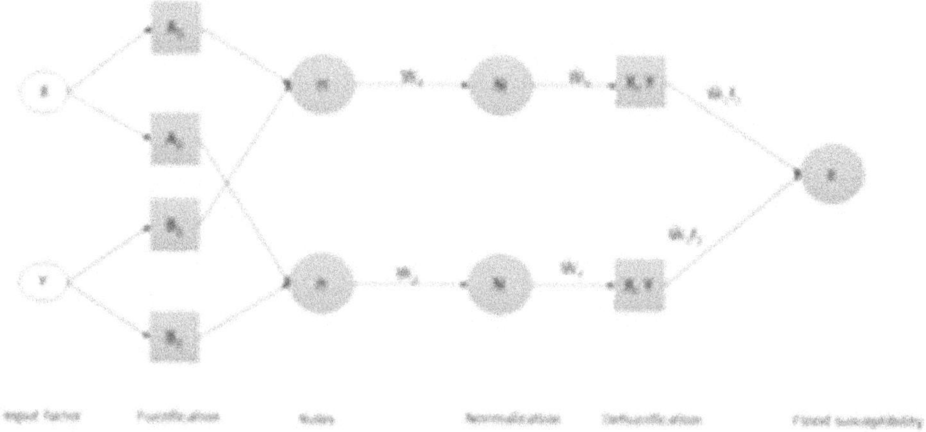

AI IN REMOTE SENSING: TRANSFORMING DATA ANALYSIS AND DECISION-MAKING

Artificial Intelligence (AI) has emerged as a game-changing technology in the field of remote sensing, enhancing our ability to analyze vast amounts of spatial data collected from various sensors, including satellites, drones, and aerial imagery. By integrating AI techniques, such as machine learning and deep learning, remote sensing applications have become more efficient, accurate, and capable of providing timely insights for diverse sectors, including agriculture, urban planning, environmental monitoring, and disaster management.

KEY APPLICATIONS OF AI IN REMOTE SENSING

1. **Land Use and Land Cover Classification**

AI algorithms, particularly supervised machine learning methods like Support Vector Machines (SVM) and Random Forest, are widely used to classify land use and land cover from remote sensing data. These models can automatically learn patterns from training datasets, allowing for precise categorization of different land types, such as forests, urban areas, and water bodies. For example, convolutional neural networks (CNNs) have shown remarkable success in accurately classifying complex land cover types from high-resolution satellite images.

2. **Change Detection**

Monitoring changes over time is critical for environmental management and urban planning. AI techniques can analyze time-series remote sensing data to detect changes in land cover, vegetation health, and infrastructure development. By applying algorithms that compare images taken at different times, AI can identify areas experiencing significant transformations, such as deforestation, urban sprawl, or natural disaster impacts.

3. **Crop Monitoring and Precision Agriculture**

In agriculture, AI-enhanced remote sensing is revolutionizing crop monitoring and management. By analyzing multispectral and hyperspectral satellite imagery, AI models can assess crop health, estimate yields, and detect pest infestations or diseases. This information enables farmers to make informed decisions regarding irrigation, fertilization, and pest control, ultimately leading to improved agricultural productivity and sustainability.

4. **Disaster Management and Response**

AI plays a vital role in disaster management by enabling rapid assessment and response during emergencies. Remote sensing data, when combined with AI techniques, can provide real-time information on the extent of damage caused by disasters such as floods, wildfires, and earthquakes. Machine learning models can analyze satellite images to map affected areas, assess the severity of damage, and support decision-making for effective emergency response and recovery efforts.

5. **Environmental Monitoring**

Remote sensing data, augmented by AI, is crucial for monitoring environmental changes and assessing the health of ecosystems. AI algorithms can analyze satellite images to track changes in land cover, monitor water quality, and assess biodiversity. This information is invaluable for conservation efforts, as it helps identify critical habitats, track endangered species, and evaluate the impacts of climate change.

6. **Urban Planning and Smart Cities**

In urban planning, AI in remote sensing facilitates the analysis of urban growth patterns, infrastructure development, and transportation networks. By processing remote sensing data, AI models can provide insights into population density, land use trends, and urban heat islands, helping city planners design more efficient and sustainable urban environments.

AI SURROGATES OF SIMULATIONS

AI surrogates have become important tools for advancing climate risk and impact assessments, providing a faster alternative to traditional simulation models. These machine learning-based models are trained to mimic the output of complex simulations by recognizing patterns between various input factors, such as weather conditions and land characteristics and outcomes like rainfall, temperature, or flood risk. This method significantly reduces computation time while preserving accuracy, making it ideal for applications where rapid decision-making is needed, such as in emergency response and disaster management, where traditional modeling approaches may be too slow. A major advantage of AI surrogates is their capacity to process large volumes of high-resolution data, which is essential for accurately forecasting local phenomena like urban heat islands or flash floods. While conventional models often struggle with fine-scale details due to limited resolution, AI surrogates can deliver precise predictions by leveraging vast datasets, including satellite imagery and historical climate records. They also support probabilistic risk assessments, where multiple scenarios are quickly evaluated to gauge the likelihood and uncertainty of various climate outcomes, guiding more effective resilience planning.

However, using AI surrogates does present certain challenges. Their accuracy depends on the quality of training data and the original simulation models they aim to replicate, making rigorous validation against real-world conditions necessary. Additionally, interpreting AI models, especially those based on deep learning techniques, can be challenging because they often function as "black boxes," making it difficult to understand the basis for their predictions. Another issue is ensuring that these models can generalize effectively to new situations, particularly for unprecedented climate events, which may not be well represented in existing training data. Looking ahead, combining AI surrogates with traditional physics-based models could enhance their predictive power, using the strengths of both approaches to achieve greater accuracy. Additionally, advancements in techniques like transfer learning may allow models trained in one geographic area to be adapted for use in another, addressing the problem of data scarcity and regional differences. As development continues, AI surrogates promise to accelerate climate modeling, providing organizations with valuable tools to better anticipate and adapt to the impacts of climate change and extreme weather events.

AI TO LEVERAGE GEOSPATIAL BIG DATA

AI has become a crucial tool for harnessing the potential of geospatial big data, enabling organizations to manage, analyze, and gain insights from the massive datasets generated by satellites, drones, and other remote sensing technologies. With the volume of geospatial data expanding rapidly, traditional data processing methods struggle to keep pace, while AI offers the ability to efficiently process and interpret these vast datasets. By applying advanced machine learning algorithms, AI can automatically detect patterns, identify objects, and classify land cover types, providing valuable insights that inform decision-making in sectors like agriculture, urban planning, environmental monitoring, and disaster management.

The application of AI in geospatial big data helps tackle challenges related to data variety and complexity. With geospatial data coming in various forms such as satellite images, LiDAR scans, and GPS records, AI algorithms are adept at integrating different data types to provide a comprehensive view of the landscape. This fusion of data sources allows for more detailed mapping and better understanding of spatial relationships, leading to accurate monitoring of environmental changes, such as deforestation, urban expansion, or coastal erosion. AI-powered models can continuously learn and improve from new data inputs, enhancing the ability to predict changes and assess the impact of different factors on the environment.

One significant advantage of using AI for geospatial data analysis is its capability to deliver near-real-time insights. In fields such as emergency response and disaster management, timely information is critical. AI algorithms can rapidly process incoming data from satellites or drones to detect areas affected by natural disasters like floods, wildfires, or earthquakes. This swift analysis enables emergency responders to act quickly, allocating resources and coordinating evacuations more effectively. Additionally, in agriculture, AI-driven geospatial analysis can optimize precision farming techniques by monitoring crop health, predicting yields, and identifying areas that require irrigation or pest control, thereby improving productivity and sustainability.

As AI technologies continue to evolve, their integration with geospatial big data is expected to become even more sophisticated. Future advancements may include the development of AI models that can predict complex geospatial phenomena, such as climate-driven changes in land use or the spread of invasive species, with greater accuracy. Moreover, as cloud computing and data-sharing platforms expand, AI's capacity to handle even larger datasets will further enhance its role in global efforts to manage natural resources, mitigate climate risks, and support sustainable development initiatives. AI's ability to transform geospatial big data into actionable insights is driving a new era of data-driven decision-making across various industries.

OPTIMIZING FLOOD RISK MANAGEMENT WITH GEOAI

Flood risk management is critical for safeguarding communities and minimizing the economic impact of floods. Traditional approaches often rely on historical data and manual analysis, which can be time-consuming and less adaptive to changing conditions. GeoAI, the fusion of geospatial technologies with artificial intelligence, offers a transformative approach to flood risk management by enhancing predictive capabilities, automating analysis, and enabling real-time decision-making.

FEATURES OF GEOAI IN FLOOD RISK MANAGEMENT

1. **Real-Time Flood Monitoring and Early Warning Systems** GeoAI integrates data from various sources, such as satellite imagery, weather stations, and social media feeds, to monitor flood conditions in real time. Machine learning algorithms process this data to detect patterns indicating rising water levels or potential flooding. When certain thresholds are met, early warning systems can be triggered to alert authorities and the public, providing valuable lead time for evacuation and preparation.
2. **Predictive Flood Modeling** Traditional flood models often struggle with accurately predicting floods due to complex variables such as rainfall intensity, terrain characteristics, and drainage systems. GeoAI enhances predictive modeling by using AI techniques to analyze large datasets, identify correlations, and simulate flood scenarios under different conditions. These models can account for climate change impacts, urban development, and other factors that traditional methods may overlook, resulting in more accurate forecasts.
3. **Automated Flood Extent Mapping** After a flood event, it is crucial to map the affected areas quickly to assess damage and coordinate relief efforts. GeoAI can automate the process of generating flood extent maps by using deep learning algorithms to analyze satellite and aerial imagery. These algorithms can detect water bodies, classify land cover, and differentiate between flooded and non-flooded areas, enabling rapid assessment of the situation and more efficient allocation of resources.
4. **Risk Assessment and Vulnerability Analysis** GeoAI can analyze spatial data to identify areas at high risk of flooding based on factors such as topography, land use, soil type, and historical flood records. By combining this information with socioeconomic data, it is possible to assess the vulnerability of communities and infrastructure. Decision-makers can then prioritize mitigation efforts, such as reinforcing flood defenses or revising zoning regulations, in the most vulnerable areas.

5. **Adaptive Flood Management Strategies** GeoAI supports the development of adaptive flood management strategies that can evolve with changing conditions. By continuously analyzing new data, GeoAI systems can update flood models and risk assessments dynamically. This adaptability allows for more flexible planning, such as adjusting floodplain boundaries, optimizing the placement of temporary barriers, or refining emergency response plans as new information becomes available.
6. **Post-Flood Damage Assessment and Recovery Planning** After a flood, GeoAI can facilitate damage assessment by analyzing post-event imagery to estimate the extent of damage to buildings, roads, and other infrastructure. This information helps prioritize recovery efforts, allocate resources efficiently, and guide the reconstruction process. Additionally, insights gained from past flood events can inform long-term resilience strategies to reduce the impact of future floods.

IMPACTFUL AI USE CASES IN GEOSPATIAL AND GIS INDUSTRY

In the geospatial and GIS industry, AI has become a game-changer, enabling transformative applications that drive smarter decision-making and more efficient operations. One of the most impactful use cases is in automated land use and land cover classification. By leveraging machine learning algorithms, AI can rapidly analyze satellite imagery to identify different types of land cover, such as forests, urban areas, or agricultural fields, and detect changes over time. This capability is particularly valuable for environmental monitoring, where continuous tracking of deforestation, urban sprawl, or wetland loss is crucial for sustainable land management and conservation efforts.

Disaster management is another area where AI's impact is significant, especially in providing real-time insights for emergency response. AI-powered models can quickly process data from various sources, such as satellites and drones, to detect and assess the extent of natural disasters like wildfires, floods, or earthquakes. These insights enable rapid response planning, helping authorities prioritize areas that need immediate attention, allocate resources more effectively, and reduce the risk of further damage. AI's ability to deliver timely and accurate information is essential for improving the speed and efficiency of disaster response efforts.

In the realm of precision agriculture, AI's use in geospatial analysis has been transformative, optimizing farming practices through data-driven decision-making. By analyzing data from satellite images, drones, and sensors, AI can monitor crop health, forecast yields, and identify stress factors such as pest infestations or nutrient deficiencies. This information helps farmers make informed decisions about irrigation, fertilization, and pest control, ultimately enhancing productivity while minimizing environmental impacts. The application of AI in agriculture supports more sustainable farming practices, which are increasingly important in the context of climate change and food security.

Urban planning and infrastructure management have also greatly benefited from AI's integration with GIS technologies. AI algorithms can process large amounts of geospatial data to optimize the planning and development of cities, identifying suitable locations for new infrastructure projects, analyzing traffic patterns, and predicting future urban growth. This leads to more efficient land use, reduced congestion, and better allocation of resources in rapidly expanding urban areas. The ability to anticipate and model changes in urban landscapes allows planners to make data-informed decisions that improve the quality of life for residents and contribute to the sustainable development of cities.

CONCLUSION

The integration of AI with geospatial technologies is revolutionizing the way we approach environmental monitoring, disaster management, agriculture, and urban planning. By leveraging advanced algorithms, AI enables the rapid analysis of vast and diverse geospatial datasets, providing valuable insights that were previously difficult to obtain. This capability has become crucial for addressing complex challenges, such as tracking land use changes, optimizing agricultural productivity, and managing the impacts of natural disasters. The use cases discussed demonstrate how AI's ability to process and interpret big data is empowering industries to make smarter, more data-driven decisions that enhance resilience and sustainability.

AI's impact in the geospatial and GIS industry extends beyond real-time data analysis to predictive modeling, where machine learning techniques can anticipate future changes in land use, climate patterns, and urban development. This foresight is invaluable for long-term planning and risk assessment, enabling organizations to prepare for potential challenges and mitigate their effects. While the deployment of AI surrogates offers efficient alternatives to traditional simulation models, their integration with physical-based methods can further improve accuracy, making AI a versatile tool for tackling both current and emerging geospatial challenges.

As AI technology continues to advance, its role in transforming geospatial data into actionable insights will only grow, driving innovation across various sectors. To fully realize its potential, ongoing efforts are needed to overcome challenges related to data quality, model interpretability, and generalization across different contexts. By addressing these issues, AI can continue to push the boundaries of what is possible in geospatial analysis, providing powerful tools for sustainable development, climate resilience, and efficient resource management in an increasingly complex world.

References

Alfieri, L., Bisselink, B., Dottori, F., Naumann, G., Roo, A. D., Salamon, P., & Feyen, L. (2018). Global projections of river flood risk in a warmer world. *Earth's Future*, 6(2), 704–717.

Cheng, Q., Zhao, J., & Wu, H. (2022). Ensemble learning for flood hazard modeling using multi-source remote sensing data. *Remote Sensing of Environment*, 276, 113043.

Hoch, J. M., Neal, J., Baart, F., & Winsemius, H. C. (2019). Data scarcity for hydrological impact studies in data-sparse regions: Challenges and potential solutions. *Frontiers of Earth Science*, 7, 257.

Huang, C., Chen, Y., & Wu, J. (2020). Mapping flood susceptibility using geospatial machine learning techniques. *Water (Basel)*, 12(4), 1026.

Huang, J., Liang, S., & Wang, L. (2021). Flood forecasting with integrated hydrological models and deep learning techniques. *Journal of Hydrometeorology*, 22(3), 577–594.

Jafarzadeh, M., Shabani, M., & Alizadeh, M. (2023). Hybrid AI-based flood risk assessment using fuzzy logic and machine learning. *The Science of the Total Environment*, 868, 161542.

Khosravi, K., Shahabi, H., & Chen, W. (2022). Transfer learning for flood susceptibility modeling: A step towards universal flood risk assessment. *Environmental Modelling & Software*, 153, 105403.

Li, Z., Xu, Q., & Tan, X. (2020). Flood depth estimation from synthetic aperture radar imagery using machine learning. *Remote Sensing*, 12(3), 531.

Liu, Y., Hu, J., & Fang, Y. (2021). Real-time flood monitoring using deep learning techniques and multi-source remote sensing data. *Remote Sensing*, 13(1), 45.

Mosavi, A., Ozturk, P., & Chau, K. W. (2018). Flood prediction using machine learning models: Literature review. *Water (Basel)*, 10(11), 1536. DOI: 10.3390/w10111536

Panahi, M., Rezaie, F., & Shirzadi, A. (2022). Flood susceptibility mapping using hybrid machine learning algorithms: A case study from Iran. *Geocarto International*, 37(3), 734–749.

Pourghasemi, H. R., Rahmati, O., & Gokceoglu, C. (2017). Modeling the impact of human activities and geomorphology on flood susceptibility. *Geocarto International*, 32(3), 244–259.

Pulvirenti, L., Chini, M., Pierdicca, N., & Guerriero, L. (2016). An algorithm for operational flood mapping from synthetic aperture radar (SAR) data using fuzzy logic. *Remote Sensing*, 8(7), 565.

Rahmati, O., Pourghasemi, H. R., & Melesse, A. M. (2019). Application of GIS and remote sensing techniques in flood risk management: A case study in Ilam Province, Iran. *Environmental Earth Sciences*, 78(1), 40.

Sameen, M. I., & Pradhan, B. (2019). Assessing landslide susceptibility using machine learning and geospatial data in the high mountain environment. *Geomorphology*, 327, 11–22.

Sampson, C. C., Smith, A. M., Bates, P. D., Neal, J. C., Trigg, M. A., & Brewer, P. A. (2015). A high-resolution global flood hazard model. *Water Resources Research*, 51(9), 7358–7381. DOI: 10.1002/2015WR016954 PMID: 27594719

Schumann, G. (2011). The need for precise global information in flood risk management. *International Journal of Remote Sensing*, 32(22), 5963–5968.

Schumann, G., & Di Baldassarre, G. (2010). The direct use of radar satellites for continuous monitoring of flood inundation. *Remote Sensing of Environment*, 115(1), 2880–2890.

Sun, W., Guo, L., & Zhang, Y. (2022). Climate-informed flood risk prediction using LSTM networks and geospatial analysis. *Water Resources Research,* 58(4), e2022WR031301.

Tarboton, D. G. (1997). A new method for the determination of flow directions and upslope areas in grid digital elevation models. *Water Resources Research*, 33(2), 309–319. DOI: 10.1029/96WR03137

Tehrany, M. S., Pradhan, B., & Jebur, M. N. (2015). Flood susceptibility mapping using a novel ensemble weights-of-evidence model. *Geocarto International*, 30(6), 660–685.

Youssef, A. M., Pradhan, B., & Pourghasemi, H. R. (2023). Urban flood susceptibility mapping using hybrid AI models and GIS: Case study of Cairo, Egypt. *Geocarto International*, 38(1), 95–114.

Zhang, J., Huang, Y., & Ma, H. (2018). Application of ensemble methods in flood prediction: A case study in the Yangtze River Basin, China. *Water (Basel)*, 10(12), 1801.

Zhang, X., Wang, Q., & Li, H. (2023). Mapping flood propagation using graph neural networks and geospatial data. *Environmental Modelling & Software*, 160, 105349.

Zhao, W., Du, S., & Qi, J. (2019). CNN-based deep learning method for mapping flood areas. *Water (Basel)*, 11(10), 2060.

Chapter 7
The Role of IoT in Shaping the Future of Geospatial AI

Rachna Rana
Ludhiana Group of Colleges, Ludhiana, India

Pankaj Bhambri
https://orcid.org/0000-0003-4437-4103
Guru Nanak Dev Engineering College, Ludhiana, India

ABSTRACT

This chapter shows about the new expertise for instance AI, ML, and IoTs which has altered the geospatial sector, which involves the collecting, analysis, and display of geographical data. These technologies drive industrial innovation and growth by allowing for more precise and efficient data collecting, analysis, and decision-making. AI & ML are especially significant in the geospatial business because they enable the analysis of enormous quantity of facts that would be too time-consuming for people to handle manually. AI and ML methods can examine and understand geographical data such as satellite imaging, aerial photography, and LiDAR scans, revealing patterns and trends that humans may miss. The IoTs is also propelling the geospatial sector forward by allowing for the capture of real-time data from sensors implanted in actual things. Weather sensors, traffic sensors, and GPS trackers are examples of IoTs devices that may provide useful geographical data for decision-making in a kind of business, including farming, transportation, and town development.

DOI: 10.4018/979-8-3693-8054-3.ch007

1.1 INTRODUCTION

The convergence of the IoTs and Geospatial Artificial Intelligence (AI) is changing the way geographical data is gathered, processed, and utilized across sectors. The Internet of Things (IoT) is a network of linked devices and sensors that create real-time data, whereas geospatial AI analyzes and interprets geographic information using Machine Learning (ML) and AI techniques. When coupled, IoTs and Geospatial AI have the potential to transform industries such as town development, ecological scrutinizing, transportation, and farming by providing actual-time insights and predictive analytics depended on location-specific data (Ahmad & Nabi, 2021).

IoT devices, such as GPS-enabled sensors, drones, and smart infrastructure, are constantly collecting massive volumes of geographical data. This data is sent into AI systems, which use it to spot trends, optimize processes, and even forecast future results. The confluence of these technologies improves decision-making processes, allowing companies and governments to better manage resources, adapt to environmental changes, and create smarter cities.

As the number of IoT devices deployed grows and AI models get more complex, the future of geospatial AI will see tremendous expansion, delivering a better knowledge of our physical environment and the forces that drive it. This chapter investigates the crucial role of IoT in enhancing geospatial AI, focusing on its applications, advantages, and future prospects (Anand et al., 2021).

The fast advancement of technology has enabled the convergence of two significant innovations: the IoTs and Geospatial AI. This combination has the potential to transform how we see and interact with our environments. The IoT is a linked set-up of sensors, devices, and structures that interact and exchange data in real time, whereas geospatial AI uses machine learning and AI techniques to analyze and understand geographical or location-based data.

When IoT and Geospatial AI are joint, tremendous synergies emerge that propel progress in a variety of domains, consisting of elegant cities, ecological monitoring, transportation, farming, and disaster management (Bhambri & Bajdor, 2024a). Satellites, drones, GPS systems, smart infrastructure, and other linked technologies create massive amounts of location-specific data on a continual basis. Geospatial AI processes and analyzes data to identify trends, optimize operations, and create more precise forecasts (Bernini et al., 2023).

The Internet of Things plays a significant job in defining the future of geospatial artificial intelligence. IoT improves geospatial models' capabilities by supplying real-time, high-resolution data from a variety of sources, allowing for faster, more informed decisions. For example, IoT-powered Geospatial AI can monitor environmental conditions in real time, enabling for faster responses to natural catastrophes or ecosystem changes. As these technologies grow, the future offers even deeper integration, with smarter, more adaptable systems that will promote sustainable development, better resource management, and improve quality of life. This article investigates how the IoT is affecting the future of geospatial AI, examining its present uses, limitations, and disruptive potential (Bisht et al., 2024).

The large-scale automation of farming in the 20th century replaced labour with machines, increasing earth efficiency and attaining financial system of level. Accuracy agricultural has the possible to considerably increase farmer incomes, get better the extrinsic and intrinsic quality of agricultural goods, and reduce farming's negative environmental consequences (Bhambri & Bajdor, 2024b).

Precision farming, while not a heal-all, can help to promote large sustainable agriculture methods. The continuing 4th industrial revolution is transforming farming, ushering in the age of Agriculture 4.0. Data-driven management, innovative tool-based manufacturing, sustainability, professionalization, and a low ecological force define this next era.

Agriculture accounts for 85% of total water handled globally, emphasizing the need of precision water management in orchards, particularly in semi-arid locations where water input represents a significant monetary speculation. Weather modify exacerbates these issues for the fruit business by generating extended dearth in some location. tiny and marginal farmers account for more than 95% of the agricultural population, with over 550 million of the world's 608 million farms being tiny (less than 2 ha) or medium-sized (less than 50 ha). As a result, increasing farmer use of precision farming technology (PFTs) is critical for ensuring long-term and healthy crop output. Precision farming (PF) technology may dramatically increase land productivity (Chakravarthy et al., 2022).

Precision farming now faces a number of issues, including indefensible source usage, extended monoculture, concentrated creature husbandry, ecological dilapidation, unequal digitalization, food safety concerns, an inefficient agri-food supply chain, and opposition to change. These issues impede the achievement of efficiency, productivity, and sustainability in agricultural production while worsening unintended consequences for ecosystems.

Furthermore, accuracy agricultural systems frequently have disadvantages, such as expensive hardware prices, strict topography necessities, and vulnerability to ecological influences. PF decreases greenhouse gas emissions significantly by using machine guidance and controlled traffic farming to avoid overlaps in farm activities, resulting in a 6% reduction in fuel usage. These advantages are considerably more substantial in large-scale fields, with reduced soil compaction, runoff, and erosion (Chandrashekhar et al., 2024).

1.2 ARTIFICIAL INTELLIGENCE

AI is revolutionizing the geospatial sector by enabling more complex analysis, decision-making, and process automation based on geographic data. The geospatial sector is concerned with the gathering, management, analysis, and interpretation of data pertaining to the Earth's surface. By adding AI, the geospatial sector is moving beyond traditional approaches, allowing for deeper insights and more dynamic applications across a variety of industries (Dhibar & Maji, 2023).

AI, namely ML and deep learning, is being used to process and analyze massive volumes of geospatial data, such as satellite imaging, aerial photography, GPS data, and geographic information system (GIS). These technologies assist in automating processes that formerly needed substantial manual effort, such as picture categorization, feature extraction, and predictive analysis (Ezhilarasan & Jeevarekha, 2023). The Figure 1 is related to Applications of AI in the Geospatial Sector.

1.2.1 Key Applications of Ai in the Geospatial Sector

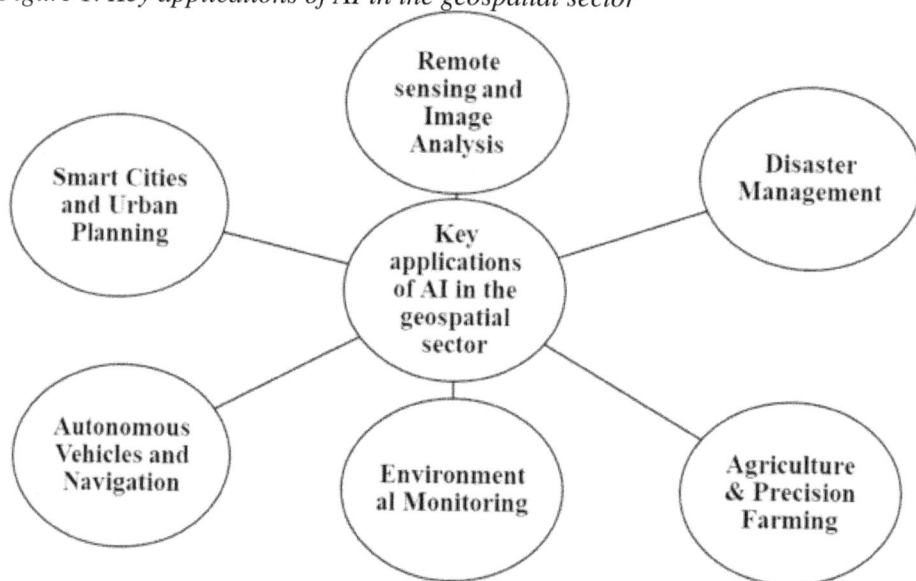

Figure 1. Key applications of AI in the geospatial sector

Remote sensing and Image Analysis: AI has transformed the analysis of satellite and drone images. Machine learning algorithms can now categorize land use, detect landscape changes, monitor deforestation, and map urban expansion with greater precision. Automated feature extraction identifies buildings, roads, and other infrastructure in real time.

Disaster Management: Artificial intelligence-powered geospatial analytics are being used to forecast natural disasters such as floods, earthquakes, and fires. AI algorithms may use historical data, weather trends, and current sensor data to anticipate danger regions, allowing authorities to plan and respond more efficiently (Bajdor & Bhambri, 2024).

Smart Cities and Urban Planning: AI algorithms use geographical data from IoT devices to optimize traffic flow, minimize energy usage, and improve public safety. AI is also utilized in urban planning to simulate various scenarios for sustainable development, which helps to improve municipal infrastructure and services.

Agriculture & Precision Farming: Using AI-powered geospatial technologies, farmers may improve crop yields by evaluating data from drones, sensors, and satellites. These systems can track soil health, forecast insect outbreaks, and evaluate crop conditions, resulting in more efficient and sustainable agricultural operations (Geetha et al., 2024).

Environmental Monitoring: Artificial intelligence models are used to track environmental changes such as climate change, pollution, and ecosystem deterioration. By evaluating geolocation data over time, AI can assist forecast environmental changes and analyze the influence of humans.

Autonomous Vehicles and Navigation: Geospatial AI is a critical component of autonomous vehicle systems, allowing them to understand their surroundings and travel safely. AI systems use real-time GPS data, street maps, and sensor input to improve navigation and avoid obstructions.

1.2.2 Obstacles and Opportunities

The geospatial industry benefits greatly from AI, but also has obstacles such as enormous datasets, high processing power, and potential biases in models. As the usage of artificial intelligence in geospatial applications develops, data privacy and ethical problems emerge.

Looking ahead, the continuing integration of AI into geospatial technology will open up new avenues for innovation, ranging from disaster response and sustainability initiatives to enhanced location-based services and personalized mapping solutions (Mohana Sundari et al., 2024). Advances in data processing, cloud computing, and the proliferation of linked devices will determine geospatial AI's future, transforming it into a vital driver of social advancement.

1.3 INTERNET OF THINGS

The IoT is changing the geospatial industry by allowing for real-time gathering, transmission, and analysis of location-based data. The IoT is a network of networked devices, sensors, and systems that can gather and share data over the internet without human interaction. In the geospatial context, IoT devices provide a constant stream of spatial data that may be utilized for monitoring, mapping, analysis, and decision-making across a variety of businesses (Garg et al., 2021).The geospatial sector may use IoT to access a wide range of data sources, including GPS-enabled devices, drones, satellites, weather stations, and smart infrastructure. This real-time data may be incorporated into Geographic Information Systems (GIS) and analyzed with geospatial methods to acquire a better understanding of patterns, trends, and spatial linkages (Gupta & Vyas, 2023).

1.3.1 Key IoT Applications in the Geospatial Sector

Figure 2 elaborates the Key IoT applications in the geospatial sector.

Figure 2. Key IoT applications in the geospatial sector

Smart City and Infrastructure: Smart cities use IoT devices to monitor traffic flow, energy usage, air quality, and public services. The data acquired by these sensors may be displayed on GIS systems, allowing city planners to optimize urban infrastructure, minimize traffic congestion, and increase sustainability (Rana & Bhambri, 2024a). For example, IoT-enabled traffic sensors may give real-time traffic information, allowing for more effective transportation system management.

Environmental Monitoring: The Internet of Things plays an important role in environmental monitoring by delivering real-time data on air and water quality, meteorological conditions, and natural resource use. Sensors installed in isolated places, rivers, woods, or seas can detect environmental changes such as pollution, deforestation, and wildlife migrations.

Precision Agriculture: In agriculture, IoT-enabled equipment such as soil moisture sensors, weather stations, and drones collect information on crop conditions, soil health, and weather patterns. Farmers may use this geographical data to make more educated planting, irrigation, and harvesting decisions. Precision agriculture based on IoT and geospatial data may boost crop yields, minimize water and fertilizer use, and encourage sustainable agricultural methods.

Disaster Management and Emergency Response: IoT devices are extremely useful in disaster management because they provide real-time monitoring of key infrastructure, weather patterns, and environmental dangers (Vigneshwari et al., 2024). For example, IoT sensors can monitor rising water levels in rivers, landslides, and seismic activity, providing early warning of floods, earthquakes, and other natural disasters. Geospatial tools may utilize this data to produce risk maps, forecast catastrophic impacts, and advise

Transport and logistics: IoT-enabled GPS devices and sensors are widely utilized in the transportation and logistics industries to monitor cars, freight, and shipments. This geographical data aids in route optimization, fuel efficiency, and supply chain management. IoT also helps smart transportation systems by giving real-time data on vehicle locations, road conditions, and traffic congestion, which may be utilized to improve traffic flow and minimize travel time.

Asset and Infrastructure Management: The Internet of Things is rapidly being used to monitor and manage infrastructure including bridges, highways, pipelines, and power grids. Sensors mounted on infrastructure assets can gather data on their state, detecting indicators of wear, damage, or failure. This real-time geographic data enables for predictive maintenance, which helps firms address issues before they become significant and save downtime.

1.3.2 Advantages of IoT in the Geospatial Sector

Real-Time Data: IoT devices generate continuous, real-time data streams, enabling for more accurate and up-to-date geographic analysis.
Improved Decision-Making: Integrating IoT data into geospatial systems allows for speedier, data-driven decisions, particularly in crucial areas such as disaster management, urban planning, and resource management.
Automation: The Internet of Things (IoT) can automate data collecting operations, eliminating the need for human data entry and allowing for real-time monitoring of geographical phenomena.
Enhanced Efficiency: IoT devices may improve processes and resource allocation by giving accurate, location-specific data, resulting in increased operational efficiency.

1.1.2 The Challenges of IoT in the Geospatial Sector

Data Privacy and Security: As IoT devices collect massive volumes of geographic data, issues about data privacy and security develop, particularly in terms of protecting sensitive location information.
Data Integration: Bringing together disparate data streams from numerous IoT devices into a single GIS system may be difficult and time-consuming.
Interoperability: The absence of defined protocols and standards for IoT devices might make it difficult for different devices and platforms to connect and share data successfully.

1.3.3 Future of IoT in the Geospatial Sector

As IoT devices become more common and geospatial technology evolve, the potential for IoT in the geospatial sector will only increase (Chithra & Bhambri, 2024). More advanced sensors are anticipated in the future, as will enhance connection via 5G networks and more integration with AI and machine learning. These developments will allow for even more precise, efficient, and real-time geospatial analysis, accelerating innovation in fields such as urban planning, environmental conservation, agriculture, transportation, and public safety.

1.4 GPS

The Global Positioning System (GPS) is one of the most essential geospatial technologies, delivering precise, real-time position data that serves as the foundation for a wide range of spatial applications. GPS is made up of a network of satellites circling the Earth that send signals to GPS receivers on the ground, allowing them to precisely calculate their position in terms of latitude, longitude, and altitude (Hassani et al., 2021).

GPS is utilized in a variety of geospatial applications, including mapping, navigation, surveying, environmental monitoring, and many more. By merging GPS data with Geographic Information Systems (GIS), remote sensing, and other geospatial technologies, businesses may acquire important insights into spatial relationships, trends, and patterns across several geographies (Hassani et al., 2021)

1.4.1 GPS's Primary Applications in the Geospatial Sector Include

Figure 3 is sharing the GPS's primary applications in the geospatial sector.

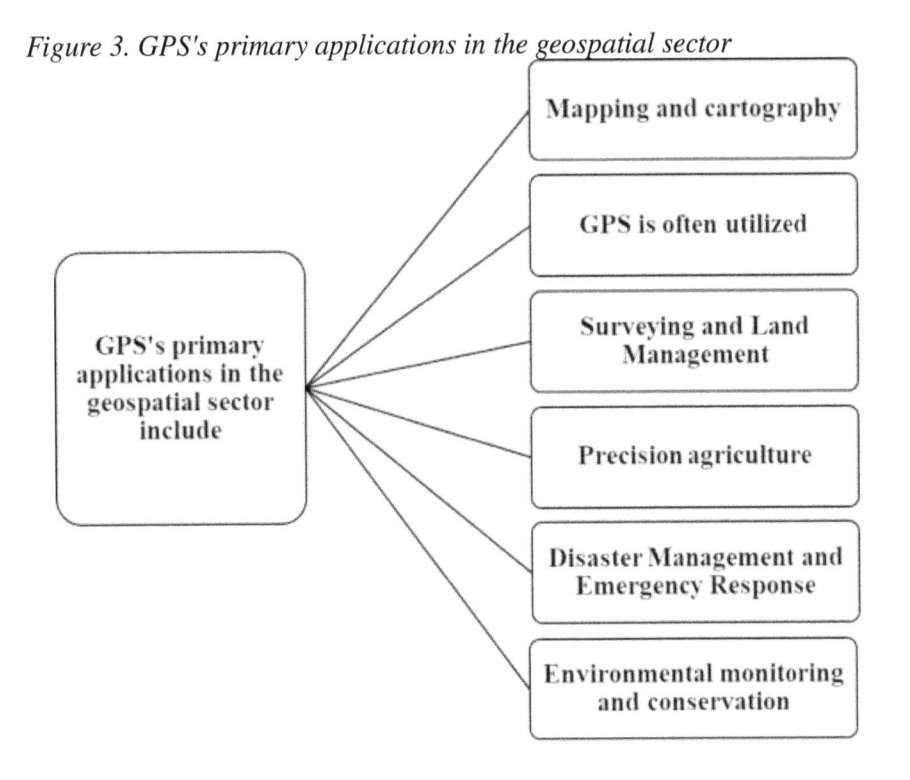

Figure 3. GPS's primary applications in the geospatial sector

GPS's primary applications in the geospatial sector include:
- Mapping and cartography
- GPS is often utilized
- Surveying and Land Management
- Precision agriculture
- Disaster Management and Emergency Response
- Environmental monitoring and conservation

Mapping and cartography: GPS is an essential tool in contemporary mapping and cartography. GPS-enabled devices capture exact position data, which is subsequently utilized to generate and update maps. GPS offers the real-time positional precision required to correctly map geographic features, whether they be comprehensive topographic maps, road maps, or land use maps. This information is critical for everything from navigation to land management.

GPS is often utilized in navigation systems for automobiles, planes, ships, and even pedestrians. In the transportation industry, GPS offers real-time position data for route planning, fleet management, and traffic monitoring. It enables more effective route planning, lowers fuel consumption, and improves overall logistics.

Surveying and Land Management: GPS revolutionized surveying by providing an efficient, precise, and dependable means of calculating coordinates and distances. Traditional land surveying techniques sometimes required substantial human labor, but GPS enables surveyors to measure land features and borders with great precision. This is critical for infrastructure development, land ownership records, real estate administration, and construction planning.

Precision agriculture uses GPS to direct farming machines, monitor crop conditions, and increase land use efficiency. Farmers may use GPS-enabled devices to properly map fields, plan planting patterns, and precisely apply fertilizers and pesticides. This geographical data enables farmers to improve their farming practices, enhance yields, decrease waste, and promote sustainability.

Disaster Management and Emergency Response: GPS is an essential tool for disaster preparedness and response. GPS-enabled systems assist in tracking the movement of people and supplies during emergencies, giving real-time position data to organize rescue operations, evacuations, and relief activities. GPS is used during natural disasters like earthquakes, floods, and wildfires to track the level of damage, identify impacted regions, and direct recovery efforts.

Environmental monitoring and conservation rely heavily on GPS technology. GPS devices, for example, are used to track animal movements, map habitats, monitor deforestation, and assess environmental changes. GPS enables conservationists and academics to better understand environmental changes and develop methods for sustainable natural resource management.

GPS is critical for gathering geographical data in a variety of businesses, including urban planning, telecommunications, and utilities. Surveyors, engineers, and planners utilize GPS to collect information on infrastructure, utilities, and natural resources. GIS systems use this geographic data to analyze and depict spatial trends, optimize resource management, and make data-driven choices.

1.4.2 Advantages of GPS in the Geospatial Sector

High Accuracy: GPS offers exact position data, enabling accurate mapping, navigation, and feature analysis.

Real-Time Data: GPS-enabled devices provide real-time position tracking, which is essential for applications such as navigation, transit, and emergency response.

GPS is widely available across the world, making it a useful tool for a variety of companies and sectors.

1.4.3 Challenges and Limitations of GPS

Signal Interference: Physical objects such as buildings, trees, or mountains can block or decrease GPS signals, reducing accuracy in some locations, notably metropolitan areas or dense woods.

Accuracy Issues: While GPS has outstanding accuracy, it may be restricted in densely populated regions where signals are bounced off buildings or in isolated places with low satellite coverage.

Power Dependency: GPS receivers require a constant power supply, therefore battery life is crucial, especially for long-duration operations in remote places.

The Future of GPS in Geospatial Sector

- As GPS technology advances, its integration with other geospatial technologies such as GIS, drones, remote sensing, and AI will provide new options (Sharma & Bhambri, 2024). The development of GPS augmentation technologies, such as differential GPS (DGPS) and real-time kinematic (RTK) positioning, is enhancing GPS data accuracy and dependability, particularly in tough settings.
- GPS will be critical for real-time navigation and spatial awareness as autonomous systems, such as self-driving automobiles and unmanned aerial vehicles (UAVs), become more prevalent. Furthermore, as new satellites are deployed and global positioning systems improve, GPS accuracy will increase, helping all businesses that rely on geospatial intelligence.

1.5 GEOSPATIAL TECHNOLOGIES FACILITATE INDUSTRIAL GROWTH

Geospatial technology have emerged as a driving factor in industrial growth, transforming how businesses operate, allocate resources, and make strategic choices. These technologies, which include GIS, remote sensing, GPS, and geospatial data analytics, enable organizations to use location-based data to boost efficiency, productivity, and competitiveness across industries (Heras et al., 2021; Hohmeier et al., 2024; Kalpana et al., 2024).Geospatial technologies have emerged as critical tools for driving innovation, enhancing decision-making, and attaining sustainable growth, as companies increasingly rely on geospatial data to manage assets, monitor operations, and evaluate market trends (Keskin & Sekerli, 2024; Khankhoje, 2024).

1.5.1 Key Ways Geospatial Technologies Promote Industrial Growth

Keyways Geospatial Technologies, which promotes the industrial growth are shown in Figure 4.

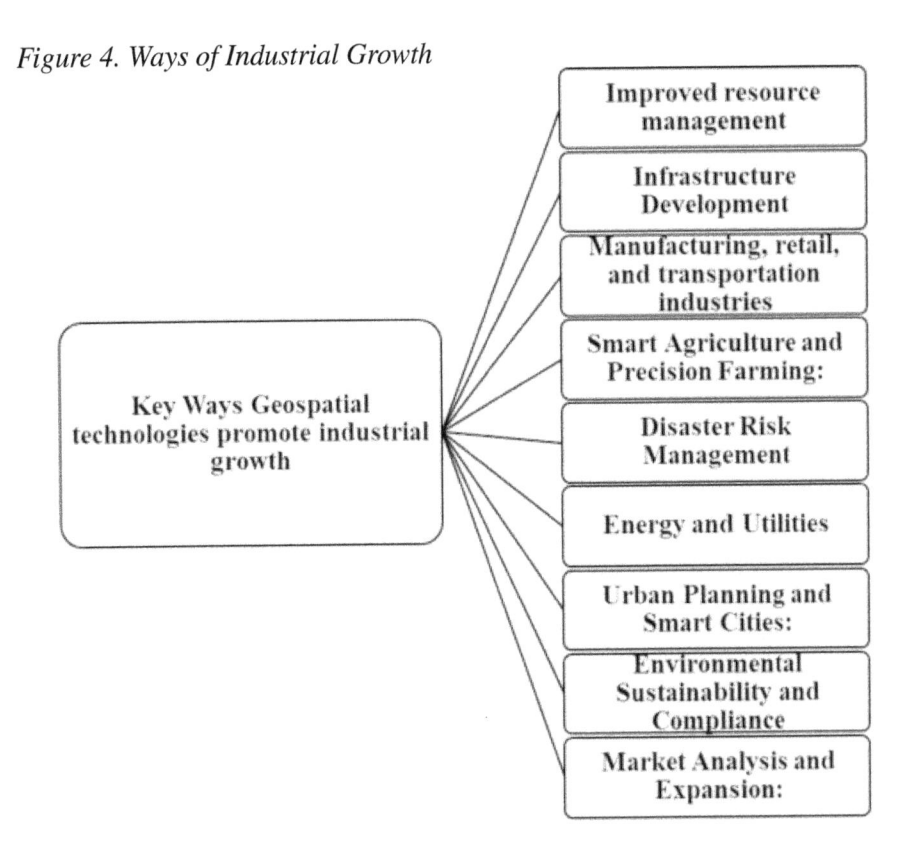

Figure 4. Ways of Industrial Growth

```
Key Ways Geospatial           ┌─ Improved resource management
technologies promote          ├─ Infrastructure Development
industrial growth       ─────┤├─ Manufacturing, retail, and transportation industries
                              ├─ Smart Agriculture and Precision Farming:
                              ├─ Disaster Risk Management
                              ├─ Energy and Utilities
                              ├─ Urban Planning and Smart Cities:
                              ├─ Environmental Sustainability and Compliance
                              └─ Market Analysis and Expansion:
```

Improved resource management: Geospatial technologies enable enterprises to map, monitor, and manage resources more precisely. In industries such as mining, oil and gas, and forestry, GIS and remote sensing technology give precise geographical data on natural resources. This enables businesses to find significant minerals, optimize extraction procedures, and reduce environmental effect. Efficient resource management reduces costs and increases production.

Infrastructure Development: Geospatial technologies are essential for planning and implementing infrastructure projects such as roads, bridges, utilities, and pipelines. Companies may use GIS and GPS to map topography, analyze environmental concerns, and determine the optimum paths for infrastructure development (Thirumalaiyammal et al., 2024). These technologies assist to decrease construction costs, eliminate dangers, and complete projects on schedule. Geospatial data helps to streamline infrastructure design and execution in industries like construction, energy, and transportation.

Manufacturing, retail, and transportation industries all rely on geospatial technology to enhance logistics and supply chain operations. GPS-enabled tracking systems give real-time information on the whereabouts of cars, shipments, and assets, allowing businesses to optimize route, minimize fuel consumption, and shorten delivery times.

Smart Agriculture and Precision Farming: Geospatial technologies are transforming the agriculture business with precision farming. GPS-guided tractors, drones, and sensors enable farmers to track soil conditions, crop health, and weather trends. GIS enables data analysis to Optimize planting schedules, irrigation, and fertilizer use, resulting in increased crop yields and lower resource use. The capacity to handle agricultural operations more efficiently increases production while reducing environmental impact, hence promoting agricultural sector growth.

Disaster Risk Management: Geospatial technology can assist organizations reduce the risks associated with natural catastrophes. Industries can predict disasters like floods, earthquakes, and hurricanes by monitoring environmental parameters such as weather patterns, terrain changes, and water levels using GIS and remote sensing.

Energy and Utilities: Geospatial technologies play an important role in resource discovery, production, and distribution. For example, GIS and remote sensing assist energy corporations in locating possible drilling sites, monitoring pipeline integrity, and mapping energy distribution networks. Geospatial data is also used by renewable energy sectors such as solar and wind to determine the best sites for solar farms and wind turbines based on parameters such as sunlight exposure and wind patterns. Geospatial technologies increase operational efficiency, save costs, and promote sustainability.

Urban Planning and Smart Cities: As cities grow and urbanize, geospatial technologies become critical for urban planning and smart city development. GIS enables urban planners to examine spatial data such as population density, land use, transportation, and utilities in order to create more efficient, livable cities (Ruby et al., 2024). Smart cities employ IoT devices to collect geographical data, which is then analyzed to improve traffic flow, cut energy usage, and increase public safety. The integration of geospatial technology with smart city projects results in higher quality of life and economic progress.

Environmental Sustainability and Compliance: Geospatial technology help companies meet environmental requirements and promote sustainability. Industries may decrease their environmental footprint by mapping environmental hazards, monitoring land use changes, and tracking pollutant levels using data analysis. GIS is, for example, utilized in the forestry and agriculture industries to guarantee sustainable land use and monitor conservation initiatives. This not only assures compliance with environmental legislation, but it also improves the reputation of organizations who prioritize sustainability.

Market Analysis and Expansion: Geospatial data is increasingly being utilized to analyze markets and find new growth prospects. GIS can map consumer demographics, rival locations, and economic trends to help choose where to grow operations. Geospatial data is used by the retail, real estate, and telecommunications industries to determine where to establish new stores, deploy networks, and target advertising. This data-driven strategy to market growth enables organizations to penetrate new areas more successfully and expand their consumer base.

1.5.2 The Advantages of Geospatial Technologies for Industrial Growth

Data-driven decision-making: Geospatial technology give companies with reliable, location-based data that helps them make better decisions. Companies may use geographical data to find patterns, optimise processes, and decrease risks.

Cost Efficiency: Industries may cut operating costs and increase profitability by leveraging geospatial data to simplify operations like logistics, resource management, and infrastructure construction.

Sustainability: Geospatial technologies assist companies in monitoring and managing their environmental effect, supporting sustainable practices that comply with legal standards and corporate social responsibility.

Increased Productivity: Geospatial solutions enable automation and exact data collecting, which improves operational efficiency and output.

Companies that use geospatial technology gain a competitive advantage by making better decisions, expanding into new areas, and responding more effectively to industry difficulties.

1.5.3 The Challenges of Geospatial Technologies in Industrial Growth

Data Complexity: Managing and interpreting large volumes of geographic data may be difficult, necessitating expert staff and sophisticated tools.

High initial costs: Implementing geospatial technology might necessitate a considerable upfront investment in hardware, software, and training.

Data Privacy Concerns: The collecting and use of geographic data raises privacy concerns, particularly the tracking of people's whereabouts.

1.5.4 The Future of Geospatial Technology in Industrial Growth

The future of geospatial technology in fostering industrial growth is bright. Geospatial data will become increasingly accurate, accessible, and incorporated into daily operations as satellite imagery, AI, machine learning, and IoT technologies progress.

As companies implement 5G networks, the volume of real-time geospatial data will expand, allowing for faster decision-making and more automation.

1.6 MACHINE LEARNING

ML, a subset of AI, has been a game changer in the geospatial industry, allowing for more efficient processing, analysis, and interpretation of massive volumes of geographical data. Geospatial technologies produce large datasets from satellites, drones, GPS, remote sensors, and GIS. Machine learning techniques assist to unleash the value of these datasets by recognizing patterns, generating predictions, and automating difficult geographical operations. Agriculture, urban planning, transportation, environmental monitoring, and disaster management industries all benefit from more accurate insights, faster decision-making, and increased efficiency as a result of combining machine learning and geographic data (Kharad & Thakur, 2023; Khardia et al., 2022; Khatri, 2023).

1.6.1 Key Applications of Machine Learning in the Geospatial Sector

Remote sensing and Image Classification: Machine learning is essential for evaluating remote sensing data, such as satellite and drone photography. ML algorithms can automatically categorize land cover (forests, urban areas, and bodies of water) by examining the spectral features of pixels in photographs. This is critical for environmental monitoring, city planning, and land use management. Deep learning approaches, such as convolutional neural networks (CNNs), have increased image classification capabilities, allowing for high accuracy in feature detection and mapping from aerial and satellite pictures.

Land Use and Change Detection: By analyzing satellite imagery and historical geospatial data, machine learning may help detect changes in land use over time. ML models can detect changes like deforestation, urbanization, desertification, and agricultural growth. By detecting trends in time-series data, machine learning assists businesses such as forestry, agriculture, and urban development in monitoring landscape changes, resulting in better planning and sustainable land management techniques.

Environmental monitoring and climate modeling use machine learning to observe and anticipate changes in ecosystems, weather patterns, and natural resources. ML models may use geographical and temporal data, such as temperature, humidity, and soil composition, to accurately anticipate climate change, water availability, and pollution levels. For example, ML models are used to forecast the effects of increasing sea levels, monitor city air quality, and track the spread of wildfires and other environmental risks.

Disaster Prediction and Management: Machine learning can evaluate geographical data to forecast the occurrence of natural catastrophes like earthquakes, floods, and hurricanes. ML algorithms use previous catastrophe data and current environmental factors to estimate risk regions and their consequences. This enables governments and companies to create early warning systems and take proactive steps to mitigate damage. ML also aids in the assessment of catastrophe damage by evaluating satellite photos and producing real-time damage reports.

Agriculture & Precision agricultural: Machine learning is used in agriculture to enhance crop management and agricultural methods by processing geospatial data from sensors, drones, and satellite imagery. ML models use characteristics such as soil health, weather, and crop development to offer planting, irrigation, and fertilizing recommendations. Machine learning enables precision farming by forecasting insect outbreaks, monitoring crop health, and optimizing resource usage, therefore boosting yields and lowering environmental impact.

Machine learning helps urban planners analyze geospatial data so they can make educated decisions regarding infrastructure development, transportation management, and land use. ML models can forecast population increase, determine the best places for new infrastructure, and simulate various urban development scenarios. Machine learning works in tandem with IoT devices in smart cities to evaluate real-time geographical data for traffic flow optimization, energy consumption control, and public safety. This results in more efficient, sustainable, and responsive urban ecosystems.

Natural resource exploration: Machine learning aids in the analysis of geographical data in industries like as mining, oil and gas, and renewable energy. ML models analyze geological, topographical, and seismic data to detect patterns that point to the presence of minerals, oil reserves, or geothermal energy sources. This allows for more efficient exploration and extraction, resulting in lower costs and environmental impact.

Autonomous Systems and Navigation: Machine learning is an essential component of self-driving automobiles and drones that use geographical data to navigate. ML models assist autonomous cars in interpreting their surroundings by using real-time GPS data, maps, and sensor inputs. This allows the cars to identify impediments, make route decisions, and travel safely. Machine learning is utilized in drone-based applications to improve efficiency and accuracy by performing activities such as aerial mapping, surveying, and monitoring.

Geographical Data Prediction and Forecasting: Machine learning algorithms, such as regression analysis and neural networks, use historical data to forecast future geographical trends. For example, machine learning algorithms can anticipate traffic congestion, population increase, urban sprawl, and disease transmission.

1.6.2 Advantages of Machine Learning in the Geospatial Sector

Automating Complex Tasks: Machine learning automates the study of huge geographic datasets, saving time and effort on tasks such as picture categorization, feature extraction, and pattern detection.

Improved Accuracy: ML models, particularly deep learning algorithms, have increased the accuracy of geospatial analysis by learning from large datasets and improving predictions over time.

Faster decision-making: Machine learning's real-time processing capabilities allow for speedier decision-making, particularly in time-sensitive circumstances such as disaster management, emergency response, and autonomous navigation.

Cost Efficiency: By automating data processing and increasing prediction accuracy, machine learning lowers operating costs and decreases the need for manual intervention in industries such as agriculture, infrastructure, and resource management.

Scalability: Machine learning is scalable, allowing industry to evaluate increasingly huge geographical information from numerous sources, such as satellites, drones, and IoT devices, while maintaining accuracy and performance.

1.6.3 The Challenges of Machine Learning in the Geospatial Sector

Data Quality and Availability: Machine learning models require a lot of high-quality data for training and validation. In some places or businesses, data availability or accuracy may be limited, affecting the effectiveness of machine learning models.

Model Interpretability: Complex machine learning models, particularly deep learning networks, can be difficult to understand and explain, thereby limiting their usage in applications that need openness and accountability.

Computer requirements: Machine learning models, particularly deep learning algorithms, necessitate extensive computer resources and data science skills, which may be prohibitive for certain enterprises.

Ethical and Privacy problems: The combination of machine learning with geographical data presents privacy problems, especially when evaluating data about people's movements or whereabouts. When making decisions based on such data, ethical aspects must be considered.

1.6.4 Future of Machine Learning in Geospatial Applications

As machine learning technologies progress, their integration with geographical data will become increasingly powerful. The convergence of ML, cloud computing, 5G, and edge computing will allow for real-time geographic data processing at new rates and sizes. Furthermore, the integration of AI-powered tools with IoT devices, such as smart sensors and drones, will improve companies' capacity to gather and analyze geographical data automatically.

As increasingly sophisticated information become accessible for study, machine learning will play a larger role in geospatial applications such as climate change modeling, urban resilience, and disaster recovery. Emerging explainable AI (XAI) solutions will solve model transparency and interpretability challenges, making machine learning more accessible to decision-makers.

1.7 LIDAR SCANS

Light Detection and Ranging (LiDAR) is a remote sensing technique that has revolutionized the geospatial industry by producing very precise and comprehensive 3D models of the Earth's surface. LiDAR works by generating laser pulses at a target and measuring how long it takes for the light to bounce back, allowing for exact distance measurements. These data are then used to generate 3D models, or point clouds, that depict landscape, vegetation, buildings, and other elements. LiDAR is being more widely used in geospatial workflows in a range of industries, including urban planning, forestry, agriculture, environmental monitoring, infrastructure construction, and disaster management. LiDAR is an indispensable tool for precise mapping, analysis, and decision-making due to its ability to penetrate dense vegetation and give high-resolution data even in difficult conditions (Kumawat et al., 2023; Nagaraj, 2023)

1.7.1 Key Applications of LiDAR in the Geospatial Sector

Topographic mapping and terrain modelling: LiDAR is commonly used to generate high-resolution Digital height Models (DEMs) and Digital Terrain Models (DTMs), which depict the height and contour of the Earth's surface. These models are critical for comprehending topographical characteristics including slopes, ridges, and valleys. LiDAR can capture minute features of both natural and man-made environments, making it an essential tool for topographic mapping in fields such as civil engineering, construction, and flood risk assessment.

Urban Planning and 3D City Modeling: LiDAR is increasingly being utilized in urban planning to generate detailed 3D city models. These models give useful information on building heights, infrastructural layouts, and urban density, which are critical for planning new buildings, transit networks, and utilities. LiDAR scans assist city planners in visualizing urban environments and simulating various growth scenarios, assuring effective land use while reducing environmental damage.

Forestry and Vegetation Monitoring: LiDAR is used to evaluate forest structure, canopy height, and biomass density. Unlike other remote sensing techniques, LiDAR can penetrate tree canopies to measure the ground underneath, making it valuable for studying both the forest floor and the vegetation above. This is crucial for sustainable forest management, carbon sequestration research, and biodiversity evaluations. LiDAR data are also used to track deforestation, forest degradation, and the effects of climate change on forest ecosystems.

Coastal and floodplain mapping: LiDAR is an effective technique for coastal management and flood risk assessment. LiDAR maps regions at risk of sea-level rise, storm surges, and floods by collecting accurate elevation data along coasts and floodplains. This information is used to estimate flood risks, build infrastructure resilience, and develop flood mitigation techniques. Bathymetric LiDAR, which measures underwater topography, maps shallow coastal regions, rivers, and lakes, giving essential data for environmental monitoring and catastrophe planning.

Infrastructure and Transportation Planning: LiDAR is used to survey transportation networks such as roads, bridges, and railroads with great precision. This information enables engineers to analyze the status of current infrastructure and build new transportation routes. LiDAR enables planners to discover the most efficient and cost-effective pathways for roads and services while avoiding environmental impact. LiDAR also aids building efforts by giving precise data for site preparation and grading.

Archaeology and Cultural Heritage: LiDAR has transformed archaeological study by allowing the finding of ancient buildings and landscapes that were previously difficult to identify using standard surveying methods. Dense vegetation and difficult terrain frequently conceal historical sites, but LiDAR's ability to penetrate foliage and collect high-resolution topography makes it an effective tool for locating buried ruins, temples, roadways, and other cultural heritage assets. LiDAR data enables archaeologists to map and preserve ancient places, especially in distant or inaccessible locations.

Disaster Management and Risk Assessment: LiDAR contributes significantly to disaster management by providing reliable data for hazard assessment, emergency response, and recovery planning. For example, LiDAR may be used to map earthquake fault lines, landslides, and flood zones. Following natural catastrophes such as hurricanes or wildfires, LiDAR scans may estimate damage by comparing pre- and post-event terrain, assisting authorities in planning recovery efforts. LiDAR data also aids in infrastructure maintenance and reconstruction by detecting structural damage and terrain changes.

Precision agriculture use LiDAR to measure soil quality, monitor crop health, and optimize irrigation systems. Farmers use LiDAR-generated elevation data to understand the terrain of their fields, allowing them to design planting patterns, drainage systems, and soil management measures. This technology also aids in the development of precision farming techniques by giving extensive information on crop and soil variability, hence increasing resource efficiency and crop yields.

Mining and Resource Exploration: LiDAR is used in the mining sector to map mining sites' topography, monitor landscape changes caused by excavation, and assess mine safety. LiDAR delivers precise terrain data for mining operations, maximizing resource extraction, and assuring environmental compliance. Furthermore, LiDAR aids in the identification of possible risks such as landslides or unstable slopes, hence improving safety at mining sites.

1.7.2 Types of LiDAR Systems (Tharayil et al., 2024)

Airborne LiDAR: Airborne LiDAR is usually installed to planes, helicopters, or drones to collect data over wide areas. Landscapes, woods, coasts, and urban areas are all commonly mapped using it. Airborne LiDAR has broad coverage, making it excellent for producing comprehensive topographic maps and 3D models of big areas.
Terrestrial LiDAR: LiDAR systems installed on tripods or vehicles for ground-based scanning. Surveying, construction, and infrastructure monitoring all use this sort of LiDAR. Terrestrial LiDAR allows for very detailed scans of specific regions, such as houses, roads, and archeological sites.
Bathymetric LiDAR measures underwater terrain. It maps the seafloor and other submerged structures using green laser light that can penetrate water. This is important for coastal and marine applications, including harbor development, underwater habitat mapping, and flood modeling.

1.7.3 Advantages of LiDAR in the Geospatial Sector

High Precision: LiDAR produces extremely accurate and comprehensive 3D data, enabling for precision mapping and study of topography, vegetation and infrastructure.
Rapid Data Collection: LiDAR systems can cover huge regions fast, making them perfect for time-sensitive applications like disaster relief and building planning.
LiDAR is versatile and can capture both ground and above-ground characteristics in a variety of situations, including thick woods and metropolitan areas.
Ability to pierce Vegetation: One of LiDAR's distinguishing features is its ability to pierce forest canopy and precisely map the ground underneath, which is particularly beneficial in densely vegetated areas.
3D Modeling: LiDAR produces comprehensive 3D models, allowing for improved display and analysis of complicated geographic data.

1.7.4 Challenges of Using LiDAR

Cost: LiDAR systems may be expensive to implement, particularly for large-scale projects requiring aerial platforms or high-resolution data collecting.
Data Processing: LiDAR generates large volumes of data that might be difficult to store, process, and evaluate. To efficiently manage LiDAR data, specialized software and skills are necessary.

Weather Dependency: Weather conditions can have an impact on LiDAR data collecting, particularly airborne LiDAR, which can be hindered by cloud cover, rain, or fog.

LiDAR technology continues to progress and integrate with other geospatial technologies, including machine learning, AI, and drones, indicating a potential future for geospatial applications. These developments are likely to increase the accuracy, efficiency, and cost of LiDAR systems, making them more useful for a wider range of applications. Autonomous cars, smart cities, and infrastructure development all stand to profit from LiDAR's capacity to deliver real-time 3D mapping and navigation. Furthermore, as LiDAR technology advances, its applications in environmental protection, disaster management, and resource exploration will become increasingly important, promoting long-term growth and resilience in companies throughout the world. LiDAR is a strong and adaptable instrument in the geospatial industry, providing precise and comprehensive 3D data that benefits a variety of businesses. From urban planning and infrastructure development to environmental monitoring and catastrophe management, LiDAR continues to be an important tool for geospatial analysis and decision-making.

1.8 GEOSPATIAL BUSINESS OVERVIEW AND OPPORTUNITIES

The geospatial business sector includes businesses and sectors that generate, process, analyze, and use geographic data to solve issues, make decisions, and drive innovation. This discipline uses GIS, remote sensing, GPS, and other spatial technologies to collect, organize, and evaluate location-based data. Agriculture, urban planning, transportation, telecommunications, real estate, natural resource management, disaster management, and environmental conservation all rely heavily on geospatial technology. As technology advances and demand for location-based services rises, the geospatial industry has grown dramatically, creating new opportunities for startups, existing businesses, and entrepreneurs.

1.8.1 Major Elements of the Geospatial Business

Geospatial Data and Analysis: Data—whether obtained by satellites, drones, IoT devices, or other remote sensing tools—is at the heart of the geospatial industry. Companies in this field specialize in acquiring and analyzing massive volumes of spatial data. This data is then examined to derive useful insights that may be used to make business choices. Examples include land-use analysis, traffic pattern prediction, and environmental monitoring.

With advances in big data analytics, AI, and machine learning, organizations can extract more important information from geographical data, enhancing accuracy and forecasting skills (Nagaraj, 2023; Nowak, 2021; "Organic farming: The way to sustainable agriculture and environmental protection," 2022; Osupile et al., 2022)

Geographic Information Systems are used to view, analyze, and interpret spatial data, and the software industry that surrounds GIS is an important part of the geospatial sector. Companies such as Esri (ArcGIS), Autodesk, and Hexagon offer strong GIS systems that assist businesses in managing geographical data and creating maps. Governments, enterprises, and non-profit organizations utilize GIS technology to make data-driven choices on land management, urban growth, and environmental conservation.

Remote sensing and imaging: Another important aspect of the geospatial industry is the collecting of data from satellites, drones, and airplanes. Companies in this field specialize in taking high-resolution photographs of the Earth's surface to track weather patterns, environmental changes, infrastructure, and land use. As the need for real-time, high-resolution photography develops, so do firms that specialize in satellite imaging, aerial surveying, and drone-based data collecting. Maxar, Planet Labs, and Airbus Defence and Space offer imaging services to a variety of businesses.

Location-Based Services (LBS): The rapid advancement of mobile technology has fueled the growth of location-based services, which provide information, entertainment, and commercial services based on users' physical locations. This includes navigation applications (Google Maps, Waze), ride-sharing platforms (Uber, Lyft), and retail apps that provide location-specific discounts or delivery alternatives. The integration of geographic data with mobile apps generates new income streams for LBS-focused enterprises, allowing them to improve user experiences and provide tailored offerings.

Geospatial Consulting and Services: Geospatial consulting organizations provide specialized services to businesses and governments seeking to use spatial data for specific purposes. These enterprises specialize in GIS installation, spatial analysis, data visualization, and project management, assisting organizations in areas such as urban planning, environmental protection, and disaster preparedness. Consultants assist firms in integrating geospatial technology into their operations, allowing them to maximize the benefits of location-based information.

Geospatial gear and Sensors: This section comprises manufacturers of data collecting gear such as GPS devices, drones, LiDAR scanners, and other geospatial sensors. As companies want more exact and comprehensive geospatial data, the demand for novel hardware solutions has grown. Businesses in this field frequently collaborate with enterprises in agriculture, mining, construction, and natural resource development (Tewary et al., 2023).

1.8.2 Major Sectors Leveraging Geospatial Business

Geospatial Technology have transformed agriculture by allowing for precise agricultural operations. Farmers utilize GIS, remote sensing, and drones to track crop health, soil conditions, and weather patterns. They can improve yields and reduce costs and environmental impact by evaluating geographical data. Companies that provide geospatial solutions to the agriculture industry have experienced substantial development as demand for sustainable farming techniques rises.

Urban planners employ geospatial technology to create efficient cities and manage resources. GIS is used to map infrastructure, design transportation networks, and optimize land use. Smart cities use geospatial data and IoT devices to control traffic, monitor air quality, and improve public safety. As cities become more linked, geospatial technology will play an increasingly important role in constructing sustainable and efficient urban settings.

The transportation and logistics industry extensively rely on geospatial data for route optimization, fleet management, and delivery tracking. Companies employ GPS, GIS, and real-time traffic data to improve the efficiency of transportation systems. Autonomous vehicle development also requires high-precision geospatial data for navigation. Businesses who offer solutions in this sector are at the forefront of technological advancements such as self-driving automobiles, drone delivery, and supply chain automation.

Governments and environmental groups employ geospatial techniques to monitor ecosystems, track deforestation, manage natural resources, and battle climate change. Remote sensing and GIS are used to evaluate land degradation, water quality, and biodiversity. As environmental concerns grow, so will the demand for precise geospatial data, spurring growth in firms that specialize in environmental monitoring solutions.

Telecommunications and infrastructure firms utilize geographic data to develop and operate networks, assuring optimal coverage and capacity. GIS is critical for installing fiber optic cables, erecting cellular towers, and managing infrastructure assets. As 5G technology becomes available, enterprises in this industry will rely more on geospatial technologies to fulfill the increased need for connection.

Real estate and location intelligence organizations use geographic data to assess property values, discover development prospects, and study market trends. Location intelligence technologies maximize site selection for new retail, offices, and industrial buildings by combining geospatial data with business intelligence capabilities. The capacity to examine spatial data allows real estate developers and investors to make better judgments regarding land use and investment.

1.8.3 Geospatial Business Trends and Opportunities (Sharma & Shivandu, 2024)

Integration with AI and machine learning. Geospatial data integration with AI and machine learning enables new predictive modeling and automation possibilities. AI-powered solutions can handle massive volumes of geographical data faster and more precisely than ever before, enabling organizations to spot patterns, forecast trends, and optimize decision-making. Machine learning algorithms are, for example, employed to evaluate satellite data for land use categorization, deforestation identification, and catastrophe risk assessment.

1.8.4 The Emergence of Cloud Computing and Big Data Analytics

The emergence of cloud computing and big data analytics is changing how geographical data is stored and analyzed. Businesses can now access and analyze enormous amounts of geographic data in real time, resulting in faster and more accurate insights. Cloud-based GIS solutions allow enterprises to work on geospatial projects remotely, avoiding the requirement for costly on-premise equipment.

Drones are becoming increasingly popular in geospatial data collection because to their capacity to acquire high-resolution pictures and LiDAR data swiftly and cost-effectively. The drone sector is quickly expanding, with applications in agriculture, construction, mining, and disaster management.

Expansion of Location-Based Services: The growing popularity of smartphones and mobile devices has spurred demand for location-based services (LBS). As 5G networks become more widely available, LBS capabilities will improve, allowing for additional real-time applications such as augmented reality, real-time navigation, and targeted marketing. Businesses that can use LBS to build unique consumer experiences will have substantial growth potential.

Sustainability and Climate Action: The geospatial business sector is well-positioned to contribute significantly to climate change mitigation and sustainability efforts. Geospatial enterprises assist governments and organizations in making educated environmental and resource management choices by providing data on land use, deforestation, water resources, and carbon emissions. As climate rules and regulations become increasingly strict, the demand for geospatial solutions to sustainability will increase.

1.8.5 Challenges of Geospatial Business (Parikh, 2024)

Data Privacy and Security: The collecting and use of geographic data raises privacy problems, particularly when tracking people's movements or sensitive information. Businesses must negotiate legal and ethical issues including data privacy and user permission.

High Technology Costs: Developing and deploying geospatial technology such as satellites, LiDAR, and drones may be costly. High initial expenditures can be a barrier to entry for small and medium-sized firms (SMEs).

Technical expertise: Managing and interpreting massive amounts of geographical data necessitates GIS, remote sensing, and data analysis expertise. Companies may struggle to locate competent individuals to manage and maintain geospatial systems.

The combination of the IoT with geospatial AI is a transformational force with the potential to change a wide range of enterprises and social functions. The combination of real-time data gathering from IoT devices with the analytical capacity of geospatial AI has the potential to transform how we manage cities, handle environmental concerns, and improve industrial processes. In this thorough conclusion, we will look at how IoT-driven geospatial AI is influencing the future of smart cities, agriculture, disaster management, and environmental monitoring, with an emphasis on the larger implications for efficiency, sustainability, and innovation across numerous industries ((Rana & Bhambri, 2024b). Smart cities are likely to benefit the most directly from the convergence of IoT and geospatial AI. Cities are relying more on location-based data to manage everything from traffic flow and public transit to energy use and garbage disposal. IoT devices such as sensors, cameras, and GPS-enabled devices are continually gathering massive volumes of geographical and temporal data, which AI systems may evaluate to enhance urban operations. For example, traffic sensors installed at major junctions collect data on vehicle movement, which geospatial AI then uses to minimize congestion, improve traffic signal timings, and even anticipate future traffic patterns. This can lead to less congestion, lower emissions, and better road safety.

IoT and geospatial AI's capacity to collaborate smoothly is critical to the development ofsmart cities. As urban populations rise, cities will confront increasing issues in transportation, housing, pollution, and resource management. IoT devices, along with geospatial AI's analytical skills, will be critical in discovering novel solutions to these problems, allowing for more livable, efficient, and sustainable urban settings. Agriculture is another area that is seeing a significant shift as a result of the confluence of IoT and geospatial AI. Precision agriculture, which uses data to enhance agricultural techniques, is gaining popularity as IoT devices such as soil sensors, weather stations, and GPS-enabled gear become more widely available. These sensors capture real-time data on soil conditions, crop health, weather patterns, and equipment usage, which geospatial AI systems may subsequently evaluate. Geospatial AI helps make sense of this data by recognizing trends and providing insights that farmers can utilize to make better decisions. For instance, AI can assess soil moisture data from IoT sensors and prescribe accurate irrigation schedules, ensuring crops receive the exact quantity of water they require.

Similarly, AI can analyze satellite photos and drone data to detect early indicators of insect infestations or agricultural illnesses, allowing farmers to take precautions before serious harm occurs. As the world's population grows, so does the need for food, putting pressure on agricultural systems to become more productive and efficient. IoT-driven geospatial AI provides a solution by allowing farmers to produce more food with less resources, hence lowering environmental impact and increasing profitability. In the future, we can expect increased integration of IoT devices and geospatial AI in agriculture to generate even greater advancements in precision farming, allowing us to feed a growing world more sustainably.

Disaster management and emergency response are crucial areas in which IoT and geospatial AI are making substantial progress. IoT sensors placed in sensitive places can offer real-time data on seismic activity, meteorological conditions, and other environmental elements, helping authorities to respond more rapidly and efficiently to natural catastrophes like earthquakes, floods, and hurricanes. Geospatial AI improves this process by evaluating the massive volumes of data created by these IoT devices in order to forecast catastrophe trends, assess risk levels, and design more effective response plans. Geospatial AI, for example, may evaluate data from flood sensors, weather predictions, and topographical maps to anticipate where floods will occur and which places are most vulnerable. This allows authorities to prioritize evacuation operations and use resources more efficiently, thereby saving lives and lessening the impact of disasters. After a disaster, geospatial AI can assist with damage assessment and recovery operations. AI algorithms can swiftly identify the most severely impacted places, estimate the level of damage, and direct rescue and recovery efforts by analyzing satellite photos and drone footage. The continuing integration of IoT and geospatial AI in disaster management will become more crucial in improving global resilience to natural catastrophes and lessening their impact on populations. IoT and geospatial AI are also playing important roles in environmental monitoring and sustainability activities. The capacity to collect real-time data on environmental conditions is critical for understanding and mitigating the effects of climate change, pollution, and other ecological issues. IoT devices, such as air quality sensors, water level monitors, and animal trackers, generate continuous data streams that geospatial AI systems can evaluate.

Geospatial AI, for example, may evaluate data from IoT-enabled air quality sensors to monitor pollution levels in cities and identify emission sources. Policymakers may then use this information to create targeted initiatives to reduce air pollution and improve public health. Similarly, IoT devices installed in rivers and seas can monitor the water quality and identify changes. In the future, integrating IoT and geospatial AI will be critical in solving global concerns such as climate change and environmental sustainability. By giving real-time insights on the health of the environment, these technologies will enable better informed decision-making and assist governments, companies, and communities in taking proactive efforts to safeguard the planet. The combination of IoT with geospatial AI has far-reaching consequences beyond the industries mentioned above. The capacity to gather and analyze real-time location-based data is driving new levels of efficiency and creativity in areas as diverse as logistics and transportation, energy, and infrastructure. For example, in logistics, IoT-enabled GPS tracking devices give real-time data on the location of items, which geospatial AI may use to improve delivery routes and save transportation costs. In the energy industry, IoT sensors can track the performance of renewable energy sources, while geospatial AI uses this information to improve energy production and distribution. The Table 1 is related to applications, challenges, and future scope of different technologies in Geospatial sector.

Table 1. Applications, Challenges, and Future Scope of Different Technologies in the Geospatial Sector

Techniques	Applications	Challenges	Future Scope
Artificial Intelligence	Disaster Management, Agriculture & Precision Farming, Environmental Monitoring, Autonomous Vehicles and Navigation, Smart Cities, and Urban Planning, Remote sensing, and Image Analysis	Enormous datasets High processing power Potential biases in models.	Advances in data processing, cloud computing, and the proliferation of linked devices will determine geospatial AI's future, transforming it into a vital driver of social advancement.
Internet of Things	Smart City and Infrastructure, Precision Agriculture, Disaster Management and Emergency Response, Transport and Logistics, Asset and Infrastructure Management, Environmental Monitoring	Data Privacy and Security Data Integration Interoperability	These developments will allow for even more precise, efficient, and real-time geospatial analysis, accelerating innovation in fields such as urban planning, environmental conservation, agriculture, transportation, and public safety.
Global Positioning System	Mapping and cartography, GPS is often utilized, Surveying and Land Management, Precision agriculture, Disaster Management, and Emergency Response, Environmental monitoring and conservation.	Signal Interference Accuracy Issues Power Dependency	new satellites are deployed and global positioning systems improve, GPS accuracy will increase, helping all businesses that rely on geospatial intelligence.

continued on following page

Table 1. Continued

Techniques	Applications	Challenges	Future Scope
Machine Learning	Remote sensing and Image Classification Land Use and Change Detection Environmental monitoring and climate modeling Disaster Prediction and Management Agriculture & Precision agricultural Machine learning helps urban planners analyze Natural resource exploration Autonomous Systems and Navigation Geographical Data Prediction and Forecasting	Data Quality and Availability Model Interpretability Computer requirements Ethical and Privacy problems	As machine learning technologies progress, their integration with geographical data will become increasingly powerful.
LiDAR Scans	Topographic mapping, and terrain modeling, Urban Planning, and 3D City Modeling, Forestry and Vegetation, Monitoring Coastal and floodplain mapping Infrastructure and Transportation Planning Archaeology and Cultural Heritage, Disaster Management, and Risk Assessment, Precision agriculture, Mining, and Resource Exploration	Cost Data Processing Weather dependency	ongoing research focused on improving its range, resolution, and real-time processing capabilities

1.9 DIGITAL TWINS AND REAL TIME ANALYTICS

Digital Twins are used with geospatial AI to stimulate and analyze the real time world conditions for planning of urban areas, management of infrastructure and environment. Virtual representations of physical things updated in real time analysis using data from IoT devices. Both are revolutionized the geospatial sector by enabling more flexible, information driven insights, and decision making across the multiple industries. A digital twin is a virtual modal which is continuously updated with real time information from IoT devices, drones, and other sources. The digital twins are used for analyzing, predicting, and monitoring the real time world. There are many applications where digital twins and real time analytics are used like Planning of Urban Areas and smart cities, managing of infrastructure, monitoring of environment, responding to the disaster. The main advantages of Digital Twins and Real Time Analytics in Geospatial Sectors are like digital twins provide 3D, real time view of environments; these can be improved by understanding and decision making. They enable simulation of future scenarios and allowing planners that they can make models of potential impacts and mitigate risks. They also maintain the schedules so that they could reduce the downtime and costs.

Real time analytics includes continuously analyzing the data which is collected in the geospatial sector and enabling the immediate insights and decisions. In real time analytics, harness information from IoT sensors, GPS devices, and satellites, to provide moveable intelligence. Traffic and Transportation management, agriculture and precision farming, monitoring environment, responding to disaster, tracking of supply chain and asset are applications in which Real Time Analytics are used. Real Time Analytics are beneficial in immediate decision making, enhancing the operational efficiency, improving the safety and risk management. Both are complemented each other and providing a high-fidelity, continuously updated model and enables organizations to simulate real world scenarios, optimizing the resource management, and enhancing the resilience and sustainability.

CONCLUSION

AI is quickly becoming a crucial tool in the geospatial sector, providing improved accuracy, efficiency, and automation in a variety of applications. As technology evolves, it will further change sectors that rely on spatial data, opening up new avenues for innovation and problem solutions. IoT is transforming the geospatial sector by delivering continuous streams of real-time data that can be examined for a variety of applications. IoT technology and geospatial systems will continue to play a vital role in determining the future of smart cities and the environment.

GPS is a critical component of the geospatial industry, delivering precise, real-time position data for a variety of applications. From mapping and surveying to navigation, agriculture, and disaster management, GPS continues to foster innovation and efficiency in businesses that rely on geographic data. The future of GPS promises significantly improved precision, integration, and usefulness, hence expanding its position in the geospatial industry. Geospatial technologies are an effective driver of industrial expansion. Geospatial solutions alter companies by optimizing resource management, boosting operational efficiency, facilitating market development, and promoting sustainability. As geospatial technologies advance, their effect on industrial growth will only rise, resulting in a more connected, efficient, and sustainable world. Machine learning is a game changer in the geospatial sector, revolutionizing how companies analyze, interpret, and exploit geographical data. As machine learning algorithms advance and geographical data becomes more plentiful, ML will play an increasingly important role in driving innovation, efficiency, and sustainability across numerous industries. The geospatial industry is active and expanding, with applications spanning a wide range of sectors. As technology advances and the need for location-based insights develops, geospatial firms will have tremendous potential for innovation and development. Geospatial solutions help enterprises solve difficult challenges, enhance operations, and improve decision-making in fields ranging from precision agriculture and smart cities to disaster management and environmental monitoring. With continuous investment in AI, big data, and cloud technologies, the geospatial industry is positioned for even greater growth in the years ahead.

References

Ahmad, L., & Nabi, F. (2021). IoT (Internet of things) based agricultural systems. *Agriculture 5.0: Artificial Intelligence, IoT, and Machine Learning*, 69-121. DOI: 10.1201/9781003125433-4

Anand, T., Sinha, S., Mandal, M., Chamola, V., & Yu, F. R. (2021). AgriSegNet: Deep aerial semantic segmentation framework for IoT-assisted precision agriculture. *IEEE Sensors Journal*, 21(16), 17581–17590. DOI: 10.1109/JSEN.2021.3071290

Bernini, G., Piscione, P., & Seder, E. (2023). AI-driven service and Slice orchestration. *Shaping the Future of IoT with Edge Intelligence*, 15-36. DOI: 10.1201/9781032632407-3

Bhambri, P., & Bajdor, P. (Eds.). (2024a). *Handbook of Technological Sustainability: Innovation and Environmental Awareness* (1st ed., p. 412). CRC Press., DOI: 10.1201/9781003475989

Bhambri, P., & Bajdor, P. (2024b). Technological Sustainability Unveiled: A Comprehensive Examination of Economic, Social, and Environmental Dimensions. In Bhambri, P., & Bajdor, P. (Eds.), *Handbook of Technological Sustainability: Innovation and Environmental Awareness* (pp. 80–98). CRC Press., DOI: 10.1201/9781003475989-8

Bisht, S., Bhardwaj, R., & Roy, D. (2024). Optimizing role assignment for scaling innovations through AI in agricultural frameworks: An effective approach. DOI: 10.1016/j.aac.2024.07.004

Chakravarthy, A. S., Sinha, S., Narang, P., Mandal, M., Chamola, V., & Yu, F. R. (2022). DroneSegNet: Robust aerial semantic segmentation for UAV-based IoT applications. *IEEE Transactions on Vehicular Technology*, 71(4), 4277–4286. DOI: 10.1109/TVT.2022.3144358

Chandrashekhar, B. N., Sanjay, H. A., & Geetha, V. (2024). Impact of hybrid [CPU+GPU] HPC infrastructure on AI/ML techniques in industry 4.0. *AI-Driven Digital Twin and Industry*, 4(0), 280–295. DOI: 10.1201/9781003395416-18

Chithra, N., & Bhambri, P. (2024). Ethics in Sustainable Technology. In Bhambri, P., & Bajdor, P. (Eds.), *Handbook of Technological Sustainability: Innovation and Environmental Awareness* (pp. 245–256). CRC Press., DOI: 10.1201/9781003475989-21

Dhibar, K., & Maji, P. (2023). Future outlier detection algorithm for smarter industry application using ML and AI. *Advances in Systems Analysis, Software Engineering, and High Performance Computing*, •••, 152–166. DOI: 10.4018/978-1-6684-8785-3.ch008

Ezhilarasan, K., & Jeevarekha, A. (2023). Powering the geothermal energy with AI, ML, and IoT. *Power Systems*, 271-286. DOI: 10.1007/978-3-031-15044-9_13

Garg, P., Chakravarthy, A. S., Mandal, M., Narang, P., Chamola, V., & Guizani, M. (2021). ISDNet: AI-enabled instance segmentation of aerial scenes for smart cities. *ACM Transactions on Internet Technology*, 21(3), 1–18. DOI: 10.1145/3418205

Geetha, K., Vigneshwari, J., Bhambri, P., & Thangam, A. (2024). Sustainable Solutions for Global Waste Challenges: Integrating Technology in Disposal and Treatment Methods. In Bhambri, P., & Bajdor, P. (Eds.), *Handbook of Technological Sustainability: Innovation and Environmental Awareness* (pp. 46–56). CRC Press., DOI: 10.1201/9781003475989-5

Gupta, S., & Vyas, S. (2023). Contemporary role of edge-AI in IoT and IoE in healthcare and digital marketing. *Edge-AI in Healthcare*, 75-84. DOI: 10.1201/9781003244592-6

Hassani, H., Amiri Andi, P., Ghodsi, A., Norouzi, K., Komendantova, N., & Unger, S. (2021). Shaping the future of smart dentistry: From artificial intelligence (AI) to intelligence augmentation (IA). *IoT*, 2(3), 510–523. DOI: 10.3390/iot2030026

Hassani, H., Amiri Andi, P., Ghodsi, A., Norouzi, K., Komendantova, N., & Unger, S. (2021). Shaping the future of smart dentistry: From artificial intelligence (AI) to intelligence augmentation (IA). *IoT*, 2(3), 510–523. DOI: 10.3390/iot2030026

Heras, J., Marani, R., & Milella, A. (2021). 39. semi-supervised semantic segmentation for grape bunch identification in natural images. *Precision Agriculture*, 21(11211), 331–337. DOI: 10.3920/978-90-8686-916-9_39

Hohmeier, K. C., Turner, K., Harland, M., Frederick, K., Rein, L., Atchley, D., Woodyard, A., Wasem, V., & Desselle, S. (2024). Scaling the optimizing care model in community pharmacy using implementation mapping and COM-B theoretical frameworks. *JAPhA Practice Innovations*, 1(1), 100002. DOI: 10.1016/j.japhpi.2023.100002

Kalpana, Y. B., Nirmaladevi, J., Sabitha, R., Ammal, S. G., Dhiyanesh, B., & Radha, R. (2024). Revolutionizing agriculture: Integrating IoT cloud, and machine learning for smart farm monitoring and precision agriculture. *Studies in Computational Intelligence*, 79-108. DOI: 10.1007/978-3-031-67450-1_4

Keskin, M., & Sekerli, Y. E. (2024). Mitigation of the effects of climate change on agriculture through the adoption of precision agriculture technologies. *Climate-Smart and Resilient Food Systems and Security*, 435-458. DOI: 10.1007/978-3-031-65968-3_20

Khankhoje, R. (2024). Future trends and ethical challenges in transforming gender healthcare using AI and ML. *Transforming Gender-Based Healthcare with AI and Machine Learning*, 239-259. DOI: 10.1201/9781003473435-14

Kharad, V., & Thakur, N. V. (2023). Analysis of decision support system for crop health management in smart and precision agriculture based on Internet of things (IoT) and artificial intelligence (AI). *2023 1st DMIHER International Conference on Artificial Intelligence in Education and Industry 4.0 (IDICAIEI)*, 25, 1-6. DOI: 10.1109/IDICAIEI58380.2023.10406812

Khardia, N., Meena, R. H., Jat, G., Sharma, S., Kumawat, H., Dhayal, S., Meena, A. K., & Sharma, K. (2022). Soil properties influenced by the foliar application of Nano fertilizers in maize (Zea mays L.) crop. *International Journal of Plant and Soil Science*, •••, 99–111. DOI: 10.9734/ijpss/2022/v34i1430996

Khatri, M. (2023). Transforming Indian business landscapes: The impact of AI, IoT, Metaverse, and emerging technologies. [IJSR]. *International Journal of Science and Research (Raipur, India)*, 12(10), 373–378. DOI: 10.21275/SR231003023311

Kumawat, H., Singh, D. P., Yadav, K. K., Khardia, N., Dhayal, S., Sharma, S., Sharma, K., & Kumawat, A. (2023). Response of fertility levels and liquid Biofertilizers on soil chemical properties and nutrient uptake under wheat (Triticum aestivum L.) crop. *Environment and Ecology*, 41(4), 2248–2256. DOI: 10.60151/envec/ZBLX2641

Mohana Sundari, V., Ganeshkumar, M., & Bhambri, P. (2024). Environmental Stewardship in the Digital Age: A Technological Blueprint. In Bhambri, P., & Bajdor, P. (Eds.), *Handbook of Technological Sustainability: Innovation and Environmental Awareness* (pp. 57–67). CRC Press., DOI: 10.1201/9781003475989-6

Nagaraj, A. (2023). Internet of things (IoT) with AI. *The Role of AI in Enhancing IoT-Cloud Applications*, 21-72. DOI: 10.2174/9789815165708123010006

Nagaraj, A. (2023). Integration of AI and IoT-cloud. *The Role of AI in Enhancing IoT-Cloud Applications*, 116-165. DOI: 10.2174/9789815165708123010008

Nowak, B. (2021). Precision agriculture: Where do we stand? A review of the adoption of precision agriculture technologies on Field crops farms in developed countries. *Agricultural Research*, 10(4), 515–522. DOI: 10.1007/s40003-021-00539-x

Organic farming: The way to sustainable agriculture and environmental protection. (2022). *International Journal of Biology, Pharmacy and Allied Sciences, 11*(1 (SPECIAL ISSUE)). DOI: 10.31032/IJBPAS/2022/11.1.1004

Osupile, K., Yahya, A., & Samikannu, R. (2022). A review on agriculture monitoring systems using Internet of things (IoT). *2022 International Conference on Applied Artificial Intelligence and Computing (ICAAIC), 7*, 1565-1572. DOI: 10.1109/ICAAIC53929.2022.9792979

Parikh, N. (2024). Unveiling the role of AI product managers: Shaping the future. DOI: 10.36227/techrxiv.172504030.01820212/v1

Rana, R., & Bhambri, P. (2024a). Environmental Challenges and Technological Solutions. In Bhambri, P., & Bajdor, P. (Eds.), *Handbook of Technological Sustainability: Innovation and Environmental Awareness* (pp. 187–200). CRC Press., DOI: 10.1201/9781003475989-17

Rana, R., & Bhambri, P. (2024b). Ethical Considerations in Artificial Intelligence for Environmental Solutions: Striking a Balance for Sustainable Innovation. In Bhambri, P., & Bajdor, P. (Eds.), *Handbook of Technological Sustainability: Innovation and Environmental Awareness* (pp. 389–396). CRC Press., DOI: 10.1201/9781003475989-31

Ruby, S., Biju, T., & Bhambri, P. (2024). Catalysing Sustainable Progress: Empowering MSMEs through Tech Innovation for a Bright Future. In Bhambri, P., & Bajdor, P. (Eds.), *Handbook of Technological Sustainability: Innovation and Environmental Awareness* (pp. 374–388). CRC Press., DOI: 10.1201/9781003475989-30

Sharma, K., & Shivandu, S. K. (2024). Integrating artificial intelligence and Internet of things (IoT) for enhanced crop monitoring and management in precision agriculture. *Sensors International, 5*, 100292. DOI: 10.1016/j.sintl.2024.100292

Sharma, R., & Bhambri, P. (2024). Digital Duplicity and the Disintegration of Trust: A Quantitative Inquiry into the Impact of Deep Fakes on Media Sustainability and Societal Equilibrium. In Bhambri, P., & Bajdor, P. (Eds.), *Handbook of Technological Sustainability: Innovation and Environmental Awareness* (pp. 273–291). CRC Press., DOI: 10.1201/9781003475989-24

Tewary, A., Upadhyay, C., & Singh, A. (2023). *Emerging role of AI, ML and IoT in modern sustainable energy management* (Vol. 2). IoT and Analytics in Renewable Energy Systems., DOI: 10.1201/9781003374121-23

Tharayil, S. M., Krishnapriya, M. A., & Alomari, N. K. (2024). How multimodal AI and IoT are shaping the future of intelligence. *Internet of Things and Big Data Analytics for a Green Environment*, 138-167. DOI: 10.1201/9781032656830-8

Thirumalaiyammal, B., Steffi, P. F., & Bhambri, P. (2024). Green Horizons: Navigating Environmental Challenges through Technological Innovation. In Bhambri, P., & Bajdor, P. (Eds.), *Handbook of Technological Sustainability: Innovation and Environmental Awareness* (pp. 292–304). CRC Press., DOI: 10.1201/9781003475989-25

Vigneshwari, J., Senthamizh Pavai, P., Maria Suganthi, L., & Bhambri, P. (2024). Eco-ethics in the Digital Age: Tackling Environmental Challenges through Technology. In Bhambri, P., & Bajdor, P. (Eds.), *Handbook of Technological Sustainability: Innovation and Environmental Awareness* (pp. 201–213). CRC Press., DOI: 10.1201/9781003475989-18

Key Terms

Artificial Intelligence: Artificial Intelligence also plays most important role in geospatial analysis with IOT.

Geo-Location: It is used to identify the device's physical location through GPS, Wi-Fi etc. It is used to real time mapping and tracking.

Geospatial Analysis: For the geospatial analysis, data includes geographical components like latitude and longitude. It is used in tracking, monitoring, and mapping the environment.

Geospatial Business: The geospatial business sector includes businesses and sectors that generate, process, analyze, and use geographic data to solve issues, make decisions, and drive innovation.

GIS (Geographic Information System): A system which is used to manage, analyze, and visualize spatial data i.e. GIS. It provides a platform where AI and ML models are used to enhance the data analysis.

Industrial Growth: The future of geospatial technology in fostering industrial growth is bright. Geospatial data will become increasingly accurate, accessible, and incorporated into daily operations as satellite imagery, AI, machine learning, and IoT technologies progress.

IOT Sensors: These are those devices which are used to collect real time information on environmental conditions, movement, location, and other factors. It is used to gather spatial data in geospatial AI.

LIDAR(Light Detection and Ranging):ML is used to analyze LIDAR data for terrain mapping: infrastructure assessment, and forestry etc.

Machine Learning: Machine learning (ML) is also being used to process and analyze massive volumes of geospatial data, such as satellite imaging, aerial photography, GPS data, and geographic information system (GIS).

Spatial Data Analysis: It is used to analyze the distribution, location, and patterns with the help of AI and ML.

Chapter 8
Traffic Flow Optimization Using AI

Rajesh Kanna Rajendran
 https://orcid.org/0000-0001-7228-5031
Christ University, India

N. R. Wilfred Blessing
University of Technology and Applied Sciences, Ibri, Oman

T. Mohana Priya
Christ University, India

ABSTRACT

Traffic flow optimization is a critical challenge in urban planning and transportation management, aimed at reducing congestion, improving travel times, and enhancing overall roadway efficiency. This paper explores the application of artificial intelligence (AI) techniques to address these challenges. Leveraging machine learning algorithms, neural networks, and advanced data analytics, AI-driven systems can dynamically adjust traffic signals, predict traffic patterns, and optimize routing in real-time. This approach utilizes historical traffic data, real-time sensors, and predictive modeling to make data-driven decisions that enhance traffic flow and reduce delays. Integrating AI with existing traffic management infrastructure, cities can achieve more responsive and adaptive traffic control and improved quality of life for commuters. This Chapter presents a review of current AI applications in traffic optimization, evaluates their effectiveness through case studies, and discusses potential future developments in this evolving field.

DOI: 10.4018/979-8-3693-8054-3.ch008

INTRODUCTION

Traffic congestion is a pressing issue in many cities around the globe, significantly affecting urban life and economic productivity. With over 55% of the world's population residing in urban areas—a figure projected to rise to 68% by 2050—traffic management has become a critical challenge. The increase in the number of vehicles, coupled with inadequate infrastructure, has led to severe congestion in major cities like Los Angeles, London, and Tokyo. In 2023, for instance, drivers in Los Angeles spent an average of 119 hours stuck in traffic, costing the city approximately $10 billion in lost productivity and fuel expenses. These issues are not confined to developed nations; developing cities are also struggling to cope with the rapid urbanization and its accompanying traffic woes.

Bangalore, often referred to as India's Silicon Valley, epitomizes the challenges faced by rapidly growing urban centers in the developing world. The city's population has ballooned from 5.1 million in 2001 to over 12 million in 2024, leading to a corresponding increase in vehicle ownership. In 2023, Bangalore was ranked as the second most congested city in the world, with commuters spending an average of 243 hours annually in traffic. The city's road infrastructure, originally designed for a much smaller population, has not kept pace with this explosive growth. Factors such as narrow roads, frequent bottlenecks, and a lack of efficient public transportation have exacerbated the problem, making traffic congestion one of the most significant challenges for Bangalore's civic administration.

The concept of smart cities offers a potential solution to these traffic woes. Smart cities utilize advanced technologies like the Internet of Things (IoT), Artificial Intelligence (AI), and big data analytics to improve urban infrastructure and services. In the context of traffic management, AI can be employed to analyze real-time traffic data from various sources such as cameras, sensors, and GPS devices. This data can then be used to optimize traffic light timings, manage traffic flow dynamically, and even predict traffic congestion before it occurs. For instance, Singapore's AI-based traffic management system has reduced traffic jams by 8% and cut down travel times by 20%. Similarly, Barcelona's smart city initiatives have led to a 21% decrease in traffic-related pollution, showcasing the potential of AI in transforming urban mobility.

Bangalore has also embarked on its journey towards becoming a smart city, with traffic management being one of the key focus areas. The city's Integrated Traffic Management System (ITMS) aims to leverage AI and IoT technologies to monitor and manage traffic in real time. The system is expected to include features like adaptive traffic signal control, which adjusts signal timings based on actual traffic conditions, and an integrated command and control center for better coordination between various civic agencies. These initiatives are part of a broader effort under the Smart Cities Mission, a government program launched in 2015 to promote sustainable and inclusive urban development. However, despite these efforts, Bangalore still faces significant challenges in implementing these technologies on a large scale, highlighting the need for continued investment and innovation in smart city solutions.

LITERATURE REVIEW

Traffic congestion has been a subject of extensive research over the years, particularly as urbanization intensifies globally. Scholars have explored various strategies to manage and alleviate traffic congestion, ranging from traditional traffic engineering methods to advanced technological interventions. The evolution of traffic management practices reflects a growing understanding of the complex interplay between urban infrastructure, human behavior, and emerging technologies.

Early approaches to traffic management focused on expanding road infrastructure and improving traffic signal systems. Mousavi et al. (2020) proposed a reinforcement learning-based approach for adaptive traffic signal control, which adjusts signal timings based on current traffic conditions. The study showed that this approach could reduce waiting times at intersections by up to 25%, making it a viable solution for cities struggling with traffic congestion.

With the advent of Artificial Intelligence (AI), traffic management has entered a new era of optimization and efficiency. AI-based systems are increasingly being deployed to analyze vast amounts of traffic data in real time, enabling dynamic adjustments to traffic signal timings and routing decisions. In a seminal study, Vlahogianni et al. (2015) reviewed AI applications in traffic flow prediction and management, highlighting the potential of machine learning algorithms to improve traffic efficiency by up to 30%. Their findings suggest that AI can not only optimize traffic flow but also reduce the environmental impact of congestion through more efficient fuel use and reduced idling times.

The integration of AI in traffic management is particularly relevant in the context of smart cities. Song et al. (2017) demonstrated the effectiveness of AI-driven adaptive traffic signal systems in reducing congestion in urban environments. Their study focused on a pilot implementation in Beijing, where an AI-based system reduced travel time by 12% and decreased the number of stops by 15%. These results underscore the potential of AI to transform urban mobility, particularly in densely populated areas.

Several case studies highlight the successful implementation of smart city traffic management systems that incorporate AI. For example, in Singapore, the Land Transport Authority (LTA) has implemented an AI-based traffic management system that uses data from GPS-equipped taxis and traffic cameras to monitor and manage traffic in real time. According to a report by Yap et al. (2020), this system has reduced traffic jams by 8% and decreased average travel times by 20%. Similarly, Barcelona's smart city initiatives, which include the deployment of AI for traffic and environmental monitoring, have resulted in a 21% reduction in traffic-related pollution (Díaz-Díaz et al., 2017).

These case studies provide valuable insights into the potential benefits of AI in traffic management. However, they also highlight the challenges associated with implementing such systems in different urban contexts. For instance, while Singapore's success is often cited as a model for other cities, the city's relatively small size and centralized administration are factors that may not be easily replicated in larger, more complex urban environments like Bangalore.

Mahadevan et al. (2023) explored the potential of Bangalore's Integrated Traffic Management System (ITMS) as part of the city's smart city initiatives. The ITMS aims to integrate AI and IoT technologies to monitor traffic in real time and adjust traffic signals dynamically. The study highlighted the system's early successes in reducing congestion at major intersections and its potential to be scaled up across the city. However, the authors also noted challenges, such as the need for substantial investment and coordination between various stakeholders, to fully realize the benefits of AI-driven traffic management in Bangalore.

Bangalore presents a unique case study in the application of AI for traffic management within the broader framework of smart city initiatives. The city's traffic congestion issues have been the subject of numerous studies, many of which emphasize the need for innovative solutions that go beyond traditional traffic engineering methods. Nagendra et al. (2018) explored the impact of Bangalore's rapid urbanization on traffic patterns and highlighted the potential role of AI in addressing these challenges. Their study called for a comprehensive approach that combines AI with improvements in public transportation and urban planning.

Despite the promise of AI, Bangalore faces significant challenges in implementing these technologies on a large scale. A study by Ranganathan et al. (2021) pointed out that while the city's Integrated Traffic Management System (ITMS) represents a step forward, issues such as data integration, funding, and coordination between various civic agencies continue to hinder progress. The study also emphasized the importance of community engagement and public awareness in ensuring the success of such initiatives.

Zhang et al. (2021) conducted a comprehensive study on predictive traffic management systems using AI and machine learning algorithms. Their research focused on large metropolitan areas in China, where rapid urbanization has led to severe traffic congestion. The study found that AI-driven predictive models, which analyze historical and real-time traffic data, can forecast traffic conditions with over 90% accuracy. These predictions allow for preemptive traffic management measures, such as dynamic route adjustments and real-time traffic signal optimization, leading to a reduction in congestion by up to 25%.

The advent of autonomous vehicles (AVs) presents a significant shift in traffic management strategies. Goodall et al. (2020) explored the impact of AVs on traffic flow in urban settings, using simulation models to assess their potential benefits. The study, conducted in New York City, demonstrated that widespread adoption of AVs could reduce traffic delays by 30% due to their ability to communicate with each other and with traffic management systems in real time. The researchers emphasized that the integration of AVs into existing traffic systems requires a robust AI infrastructure to manage the complex interactions between human-driven and autonomous vehicles.

Public transportation plays a crucial role in alleviating traffic congestion, and AI has been increasingly applied to enhance its efficiency. An influential study by Bianchini et al. (2019) examined the application of AI in optimizing bus schedules and routes in Rome. The research highlighted that AI-driven systems could adjust bus frequencies in real time based on passenger demand and traffic conditions. This approach not only improved the punctuality of buses by 15% but also increased overall public transportation usage by 10%, thus reducing the number of private vehicles on the road.

Recent advancements in smart traffic light systems have been instrumental in managing traffic in real-time. A study by Papageorgiou et al. (2022) in Athens explored the deployment of AI-based adaptive traffic light systems. These systems use real-time traffic data to adjust signal timings dynamically, optimizing traffic flow at intersections. The results showed a 12% reduction in average waiting times and a 20% decrease in fuel consumption across the city. The study underscores the potential of AI to significantly enhance urban mobility through smarter traffic signal management.

A review by Thomas et al. (2023) analyzed the deployment of AI-based traffic systems in various cities worldwide, including those in developing regions. The study identified key barriers such as data privacy concerns, the high cost of implementation, and the lack of standardized infrastructure. In particular, the research highlighted that in cities like Lagos and Jakarta, the effectiveness of AI systems is often limited by poor data quality and inadequate technological infrastructure. The authors called for greater international collaboration to address these challenges and to develop scalable solutions that can be adapted to different urban contexts.

Kramers et al. (2018) examined the implementation of smart city technologies in Stockholm, focusing on AI's role in traffic management. The study found that AI applications, such as predictive analytics for traffic forecasting and adaptive traffic control, reduced congestion by approximately 20% and enhanced the overall efficiency of the city's transportation network . These findings are consistent with similar initiatives in cities like Barcelona and Amsterdam, where smart city technologies have led to measurable improvements in traffic conditions.

As Hossain et al. (2021) point out, AI-driven traffic management systems are a cornerstone of smart city projects, enabling real-time monitoring and management of traffic flows. Their study on Singapore's smart city initiatives illustrates how AI-based systems can significantly reduce traffic congestion and improve overall urban mobility .

Chen et al. (2020), issues such as data privacy, the high cost of implementation, and the need for robust infrastructure pose significant barriers to the widespread adoption of AI-based traffic management systems . Moreover, the success of these systems depends on the availability of high-quality, real-time data, which can be difficult to obtain in cities with limited technological infrastructure.

TRAFFIC FLOW MODELS AND REINFORCEMENT LEARNING

Traffic flow models have long been a critical component of urban traffic management, providing a framework to understand and predict the movement of vehicles through road networks. These models are essential for planning infrastructure, optimizing traffic signal timings, and mitigating congestion. Traditionally, traffic flow models were based on mathematical and statistical techniques that attempted to capture the relationship between traffic density, speed, and flow. However, these models often struggled to account for the complex and dynamic nature of real-world traffic, particularly in rapidly urbanizing areas like Bangalore, where unpredictable patterns and diverse driving behaviors add layers of complexity.

With the advent of Artificial Intelligence (AI) and, more specifically, Reinforcement Learning (RL), traffic flow modeling has entered a new era of precision and adaptability. Reinforcement Learning, a subset of machine learning, is particularly well-suited to traffic management because it mimics the trial-and-error learning process of human decision-making. In the context of traffic management, RL algorithms can learn from the environment—such as traffic patterns, signal timings, and vehicle movements—and continuously adapt their strategies to optimize flow and reduce congestion.

One of the most promising applications of RL in traffic flow models is in adaptive traffic signal control. Traditional traffic lights operate on pre-set timing plans, which are often inefficient during peak hours or in the face of unexpected traffic conditions. RL-based systems, on the other hand, can dynamically adjust signal timings in real-time based on the current state of traffic at an intersection. For example, if the algorithm detects an unusual build-up of vehicles on one approach, it can increase the green light duration for that direction to alleviate congestion. Studies, such as those by Mousavi et al. (2020), have shown that RL-based traffic signal control systems can significantly reduce waiting times at intersections, sometimes by as much as 25%, compared to traditional methods.

Beyond traffic signal control, RL is also being explored for broader traffic flow management tasks, such as route optimization and congestion pricing. In these applications, RL algorithms can learn to predict traffic bottlenecks and suggest alternative routes or pricing strategies to distribute traffic more evenly across the network. This is particularly relevant in smart cities, where the integration of RL with Internet of Things (IoT) devices—such as connected vehicles and smart sensors—can provide a holistic and real-time view of traffic conditions. This interconnected ecosystem allows RL models to not only react to traffic conditions but also anticipate them, leading to more proactive and efficient traffic management.

The integration of RL into traffic flow models is especially pertinent in cities like Bangalore, where traffic congestion is a daily challenge. The city's ongoing efforts to implement smart city technologies, including the Integrated Traffic Management System (ITMS), provide a fertile ground for RL applications. By leveraging RL, Bangalore could potentially transform its traffic management, moving from reactive measures to a more anticipatory and adaptive system that can cope with the city's rapid urbanization and diverse traffic patterns.

Reinforcement Learning offers a transformative approach to traffic flow modeling, providing the ability to adapt and optimize traffic management in real-time. As urban areas continue to grow and traffic becomes more complex, the role of RL in traffic management is likely to expand, offering new solutions to one of the most persistent challenges of modern urban life.

ROLE AND CHALLENGES OF AI IN DEVELOPING TRAFFIC EFFICIENCY

Role of AI in Traffic Efficiency

1. **Real-Time Traffic Monitoring and Management**

AI systems leverage vast amounts of data from traffic cameras, sensors, and GPS devices to monitor traffic conditions in real-time. This enables dynamic traffic management by adjusting traffic signal timings, managing traffic flows, and detecting congestion before it becomes a major issue. For example, AI algorithms can analyze live traffic data to optimize traffic signals, reducing waiting times and improving traffic flow.

2. **Predictive Traffic Management**

AI models, particularly those using machine learning and deep learning techniques, can predict traffic patterns and potential congestion based on historical data and real-time inputs. This predictive capability allows for proactive measures, such as adjusting signal timings or rerouting traffic before congestion peaks. Studies have shown that predictive models can enhance traffic efficiency by up to 30% (Vlahogianni et al., 2015).

3. **Adaptive Traffic Signal Control**

Traditional traffic signal systems often operate on fixed schedules, which may not reflect current traffic conditions. AI-powered adaptive traffic signal control systems can adjust the timing of lights in response to real-time traffic data, optimizing traffic flow and reducing delays. For instance, reinforcement learning-based systems can learn from traffic patterns and adjust signals dynamically, improving intersection efficiency (Mousavi et al., 2020).

4. **Integration with Autonomous Vehicles (AVs)**

AI plays a crucial role in the development of autonomous vehicles, which can communicate with traffic management systems to optimize their routes and reduce congestion. By integrating AVs into traffic management systems, cities can achieve smoother traffic flow and reduce delays. Simulation models have demonstrated that AVs could reduce traffic delays by up to 30% (Goodall et al., 2020).

5. **Enhanced Public Transportation Efficiency**

AI can optimize public transportation systems by adjusting schedules and routes based on real-time passenger demand and traffic conditions. AI-driven systems can improve bus punctuality and increase overall public transport usage, thus reducing the number of private vehicles on the road (Bianchini et al., 2019).

Challenges of AI in Traffic Efficiency

1. **Data Privacy and Security**

The collection and analysis of large amounts of traffic data raise significant privacy and security concerns. Ensuring that personal data is anonymized and protecting it from cyber threats are critical challenges. Regulatory frameworks and robust security measures are needed to address these concerns.

2. **High Implementation Costs**

Developing and deploying AI-based traffic management systems requires significant investment in technology and infrastructure. For many cities, especially in developing regions, the cost of implementation can be a major barrier. Funding and resource allocation are crucial to overcoming this challenge.

3. **Integration with Existing Infrastructure**

Integrating AI systems with existing traffic infrastructure can be complex and requires careful planning. Compatibility issues and the need for system upgrades can hinder the smooth implementation of AI solutions. Ensuring that new technologies work seamlessly with legacy systems is a key challenge.

4. **Data Quality and Availability**

AI models rely on high-quality, real-time data to function effectively. In many cities, data quality and availability can be inconsistent, affecting the accuracy and reliability of AI-driven traffic management systems. Ensuring the availability of accurate and comprehensive data is essential for effective AI applications.

5. **Public Acceptance and Trust**:

The success of AI in traffic management depends on public acceptance and trust in these systems. Educating the public about the benefits of AI and addressing any concerns about its impact on daily life are important for gaining support and ensuring the successful adoption of AI technologies.

6. **Ethical and Bias Issues**

AI systems can inadvertently perpetuate biases present in the data they are trained on. Ensuring that AI models are fair and do not discriminate against certain groups is crucial. Addressing ethical issues and ensuring transparency in AI decision-making processes are essential for the responsible deployment of AI in traffic management.

Addressing these challenges requires a collaborative effort from technology developers, policymakers, and urban planners to ensure that AI solutions are effective, equitable, and sustainable.

TRAFFIC FLOW PREDICTION FOR SMART TRAFFIC LIGHTS USING MACHINE LEARNING ALGORITHMS

As urban areas grow increasingly congested, the need for intelligent traffic management solutions becomes more pressing. Traditional traffic light systems, which operate on fixed schedules, often fail to adapt to real-time traffic conditions, leading to inefficiencies and increased congestion. Machine learning (ML) algorithms, however, offer a transformative approach to traffic flow prediction, enabling smart traffic lights to dynamically adjust signal timings based on real-time data. This capability not only optimizes traffic flow but also enhances overall urban mobility and reduces congestion.

1. Machine Learning Algorithms for Traffic Flow Prediction

Machine learning algorithms, particularly supervised learning techniques, are increasingly being used to predict traffic flow and optimize traffic light management. These algorithms can analyze historical traffic data, such as vehicle counts, speeds, and travel times, to forecast future traffic conditions.

Regression Models: Linear and polynomial regression models can predict traffic flow based on historical data. These models establish relationships between traffic variables and predict future conditions, helping in setting appropriate traffic light timings.

Time Series Analysis: Techniques such as Autoregressive Integrated Moving Average (ARIMA) and Long Short-Term Memory (LSTM) networks are used to analyze temporal patterns in traffic data. Time series models can forecast traffic volumes and identify trends, which are crucial for adjusting traffic light schedules.

Classification Models: Algorithms like Decision Trees, Random Forests, and Support Vector Machines (SVMs) classify traffic conditions into categories such as high, moderate, or low congestion. These classifications help in making real-time decisions about traffic light adjustments.

2. Adaptive Traffic Light Control Systems

Adaptive traffic light systems utilize ML algorithms to continuously adjust signal timings based on real-time traffic data. These systems integrate with traffic sensors, cameras, and GPS data to monitor traffic conditions at intersections.

Reinforcement Learning (RL): RL algorithms learn from the environment through trial and error. In traffic management, RL can dynamically adjust traffic light timings to optimize flow and reduce delays. By receiving feedback on traffic conditions, RL models refine their strategies over time. Studies have shown that RL-based traffic signal control systems can significantly improve intersection efficiency.

Deep Learning: Deep learning models, particularly Convolutional Neural Networks (CNNs) and Recurrent Neural Networks (RNNs), can process large volumes of traffic data from cameras and sensors. These models extract features and patterns from traffic images and videos, enabling more accurate predictions and adjustments to traffic light operations.

Multi-Agent Systems: These systems use ML algorithms to manage traffic at multiple intersections simultaneously. Each agent represents a traffic light, and they work together to coordinate traffic flow across the network. This approach enhances the overall efficiency of traffic management by ensuring that changes at one intersection do not adversely affect others.

3. Integration with Smart City Technologies

The integration of ML-based traffic flow prediction with smart city technologies is transforming urban traffic management. Smart traffic lights, powered by AI and IoT devices, can communicate with other elements of the smart city infrastructure, such as connected vehicles and smart road sensors.

Real-Time Data Analysis: Smart traffic lights can process data from various sources in real-time, including vehicle-to-infrastructure (V2I) communications. This enables immediate adjustments to traffic light timings based on current traffic conditions and congestion levels.

Predictive Adjustments: By forecasting traffic patterns using ML algorithms, smart traffic lights can proactively adjust their schedules to prevent congestion before it occurs. This predictive capability is particularly useful in managing traffic during peak hours and special events.

Enhanced User Experience: Integrating smart traffic lights with navigation apps and real-time traffic information services can provide drivers with optimal route suggestions, reducing travel time and improving overall traffic flow.

Machine learning algorithms offer significant potential for enhancing traffic flow prediction and optimizing smart traffic lights. By leveraging real-time data and advanced analytics, these systems can improve traffic management, reduce congestion, and contribute to more efficient urban mobility.

IMPACT OF WEATHER CONDITIONS ON TRAFFIC FLOW PREDICTIONS USING MACHINE LEARNING

Weather conditions play a significant role in influencing traffic flow, and incorporating these factors into machine learning models is crucial for accurate traffic predictions. Adverse weather such as rain, snow, or fog can significantly impact road conditions, reduce visibility, and alter driving behavior, leading to increased congestion and accidents. Machine learning models that integrate weather data with traffic analytics can enhance their predictive accuracy by accounting for these variables. For example, algorithms can use historical weather data and real-time weather reports to adjust traffic flow forecasts, predicting how weather-related factors will influence traffic patterns and adjusting recommendations for route planning accordingly.

Machine learning techniques, such as regression analysis and ensemble methods, can analyze complex interactions between weather variables and traffic data to improve prediction accuracy. This integration enables traffic management systems to issue timely warnings and advisories to drivers, adjust traffic signal timings, and optimize public transportation schedules in response to changing weather conditions. As climate patterns become more unpredictable and extreme weather events more frequent, the ability to accurately incorporate weather effects into traffic predictions will be increasingly essential for maintaining safe and efficient transportation networks.

A TRAFFIC-WEATHER GENERATIVE ADVERSARIAL NETWORK FOR TRAFFIC FLOW PREDICTION

Generative Adversarial Networks (GANs) have revolutionized data prediction by generating realistic synthetic data through adversarial training of two neural networks—the generator and the discriminator. In the realm of traffic flow prediction, integrating weather data with traffic information through a Traffic-Weather GAN enhances prediction accuracy. This model combines historical traffic patterns with real-time weather conditions, such as precipitation and temperature, which can significantly impact traffic flow. By incorporating these factors, the GAN provides more accurate and reliable traffic forecasts.

The Traffic-Weather GAN consists of a traffic data generator that simulates traffic conditions and a weather data generator that models various weather scenarios. The discriminator evaluates the synthetic data, ensuring it closely matches real-world conditions. This setup allows for dynamic and real-time adjustments in traffic management strategies, such as optimizing traffic signal timings and adjusting route recommendations based on current weather and traffic data. As a result, the GAN improves urban mobility and helps in effective congestion management.

The implementation of a Traffic-Weather GAN faces challenges, including the need for high-quality data, significant computational resources, and integration with existing traffic management systems. Ensuring the accuracy of the traffic and weather data, addressing computational demands, and achieving seamless integration are crucial for the successful deployment of this technology. Future developments in GANs and their integration with smart city infrastructure hold the promise of more responsive and efficient traffic management systems.

RECENT DEVELOPMENTS IN URBAN TRAFFIC MANAGEMENT SYSTEMS

Recent advancements in urban traffic management systems have significantly transformed how cities address traffic congestion and optimize mobility. One of the most notable developments is the integration of artificial intelligence (AI) and machine learning (ML) into traffic management strategies. These technologies enable real-time analysis of traffic data, leading to more adaptive and efficient traffic control. For instance, AI algorithms can predict traffic patterns and adjust traffic signal timings dynamically, reducing delays and improving overall traffic flow. This approach allows cities to respond swiftly to changing traffic conditions and mitigate congestion more effectively.

Another significant advancement is the deployment of smart traffic signals and adaptive signal control systems. These systems use data from traffic sensors and cameras to adjust signal timings based on current traffic volumes and patterns. Recent implementations in cities like Los Angeles and Singapore have demonstrated substantial improvements in traffic efficiency, with reductions in average travel times and vehicle emissions. The ability to adapt signal timings in real time helps minimize stop-and-go traffic, leading to smoother and more predictable travel experiences for commuters.

The rise of connected and autonomous vehicle technologies is also reshaping urban traffic management. These vehicles communicate with each other and with traffic management systems, providing valuable data that enhances traffic flow and safety. For example, autonomous vehicles can coordinate with traffic signals to optimize their travel routes, reducing the likelihood of congestion. Additionally, the integration of connected vehicle data into traffic management systems allows for more accurate and timely traffic predictions, further enhancing the effectiveness of traffic control measures. As these technologies continue to evolve, they hold the potential to significantly improve urban mobility and reduce traffic-related challenges.

AI-BASED SOLUTIONS FOR PEAK HOUR TRAFFIC PREDICTION AND NOTIFICATION

Artificial Intelligence (AI) has become a game-changer in urban traffic management, especially for predicting and managing peak hour traffic. Traditional methods of traffic forecasting often fall short in addressing the complexities of real-time traffic dynamics, particularly during peak hours. AI-based solutions offer a more sophisticated approach by leveraging large datasets and advanced algorithms to predict traffic conditions with greater accuracy. These solutions not only enhance traffic flow during congested periods but also provide timely notifications to commuters, helping them make informed travel decisions.

AI-based traffic prediction models utilize machine learning algorithms to analyze historical and real-time traffic data. These models can forecast traffic conditions by recognizing patterns and trends from various data sources, including traffic sensors, GPS devices, and social media reports. For example, deep learning techniques such as Long Short-Term Memory (LSTM) networks and convolutional neural networks (CNNs) are employed to capture temporal and spatial dependencies in traffic data. Studies have shown that these models can predict traffic conditions with high accuracy, even during peak hours, by considering factors such as traffic volume, speed, weather conditions, and special events.

AI-based systems can deliver real-time notifications to commuters about peak hour traffic conditions. These systems leverage data from traffic prediction models and integrate with mobile apps and navigation platforms to provide timely alerts and alternative route suggestions. For instance, AI-powered apps can notify users about upcoming traffic jams, accidents, or road closures, allowing them to adjust their routes or departure times accordingly. By incorporating real-time data and user preferences, these systems enhance the overall travel experience and reduce the impact of peak hour congestion.

While AI-based solutions offer significant benefits, there are challenges to address. Ensuring data accuracy and handling the vast amounts of data required for accurate predictions can be demanding. Additionally, integrating these systems with existing traffic management infrastructure and addressing privacy concerns related to data collection are crucial for successful implementation. Future developments in AI, such as improvements in predictive algorithms and increased integration with smart city infrastructure, hold the promise of further enhancing peak hour traffic management. By continually refining these technologies, cities can better manage congestion and improve urban mobility.

ASSESSING THE ACCURACY OF AI MODELS FOR PREDICTING TRAFFIC CONGESTION IN MIXED TRAFFIC ENVIRONMENTS

The rapid advancement of autonomous and connected vehicle technologies has introduced a new dimension to urban traffic management. In mixed traffic environments, where autonomous vehicles (AVs) and human-driven vehicles coexist, predicting traffic congestion presents unique challenges. Assessing the accuracy of AI models designed to forecast traffic conditions in such environments is crucial for developing effective traffic management solutions. This assessment involves evaluating how well these models handle the complexities introduced by the interaction between different types of vehicles and varying driving behaviors.

Mixed traffic environments pose several challenges for AI-based traffic prediction models. Autonomous vehicles, equipped with advanced sensors and algorithms, operate differently from human-driven vehicles, which exhibit a wider range of driving behaviors and decision-making processes. AI models must account for these differences to provide accurate predictions. Key challenges include:

1. Behavioral Variability

Human drivers exhibit diverse driving behaviors influenced by factors such as experience, mood, and reaction to traffic conditions. In contrast, autonomous vehicles follow predefined algorithms and protocols. AI models must integrate these behavioral variations to predict traffic patterns accurately.

2. Data Integration

Collecting and integrating data from both autonomous and human-driven vehicles is complex. The data must be harmonized to reflect real-time traffic conditions and interactions. Ensuring data accuracy and consistency across different vehicle types is crucial for reliable predictions.

3. Interaction Dynamics

The interaction between autonomous and human-driven vehicles affects traffic flow and congestion. For example, the presence of AVs may influence human drivers' behavior, such as changing lane usage or driving speed. AI models must account for these dynamic interactions to predict congestion effectively.

Assessing the accuracy of AI models for predicting traffic congestion in mixed traffic environments is essential for optimizing traffic management and improving urban mobility. By addressing the challenges associated with behavioral variability, data integration, and interaction dynamics, and by employing advanced AI techniques and real-time data, these models can provide more accurate and reliable predictions. As autonomous and connected vehicles become more prevalent, ongoing research and development will be crucial in refining these models and ensuring their effectiveness in managing complex traffic scenarios.

AI-ENHANCED TRAFFIC FLOW PREDICTION MODELS FOR LARGE-SCALE EVENTS AND EMERGENCIES

AI-enhanced traffic flow prediction models are increasingly being recognized for their potential to manage and mitigate congestion during large-scale events and emergencies. These models leverage advanced machine learning algorithms to analyze vast amounts of data and forecast traffic conditions in real time. During major events, such as sports games, concerts, or public festivals, and in emergencies like natural disasters or accidents, accurate traffic predictions are critical for effective crowd management, route planning, and ensuring public safety. By harnessing the power of AI, cities can better prepare for and respond to these high-demand situations.

AI-based models utilize a range of advanced techniques to predict traffic flow during large-scale events and emergencies. Machine learning algorithms, such as neural networks and ensemble methods, can analyze historical traffic data, real-time sensor inputs, and event-specific variables to forecast congestion patterns. For example, during a major sports event, AI models can predict increased traffic volume around the venue by analyzing data from previous events, weather conditions, and social media activity. Similarly, in emergencies, these models can integrate data from emergency response systems and real-time traffic updates to provide dynamic predictions and optimize evacuation routes.

AI-enhanced traffic flow prediction models face several challenges. One significant issue is the need for high-quality, real-time data to ensure accurate predictions. In large-scale events, the influx of data from various sources must be efficiently processed and integrated. Additionally, these models must be adaptable to rapidly changing conditions, such as sudden road closures or unexpected crowds. Solutions include implementing robust data collection systems, utilizing scalable cloud-based infrastructure, and continuously refining algorithms to improve their adaptability. Effective communication with the public, through apps and notifications, also plays a crucial role in ensuring that traffic predictions are actionable and beneficial during critical times.

AI-enhanced traffic flow prediction models offer valuable tools for managing traffic during large-scale events and emergencies. By leveraging advanced algorithms and real-time data, these models help cities navigate the complexities of high-demand situations, improving traffic flow, safety, and overall event management. Continued development and refinement of these technologies will be essential for addressing the evolving challenges of urban traffic management in dynamic and high-pressure scenarios.

PREDICTIVE TRAFFIC ANALYTICS USING HISTORICAL AND REAL-TIME DATA FUSION

Predictive traffic analytics has been revolutionized by the fusion of historical and real-time data, offering a comprehensive approach to understanding and managing traffic patterns. By integrating historical traffic data, which provides insights into long-term trends and recurring congestion points, with real-time data from sensors, GPS, and traffic cameras, AI models can deliver highly accurate predictions of traffic conditions. This fusion allows for a more nuanced understanding of how current events, such as accidents or weather changes, impact traffic flow compared to historical patterns. For instance, machine learning algorithms can analyze past traffic patterns to predict peak congestion times and adjust real-time traffic management strategies accordingly, leading to more efficient routing and reduced congestion.

This data fusion enhances the ability to respond dynamically to sudden changes in traffic conditions. Real-time data provides up-to-the-minute information on traffic volume, road conditions, and incidents, while historical data offers context and background for these changes. The combined analysis helps in forecasting traffic bottlenecks and potential disruptions before they escalate. This proactive approach enables better resource allocation, such as adjusting traffic signal timings or deploying traffic management personnel, to mitigate congestion and improve overall traffic flow. As urban areas continue to grow and traffic patterns become increasingly complex, the integration of historical and real-time data will be crucial for developing more effective and responsive traffic management systems.

AI AND IOT TECHNOLOGIES FOR REAL-TIME TRAFFIC FLOW PREDICTION

The integration of Artificial Intelligence (AI) and Internet of Things (IoT) technologies has revolutionized real-time traffic flow prediction, offering a sophisticated approach to managing urban mobility. IoT devices, such as traffic cameras, sensors embedded in roads, and GPS-equipped vehicles, collect vast amounts of data on traffic conditions, vehicle speeds, and road usage in real time. This data serves as the foundation for AI algorithms, which analyze and interpret the information to provide accurate traffic predictions and insights. By leveraging the real-time data captured by IoT devices, AI models can dynamically adjust traffic forecasts based on current conditions, helping to optimize traffic management and improve overall road safety.

AI-powered traffic management systems utilize machine learning algorithms to process and analyze the data collected by IoT devices. These algorithms can detect patterns and anomalies, predict congestion, and provide recommendations for traffic routing and signal control. For example, AI models can forecast traffic flow based on historical data and real-time inputs, such as sudden spikes in traffic volume or changes in weather conditions. The real-time capabilities of IoT devices enhance the accuracy and timeliness of these predictions, allowing for proactive management of traffic congestion and more efficient use of road infrastructure. As urban areas continue to grow and traffic conditions become more complex, the synergy between AI and IoT technologies will be crucial for developing responsive and adaptive traffic management solutions.

EFFECTIVE TRAFFIC MANAGEMENT SYSTEM USING TECHNOLOGIES IN REAL-TIME SCENARIOS

An effective traffic management system harnesses advanced technologies to monitor, predict, and control traffic flow in real time, optimizing urban mobility and enhancing road safety. Modern traffic management systems integrate various technologies, including IoT sensors, AI algorithms, and real-time data analytics, to address the dynamic challenges of urban transportation. By utilizing these technologies, cities can develop adaptive systems that respond to changing traffic conditions, reducing congestion and improving overall efficiency.

One notable example of such a system is the intelligent traffic management framework implemented in Singapore. The Land Transport Authority (LTA) of Singapore has deployed a network of IoT sensors and traffic cameras across the city to continuously monitor traffic conditions. These sensors collect data on vehicle counts, speeds, and traffic density, which is then processed by AI-driven algorithms to predict congestion and manage traffic flow dynamically. The system can adjust traffic signal timings in real time based on current traffic patterns and predicted congestion, improving the flow of vehicles and minimizing delays.

Singapore's system also integrates real-time data from GPS-equipped taxis and public transport vehicles to provide a comprehensive view of urban mobility. This data enables the system to issue timely traffic advisories and route suggestions to drivers through mobile apps, helping them avoid congested areas and optimize their travel routes. By leveraging these technologies, Singapore has successfully reduced traffic congestion, improved travel times, and enhanced the efficiency of its transportation network, setting a benchmark for other cities to follow in the pursuit of effective traffic management.

AI IN FUTURE TRAFFIC FLOW PREDICTION: FUTURE PERSPECTIVES

As urban areas continue to expand and the complexity of traffic systems grows, the role of Artificial Intelligence (AI) in traffic flow prediction is poised to become increasingly pivotal. Looking forward, AI is expected to revolutionize how cities manage traffic, providing advanced solutions that enhance efficiency and address emerging challenges. Future AI technologies will leverage increasingly sophisticated algorithms and integrate a broader range of data sources to deliver more accurate and actionable traffic predictions.

One significant development on the horizon is the integration of AI with autonomous vehicle technology. As autonomous vehicles become more prevalent, AI systems will be able to access real-time data from these vehicles, offering unprecedented insights into traffic conditions and vehicle interactions. This data will enable AI models to predict traffic patterns with higher precision, optimize traffic signal timings, and manage congestion more effectively. The collaboration between AI-driven traffic management systems and autonomous vehicles could lead to the development of dynamic traffic control systems that adapt instantaneously to changing conditions, such as sudden traffic spikes or accidents.

Another promising area is the use of AI for predictive analytics in conjunction with smart infrastructure. Future traffic management systems will likely incorporate advanced sensors, connected roadways, and real-time data from various sources, such as social media and environmental sensors. AI algorithms will analyze this vast array of data to predict traffic flow, manage road usage, and optimize public transportation routes. For example, AI could forecast traffic congestion based on event schedules, weather patterns, and historical data, enabling proactive measures to mitigate potential delays. The integration of AI with smart city infrastructure will facilitate a more holistic approach to traffic management, improving overall urban mobility and enhancing the quality of life for city residents.

The future of AI in traffic flow prediction promises to bring transformative changes to urban transportation systems. By harnessing advanced AI technologies and integrating them with emerging trends such as autonomous vehicles and smart infrastructure, cities will be able to develop more responsive, efficient, and adaptive traffic management solutions. These advancements will not only enhance the accuracy of traffic predictions but also contribute to safer, more sustainable, and well-organized urban environments.

CONCLUSION

The integration of Artificial Intelligence (AI) and advanced technologies into traffic management represents a significant leap forward in addressing the complexities of urban mobility. As cities grapple with growing populations and increasingly congested roadways, the adoption of AI-driven solutions and IoT-based systems offers promising strategies for optimizing traffic flow and enhancing overall transportation efficiency. Real-time traffic prediction models, empowered by the fusion of historical and real-time data, enable dynamic adjustments that respond to immediate conditions and long-term trends, leading to more efficient traffic management and improved safety. Future perspectives highlight the potential of AI to further revolutionize traffic management through the incorporation of autonomous vehicles, smart infrastructure, and comprehensive data analytics. As autonomous vehicles become more widespread, AI systems will gain deeper insights into traffic patterns and vehicle interactions, allowing for even more precise predictions and adaptive traffic control.

The integration of AI with smart city infrastructure and advanced sensors will enable a holistic approach to managing urban traffic, addressing challenges such as congestion and environmental impact with greater efficacy. The ongoing advancements in AI and related technologies are set to transform traffic flow prediction and management, paving the way for smarter, more responsive urban transportation systems. As these technologies continue to evolve, they promise to deliver enhanced accuracy, efficiency, and safety in traffic management, contributing to the development of more sustainable and well-organized cities. The future of traffic management lies in the seamless integration of AI and IoT solutions, which will drive innovations in urban mobility and improve the quality of life for city residents.

References

Bianchini, D., De Antonellis, V., & Melchiori, M. (2019). AI-based bus scheduling and routing optimization in smart cities: The case of Rome. *IEEE Transactions on Smart Cities*, 1(1), 56–68.

. (●●●). Challenges. *Transportation Research Part C, Emerging Technologies*, 131, 103458.

Chen, C., Jiang, L., & Yuan, Z. (2020). Challenges and prospects of AI in smart city traffic management. *Sustainable Cities and Society*, 63, 102–114.

Chen, Z., & Sun, Y. (2022). Smart Cities and Traffic Management: Innovations and Future Prospects. *Urban Planning and Development*, 148(3), 123–136.

Díaz-Díaz, R., Muñoz, L., & Pérez-González, D. (2017). Business model analysis of public services operating in the smart city ecosystem: The case of smart grid and smart mobility. *Future Internet*, 9(3), 24.

Gao, J., & Li, X. (2023). The Role of Autonomous Vehicles and AI in Future Traffic Management. *Journal of Intelligent Transport Systems*, 27(4), 362–378.

Goodall, N. J., Smith, B. L., & Park, B. B. (2020). Traffic signal control with connected and autonomous vehicles: A review of potential benefits and challenges. *Transportation Research Part C, Emerging Technologies*, 106, 1–18.

Hossain, M. S., Muhammad, G., & Amin, S. U. (2021). Smart cities with artificial intelligence, big data, and internet of things. *IEEE Communications Magazine*, 59(1), 40–46.

Kramers, A., Höjer, M., Lövehagen, N., & Wangel, J. (2018). Smart cities and climate targets: An exploration of smart city implementation in Stockholm. *Journal of Cleaner Production*, 172, 4039–4046.

Kumar, V., & Patel, A. (2023). Advances in AI-Driven Traffic Management Systems: A Comprehensive Review. *Transportation Research Part B: Methodological*, 158, 334–355.

Mahadevan, R., Bhat, V., & Rao, S. (2023). Bangalore's Integrated Traffic Management System: A smart city initiative for urban mobility. *Journal of Urban Technology*, 30(2), 150–169.

Mousavi, S. S., Schukat, M., & Howley, E. (2020). Traffic light control using deep policy-gradient and value-function-based reinforcement learning. *Engineering Applications of Artificial Intelligence*, 85, 565–574.

Nagendra, S. M. S., Khare, M., & Goyal, R. (2018). Urban air quality management in India: A review. *Atmospheric Environment*, 172, 209–225.

Papageorgiou, M., Diakaki, C., & Aboudolas, K. (2022). Real-time traffic signal control for urban networks: The potential of AI-based adaptive systems. *IEEE Transactions on Intelligent Transportation Systems*, 23(2), 512–523.

Ranganathan, A., Ramachandran, R., & Patel, V. (2021). Challenges and opportunities in implementing AI-based traffic management systems in Indian smart cities. *International Journal of Urban Sciences*, 25(4), 549–565.

Song, H., Srinivasan, D., & Choy, M. C. (2017). Hybrid cooperative traffic signal control using a novel multi-agent reinforcement learning. *IEEE Transactions on Intelligent Transportation Systems*, 18(8), 2143–2157.

Thomas, J. P., & Frantz, R. (2023). AI in traffic management: A global review of implementations, challenges, and future directions. *Journal of Urban Planning and Development*, 149(1), 04022035.

Vlahogianni, E. I., Karlaftis, M. G., & Golias, J. C. (2015). Short-term traffic forecasting: Where we are and where we're going. *Transportation Research Part C, Emerging Technologies*, 43, 3–19. DOI: 10.1016/j.trc.2014.01.005

Wang, Y., & Zhang, Q. (2021). Real-Time Traffic Flow Prediction with Machine Learning and IoT: A Survey. *IEEE Access : Practical Innovations, Open Solutions*, 9, 214674–214688.

Yap, J., Tay, R., & Ong, K. (2020). Smart cities and urban transport: Opportunities and challenges in the deployment of artificial intelligence for real-time traffic management. *Journal of Transport and Land Use*, 13(1), 25–45.

Zhang, Y., Liu, Y., & Zheng, Y. (2021). Predictive traffic management using machine learning: A case study of large metropolitan areas in China. *Journal of Transportation Systems Engineering and Information Technology*, 21(4), 234–246.

Chapter 9
Current Trends, Opportunities, and Futures Research Directions in Geospatial Technologies for Smart Cities

Vijaya Kittu Manda
https://orcid.org/0000-0002-1680-8210
PBMEIT, India

Veena Christy
https://orcid.org/0000-0001-9987-6253
SRM Institute of Science and Technology, India

Arbia Hlali
https://orcid.org/0000-0002-1850-4579
Taibah University, Saudi Arabia

ABSTRACT

Studying current trends and opportunities in geospatial technologies for Smart cities helps city planners and administrators understand technology advancements and aids in better implementation practice. Similarly, understanding future research directions enables researchers and policymakers to harness these technologies fully and anticipate upcoming developments. Overall, this approach supports the creation of more livable, sustainable, and equitable cities for all. This chapter explores the

DOI: 10.4018/979-8-3693-8054-3.ch009

current trends, emerging opportunities, and future research directions. Geospatial AI is supported by several cutting-edge technologies such as IoT, digital twins, and 3D/4D urban modeling. Future geospatial research should use advanced AI models, real-time analytics, and privacy-preserving technologies. Understanding the technologies' ethical and inclusive implementation is essential to support long-term urban sustainability and citizen well-being. Such Smart cities can ensure more sustainable, resilient, and equitable urban environments for future generations.

1. INTRODUCTION

Smart Cities (SC) are considered the future of urban living. Numerous computing and geospatial technologies are used in cohesion to help city administrators get a unified view and help them make effective decisions and formulate strategies for SCs (Mishra & Singh, 2023; Smékalová & Kučera, 2020). Urban environments generally involve lots of complexities, mostly of technologies and data. Further, the spatial data and related technologies are evolving, offering immense opportunities as Smart cities adopt geospatial technologies.

Further, the SC agenda has increased regional cooperation between cities, increased digitalization, and is meeting citizen service requirements much better than before (Masik et al., 2021). This chapter attempts to explain several challenges and opportunities that emerge with the increasing adoption of intelligent solutions for urban management. Both SCs and geospatial technologies are rapidly evolving topics. Exploring future research trends helps stakeholders stay on top of research and use it for the next phase of urban innovation. The subject is multidisciplinary and is a topic of interest for researchers in multiple fields. It spans urban planning, GIS (Geographic Information Systems), data science, computer science, and sustainability. The application and culmination of these studies are evolving into newer sub-studies. Spatial analysis primarily deals with "location information" and "attribute information". Statistics deals with uncertainty quantification. Spatial Statistics is an evolving field in which spatial analysis meets statistics (Cressie & Moores, 2023). Along similar lines is Geoinformation and Communication Technology (GeoICT), another evolving discipline that culminates in ICT implementations involving geographic information science (GIS) and systems in SCs (Ang et al., 2022). Figure 1 shows the Geospatial Data Flow in Smart Cities.

Figure 1. Geospatial Data Flow in Smart Cities

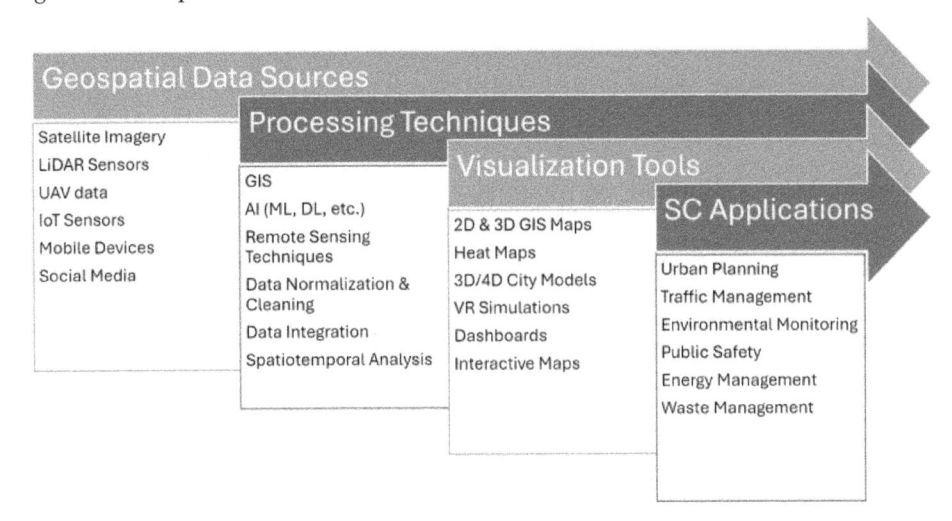

Geospatial Data Sources	Processing Techniques	Visualization Tools	SC Applications
Satellite Imagery LiDAR Sensors UAV data IoT Sensors Mobile Devices Social Media	GIS AI (ML, DL, etc.) Remote Sensing Techniques Data Normalization & Cleaning Data Integration Spatiotemporal Analysis	2D & 3D GIS Maps Heat Maps 3D/4D City Models VR Simulations Dashboards Interactive Maps	Urban Planning Traffic Management Environmental Monitoring Public Safety Energy Management Waste Management

1.1. Chapter Objectives and Structure

This chapter intends to apprise the research community of the current trends in geospatial technologies for use in Smart cities. Further, this chapter gives practical and academic relevance attempts to link the gap between academic research and real-world uses. Urban planners, government officials, and tech developers will benefit from understanding how to implement these technologies and determining the hurdles in the journey.

Section 2 of the chapter focuses on current trends in the area. It begins with an appraisal of the historical development and evolution of the technologies that shaped the technology. The section then focuses on Geospatial AI, which blends Artificial Intelligence (AI) with Geospatial technologies. The section also discusses other cutting-edge technologies that are fast emerging, including IoT, Digital Twins, and Urban Simulation.

Section 3 discusses the opportunities and advantages of using these technologies in Smart cities. The section highlights critical urban planning, land use applications, environmental monitoring, and sustainable development. It also explains traffic and transportation management, citizen engagement and public safety, and using technologies for social good.

Section 4 discusses potential future research directions that can direct the attention of researchers in upcoming areas. The section begins by listing areas for future research involving advanced AI techniques and models. The following section discusses research ideas related to 3D and 4D modeling. Since Smart cities generate an enormous quantum of data, researchers pay attention to data security and privacy. The section concludes by discussing research trends in geospatial analytics and decision-making systems.

2. CURRENT TRENDS IN GEOSPATIAL TECHNOLOGIES FOR SMART CITIES

2.1. Historical Context and Progression of Smart Cities and Geospatial Technologies

The evolution of Smart Cities is not linear and exponential. However, a complex interaction of urban planning systems, technological innovations, and socio-economic factors reflects the broader trends in urban growth and technological progress throughout history. The concept emerged because cities struggle with challenges due to rapid urban growth, population density, resource management, and environmental sustainability (Campbell, 1999). Los Angeles was the first city to consider the urban Big Data project during the 1970s. Only a year ago, in 1969, Jay W. Forrester prepared his research titled *Urban Dynamics* with MIT's Urban Systems Laboratory. Cybersyn was a project developed in Chile in the 1970s to create a nationwide network of computers and sensors to monitor and control the country's economy. Stafford Beer, a British cyberneticist, led the project (Verma et al., 2024). The project was probably one of the first attempts to use computers to manage a complex system in real time (Klumbytė & Athanasiadou, 2022). IBM was the first to coin the term "Smart city," referring to the use of technologies by cities to improve their efficiency and offer sustainable urban services. Cisco also made inspirational contributions with its networking infrastructure.

The correlation between city size and investment intensity is evident, with larger cities attracting more investment activity (Smékalová & Kučera, 2020). The end of the twentieth century saw the expansion of Information and Communication Technologies (ICT). The availability of several ICT applications offered new hope for addressing several rapid urbanization issues (Z. Cai et al., 2020). Consequently, the Smart City concept witnessed three branches - governance, sustainability, and decent urbanization (Park & Yoo, 2023). Technology integration helped lay the initial initiatives and directed the Smart city concept toward achieving the following objectives:

1. Improving urban efficiency, such as urban livability
2. Sustainability living standards

3. Social and economic prosperity

The evolution of Smart cities accelerated as the twenty-first century entered. Several technology developments happened at around the same time, including:

1. Penetration of the internet, especially high-speed internet
2. Increased deployments of sensor networks (Channi & Kumar, 2022)
3. Mobile networks and Internet of Things (IoT) infrastructure are accessible everywhere (Rosa et al., 2021)

The technological developments helped build early real-time data-driven solutions for SC, such as traffic management, hotspot congestion, energy distribution, and waste collection. As time progressed, the field evolved, and the focus shifted towards a more holistic and citizen-centric approach. Cities began using geospatial data for urban planning, transport management, and environmental monitoring. The launch of Google Earth in 2005 democratized access to geospatial information. In 2000, US President Clinton signed a bill that made the US government make Global Positioning Systems (GPS) freely available for civilian use. Open GPS for civil use has led to the advancement of mapping and related smartphone applications, which have been an enormous success. Only the US military used it earlier. The Soviet Union and the European Union subsequently began using it. GPS became an enormous success in the 1990s and useful for non-military applications (Ceruzzi, 2021).

Advancements in remote sensing and crowdsourced geographic information aided subsequent developments. The 2010s saw an integration with modern technologies. There is substantial progress in the Internet of Things (IoT), Big Data Analytics, and Artificial Intelligence (AI) (Javed et al., 2022). This evolution reflects a shift towards using geospatial data beyond urban planning. During this time, the focus of policymakers shifted towards promoting innovation, improving citizen engagement, and creating sustainable ecosystems. The growing importance of geospatial AI has further helped Smart cities enter the next phase. Table 1 presents a Chronology of Geospatial Technologies and Smart Cities Development.

Table 1. Chronology of Geospatial Technologies and Smart Cities Development

Year	Key Development
1854	Dr. John Snow used spatial analysis to map Cholera cases in London.
1963	Roger Tomlinson (Canada) coined the term Geographic Information System (GIS) while working on a national land use management program for the Canadian government.
1969	Jack and Laura Dangermond founded the Environmental Systems Research Institute (ESRI) in Redlands, California, USA. The company became a global major in GIS software, web GIS, and geodatabase management applications.
1972	The first Landsat satellite, Landsat 1, was launched for Earth observation.
1974	Los Angeles created the first urban big data project: "A Cluster Analysis of Los Angeles."
1978	The first GPS satellite, Navstar I, was launched.
1981	The first Esri User Conference was held.
1991	The term "Smart City" began gaining attention, mainly in academic and urban planning circles, as cities started exploring digital technologies to improve urban services.
1994	Amsterdam created a virtual 'digital city' to promote Internet usage.
1994	The introduction of commercial GPS for civilian use became widespread, improving urban navigation and geospatial analytics.
1996	The formation of OGC (Open Geospatial Consortium), which set spatial data interoperability standards, boosted the GIS industry.
1999	Cities like Singapore began adopting large-scale urban management systems, such as the Land Transport Authority's Intelligent Transport System, as part of early Smart city initiatives.
2005	Cisco invested $25 million over five years for research into smart cities.
2008	IBM launched the Smarter Planet project to apply sensors, networks, and analytics to urban issues.
2000	Launch of Google Earth and subsequent popularization of geospatial tools in everyday life, providing urban planners and Smart city projects with new ways to visualize data.
2003	Barcelona started developing a comprehensive Smart city strategy, becoming one of the pioneers in the Smart city movement.
2005	Wireless Sensor Networks (WSN) began being deployed in urban environments, marking the start of using IoT (Internet of Things) devices in smart cities to monitor and manage urban services.
2007	The term "smart grid" became a prominent feature of Smart city discussions, focusing on efficient energy management within cities using ICT and geospatial technologies.
2008	IBM launched its Smarter Planet initiative, positioning the company as a leader in Smart City projects by promoting intelligent systems across multiple domains such as transport, utilities, and healthcare.
2011	Barcelona hosts the first Smart City Expo World Congress.
2011	The European Commission launched the European Innovation Partnership on Smart Cities and Communities (EIP-SCC)
2015	India launched the "Smart Cities Mission" to upgrade 100 Indian cities at a funding of $24 billion.
2015	IoT platforms rise for Smart cities. The Internet of Things Security Foundation (IoTSF) was launched.
2016	The United Nations adopted the New Urban Agenda, which focuses on sustainable urbanization and Smart technologies.
2017	The UK government opened the 5G testbeds and trials program.

continued on following page

Table 1. Continued

Year	Key Development
2018	Google's Sidewalk Labs launched its Quayside project in Toronto
2019	Digital Twins gained popularity in Smart city initiatives. The inaugural IMD Smart City Index ranks Singapore as the World's Smartest City.
2020	The COVID-19 pandemic accelerated the adoption of Smart city technologies.
2021	The Smart Cities Marketplace (rebranded from the European Commission's EIP-SCC) facilitated over 300 projects globally, integrating AI, GIS, IoT, and 5G for urban innovation.
2022	Artificial Intelligence (AI) and Machine Learning (ML) integration with Geospatial analytics in Smart city platforms is seen.
2023	Emergence of autonomous urban systems.
2024	Quantum computing is integrated with geospatial technologies and smart cities.

2.2. Geospatial AI: AI Integration in Geospatial Applications

Geospatial AI (GeoAI) integrates Artificial Intelligence (AI) techniques with Geospatial Data and Technologies. Each AI technique brings in specific benefit:

1. **Machine Learning (ML) and Deep Learning (DL)**: These models help in land use classification, object detection, and spatial pattern prediction.
   a. Convolutional Neural Networks (CNNs) have excellent feature extraction and pattern recognition capabilities. So, they are frequently employed to process satellite images for land use and land cover classification, object detection (e.g., buildings, roads), and change detection over time. Advanced architectures like ResNet and DenseNet have improved model depth and feature reuse, enhancing performance on complex geospatial datasets (Al-qudah et al., 2022).
   b. Recurrent Neural Networks (RNNs) and Long Short-Term Memory (LSTM) capture temporal dependencies. These are needed for time-series analysis and have applications like traffic prediction, weather forecasting, and urban growth modeling. They process sequential data, accounting for temporal correlations that static models might overlook.
   c. Transformer models such as Vision Transformers (ViT) can provide high-resolution image analysis. They provide state-of-the-art results in segmentation and classification tasks.
   d. Multimodal Deep Learning (MDL) deals with heterogeneous data sources or modalities. Integrating such data provides complementary information and improves the robustness of the model. Urban Computing is emerging as a new discipline in sustainability development with the fusion of multiple

sources and modalities. The chemistry between Large Language Models (LLMs) and Urban Computing will be observed in the coming times (Zou et al., 2025). For example, city planners can map urban poverty levels by combining satellite imagery with socio-economic data.
2. Natural language processing (NLP) for geospatial has recently evolved to deal effectively with end-user queries so much that GeoQA (Geographic Question Answering) is now an emerging research field (Feng et al., 2023). GeoGPT brings ChatGPT to geospatial tasks for geospatial professionals to use (Y. Zhang et al., 2024). They allow urban planners and decision-makers to interact with complex geospatial data using natural language, democratizing access to spatial insights.
3. Computer Vision for 3D Urban Modeling to analyze geospatial data and extract meaningful insights, patterns, and relationships. Both computer vision and 3D modeling are expensive and resource-intensive but vital for capturing building rooftop data. GeoAI frameworks to address this requirement were developed (Z. Zhang et al., 2022).
4. Predictive Analytics for urban phenomena, such as traffic flow prediction, urban growth modeling, and crime hotspot forecasting, enable proactive urban management strategies. Studies report improved prediction accuracy with depth of bearing layers with Ensemble learning methods (Cong & Inazumi, 2024).
5. Graph Neural Networks (GNNs) are a type of neural network. They are devised to handle data that can be represented as a graph. For example, a GNN for a city can comprise of
    a. **Graphs**: The city itself is represented as the graph
    b. **Nodes**: Objects in the city, such as buildings, roads, and vehicles, can be nodes.
    c. **Edges**: The connections or relationships between these nodes, such as roads connecting buildings or vehicles traveling on roads.

ML models can work efficiently with graphs using GNN. Generally, GNNs have three features that help in Smart city situations (P. Liu et al., 2024):

a. Spatial classification (such as in identifying hotspots)
b. Regression (such as in predicting pollution levels)
c. Clustering (such as in the identification of socio-economic zones).

GNN's involve:

a. Graph Convolutional Networks (GCNs) are used to graph domains and aggregate feature information from neighboring nodes, making them suitable for image classification (Hong et al., 2021). They can guess traffic flow, analyze social networks, and prepare models of energy consumption patterns. The Spatial Graph Convolutional Network (SGCN) enhances traditional GCN. It effectively incorporates spatial features to learn from graphs from natural locations in space are now available (Danel et al., 2020).
b. Graph Attention Networks (GATs) have attention mechanisms and prioritize certain relationships within graph-structured data. They allow the models to weigh the importance of neighboring nodes differently. Urban settings can use this feature because they involve certain connections (e.g., major highways vs. minor roads, intra- and inter-city health disparity) that reflect varying significance (C. Liu et al., 2024). Improvements such as relation-augmented embedded graph attention network (EGAT) enable the complete usage of the underlying spatial and semantic relations among objects to improve detection performance regarding accuracy and context-aware understanding of the urban environment(Tian et al., 2022).
c. Spatiotemporal GNNs combine GNNs with temporal modeling techniques to address dynamic urban processes. Epidemic spread modeling, real-time navigation systems, and adaptive infrastructure management are a few applications of this model (Jin et al., 2024).
d. Graph Sample and Aggregation (GraphSAGE) is a multiscale graph sample and aggregate network. It can flexibly aggregate the new neighbor node among arbitrarily structured non-Euclidean data and capture long-range contextual relations (Ding et al., 2022). Deep GraphSAGE is also gaining attention (El Alaoui et al., 2022).

6. By learning from real-time geospatial data streams, reinforcement Learning can optimize complex urban systems, such as traffic light timing and public transportation routing. Advanced RL techniques enhance the agent's ability to handle high-dimensional state spaces typical in geospatial applications. Deep Q-Networks (DQNs) and Policy Gradient Methods are a few examples.
7. Unsupervised and Semi-Supervised Learning for Geospatial Data: Comprises of:
   a. Autoencoders and Variational Autoencoders (VAEs): These models learn compressed data representations, facilitating anomaly detection, feature extraction, and data denoising in geospatial datasets.
   b. Generative Adversarial Networks (GANs): GANs generate synthetic geospatial data for data augmentation and fill incomplete data gaps. They are employed in super-resolution mapping, simulating high-resolution images from lower-resolution inputs.

    c. Self-Supervised Learning: Leveraging inherent data structures (e.g., spatial continuity) and self-supervised techniques pre-train models on large unlabeled datasets, improving performance on downstream tasks with limited labeled data.

    These AI-driven techniques enable dynamic urban planning, real-time traffic monitoring, infrastructure management, and environmental assessment within Smart cities.

    Edge AI allows data processing locally on devices such as IoT sensors. They reduce latency and bandwidth requirements apart from allowing faster decision-making locally. Federated Learning has evolved and provides attractive, practically feasible privacy preservation models. They allow the training of AI models across multiple decentralized devices without sharing raw data. They provide a solution to the privacy and data security concerns arising from centralized data processing in Smart cities (Pandya et al., 2023). Developers can thus build geospatial AI applications that respect data sovereignty. It also promotes large-scale data integration, privacy protection, and sensitive information protection. FL with digital twins is explored in research circles (Pang et al., 2021). Fog-assisted Federated Learning (FogFL) for resource-constrained IoT devices and delay-sensitive applications is emerging. FogFL framework reduces resource-constrained edge devices' communication latency and energy consumption without touching the global model's convergence rate (Saha et al., 2021).

    Distributed learning frameworks, such as parameter servers and decentralized optimization algorithms, address computational challenges posed by large-scale geospatial datasets. Python libraries are now available to standardize the geospatial AI domain toolset (Gramacki et al., 2023). Though still in its early stages of development, Geospatial AI can change how geospatial data is collected, analyzed, and used.

2.3. Other Cutting-Edge Technologies

    IoT integrates various sensing devices with geospatial frameworks. They enable real-time data acquisition and analysis in urban environments. Sensor networks usually comprise RFID tags, GPS-enabled devices, and environmental monitors. These devices generate spatiotemporal data streams. These data streams provide inputs for the Geographic Information Systems (GIS). GIS are often centralized repositories where data analysis takes place. They allow city planners to generate dynamic maps, perform spatial analysis, and develop context-aware services.

    Countries are prioritizing the creation of Digital twins in their policies. The Korean New Deal Digital Twin exemplifies this (Lee, 2021). Digital twins generate a virtual replica of the urban environment using geospatial data (Gkontzis et al.,

2024). Metaverse tries to tie the virtual and real world together (Lv et al., 2022). The generated digital twins can become an input to a Metaverse, bringing in new applications, such as tourism and defense. Citizens of the city can use the Digital twin-Metaverse shopping mall to shop and make payments with cryptocurrencies directly from their headsets.

Several other technologies also provide vital inputs. Edge computing facilitates low-latency processing. Cloud platforms enable scalable storage and advanced analytics. Machine learning algorithms extract patterns from heterogeneous data sources, enhancing situational awareness. Applications include traffic management, air quality monitoring, and infrastructure maintenance. IoT-driven geospatial solutions follow interoperability standards (e.g., OGC SensorThings API) to create a unified SC ecosystem for data-driven decision-making and urban optimization.

Urban simulations use computational models to emulate complex city dynamics within a geospatial context. Agent-based models simulate individual entity behaviors, while cellular automata represent spatial interactions. These models integrate with 3D city models (CityGML) and digital twins for enhanced realism. Simulations incorporate multiscale data, from building-level information to regional demographics, enabling scenario analysis for urban planning. Coupling with machine learning techniques allows for predictive modeling of land use changes, traffic patterns, and environmental impacts. High-performance computing facilitates large-scale simulations, while WebGL and virtual reality technologies enable immersive visualization. Applications span urban heat island modeling, evacuation planning, and sustainable development assessment. These simulations augment traditional GIS analysis, providing dynamic insights for evidence-based urban governance and policy formulation. Urban simulations support sustainable development. They are a virtual testing ground for innovative solutions and facilitate data-driven decision-making.

3. OPPORTUNITIES AND ADVANTAGES OF GEOSPATIAL TECHNOLOGIES FOR SMART CITIES

Rapid urbanization is a global challenge. It leads to social issues, such as increased traffic congestion, air pollution, water scarcity, and loss of green space. Geospatial technologies can address these challenges by explaining the spatial relationship between the urban elements. The technologies deal with data and provide insights to administrators to plan for and manage growth. For example, geospatial data can:

1. Identify areas with flooding risk or other natural disasters and suggest proactive measures such as early warning systems (EWSs), evacuation planning, and resource allocation. AI systems for natural disasters are already well-researched.

The taxonomy for natural disaster-based AI comprises categories such as hydrological, geophysical, climatological, meteorological, biological, or hybrid (Albahri et al., 2024).
2. Plan for developing, maintaining, and optimizing large-scale transportation infrastructure (P. Zhang et al., 2024). GeoICT applications related to transportation are a large study. It comprises intelligent transportation, transportation system architectures, traffic monitoring and management, social transportation, crowdsourced transportation data, platooning for sustainable transportation, UAV-enabled transport, ridesharing, multi-station vehicle sharing, and waste transportation (Ang et al., 2022).
3. Create green amenities such as park green spaces (PGS) (Zheng et al., 2020) and landscape patterns (Kowe et al., 2021) to make green infrastructure such as parks and promenades more accessible. Green urbanism is an inspiration that attracts practitioners from urban design and planning (Tannous et al., 2021).

Precise monitoring of urban heat islands, air quality, and vegetation health can be done with advanced RS techniques and ML algorithms. LiDAR-based 3D city models give inputs for optimal placement of renewable energy infrastructure, such as solar panels and wind turbines. Spatial analysis of urban metabolism can identify inefficiencies in resource flows, informing circular economy strategies. Geospatial Big data analytics supports the development of green corridors and the mapping of ecosystem services. Integrating IoT sensors with GIS platforms allows real-time water consumption and waste management tracking, enabling adaptive resource allocation. Spatiotemporal analysis of urban mobility patterns can optimize public transportation routes, reduce carbon emissions, and promote sustainable urban mobility. The technologies provide a platform for data sharing and analysis in collaboration amongst SC stakeholders. With this, the decision-makers will have sufficient information to make more informed decisions regarding resource usage and plans.

3.1. Enhanced Urban Planning and Land Use Management

Geospatial technologies are a toolset with several tools that support proper urban planning and land use management. GIS and RS, in particular, provide spatial analysis and visualization tools to deal with land use, demographic, transportation, and environmental data. They provide a thorough understanding of the spatial relationships between various urban elements. In this context, geospatial technologies can help SCs in three significant ways:

1. **Plan for sustainable growth**: The technologies help identify environmentally sensitive areas. City planners can use this data to plan for sustainable development. They promote compact and walkable tracks for the communities (Wheeler, 2013).
2. **Improve land use efficiency**: They optimize land use patterns, reduce sprawl, and use existing infrastructure better (G. Cai et al., 2020).
3. **Facilitate public participation**: They create interactive maps and other visualizations. They allow and encourage the public to participate in the planning process. These collaborations and interactions can reach newer levels when AR/VR tools are used (Postert et al., 2022).

Some notable examples of technology in practice are:

1. High-resolution remote sensing data and advanced image processing algorithms enable precise 3D urban modeling and change detection (Buyukdemircioglu & Kocaman, 2020).
2. LiDAR-derived point clouds facilitate accurate building footprint extraction and volumetric analysis.
3. ML techniques can automate land use classification and detect urban sprawl patterns when applied to multi-temporal satellite imagery.
4. Digital twins and augmented reality, powered by real-time geospatial data streams, provide dynamic visualizations for urban planners. Similarly, digital twins with AI improve physical detail capture at the neighborhood level (Gkontzis et al., 2024).
5. Blockchain-based land registries work closely with geospatial systems. The resulting solution ensures data is not tampered with, providing transparency in property management and trust amongst the parties. Some SCs are moving towards instantaneous land registrations, reducing the overall time (Ozcelik, 2024).

Some specific examples of geospatial technologies usage to enhance urban planning and land use management include:

1. In New York City, geospatial technologies are helping develop a new zoning code to promote sustainable development. The city uses a geospatial platform to analyze the potential impacts of zoning scenarios. The zoning is based on affordability, environmental quality, and transportation efficiency (Balk et al., 2024).

2. In London, geospatial technologies helped create a digital twin of the city. This digital twin simulates the impact of different planning decisions on factors such as traffic flow, air quality, and energy consumption.
3. In Singapore, geospatial technologies have allowed for an intelligent land-use planning system. This system uses real-time data to monitor land use changes and identify opportunities for more efficient and sustainable land use.
4. In the European Union, geospatial technologies monitor land use changes. The EU has developed a land use monitoring system that collects data on land use changes across the continent. This data is used to identify land-use change trends and develop policies to promote sustainable land-use practices (d'Andrimont et al., 2020).

3.2. Geospatial Technologies for Improved Environmental Monitoring and Sustainability

Geospatial technologies can help improve urban lives by promoting inclusivity, safety, and resilience, making cities sustainable (Pérez Del Hoyo et al., 2021). Smart cities can use them to capture and keep track of environmental factors:

1. **Monitor air and water quality**: Geospatial technologies can collect and analyze air and water quality data. Urban computing technologies provide a potential solution to urban air quality management (UAQM) (Kaginalkar et al., 2021). This data helps identify pollution sources, track the spread of pollutants, and develop strategies for improving environmental quality. SCs can build a governance framework to deal with this (Kaginalkar et al., 2022).
2. **Monitor land use changes**: Geospatial technologies help monitor changes in land use over time (Yin et al., 2021). This data helps identify trends in deforestation, urbanization, and other land use changes. Further, it helps develop policies to protect sensitive environmental areas and promote sustainable land use practices.
3. **Environment Monitoring & Ecological Data**: Geospatial technologies help assess the impact of climate change on different environmental factors. Environmentalists use its data to develop adaptation and mitigation strategies to reduce the risks associated with climate change (Reutov et al., 2023). They allow monitoring of urban heat islands, biodiversity, and green spaces. The decisions taken related to these aspects support climate resilience and ecological balance.

Application of the technologies include:

1. Hyperspectral remote sensing enables precise identification of urban vegetation species and health assessment.
2. Interferometric Synthetic Aperture Radar (InSAR) techniques facilitate millimeter-scale ground deformation monitoring, which is crucial for urban subsidence detection. Sensor networks are integrated with IoT platforms and provide real-time air quality data at satisfactory spatial resolutions.
3. Machine learning algorithms applied to multi-source Earth Observation data enable automated detection of urban heat islands and microclimate modeling. LiDAR-derived 3D city models, combined with computational fluid dynamics, allow for high-resolution urban wind flow simulations.
4. Unmanned Aerial Vehicles (UAVs) are equipped with multispectral sensors to capture high-resolution data on demand for use in urban ecosystem assessments.

Several Smart cities have already begun taking advantage of geospatial technologies and become Sustainable Smart Cities:

1. In San Francisco, geospatial technologies track the city's progress towards its zero waste goal. The city has developed a geospatial data platform that allows residents and businesses to track the amount of waste they generate. Opportunities for reducing waste can thus be identified.
2. In Copenhagen, geospatial technologies help plan for a more sustainable transportation system. The city has developed a geospatial model that simulates the impact of different transportation policies on traffic flow, air quality, and greenhouse gas emissions (Ogryzek et al., 2020). Copenhagen has a highly centralized governance structure. Researchers contract smart mobility implementation with Barcelona, which has a more decentralized approach (Wolniak, 2023).
3. Geospatial technologies in Singapore allowed for a more sustainable urban environment. The city transformed from a traditional city to a smart city (Shamsuzzoha et al., 2021). It developed a geospatial platform that updates residents on air quality, water quality, and other environmental indicators.

3.3. Optimized Traffic and Transportation Systems

Geospatial technologies can help cities by understanding spatial relationships between traffic and transportation. The technologies can help in data collection and analysis, traffic modeling, and visualizations. Three key benefits can arise out of this:

1. **Improve traffic flow**: This is possible by identifying and addressing traffic bottlenecks, optimizing traffic signal timing, and developing new traffic management strategies at the street level (Y. Zhang & Raubal, 2022).

2. **Plan for future transportation needs**: This involves forecasting future traffic demand and modeling for active travel infrastructure (Hill et al., 2024). Plans for new transportation infrastructure, such as roads, bridges, and public transit systems, can be made.
3. **Promote sustainable transportation**: They promote sustainable transportation modes, such as walking, biking, and public transit.

Analyzing spatiotemporal data on passenger demand and mobility patterns allows optimization of public transportation. Geospatial analytics facilitate the design of efficient public transit routes, schedules, and service frequency. Transportation agencies can predict peak travel times and optimize resource allocation by incorporating Smart card transactions, mobile applications, and social media data.

Intelligent Transportation Systems (ITS) employ geospatial technologies to enhance the connectivity and interoperability of various transportation modes. For instance, GIS-based multimodal transportation planning integrates data from buses, trains, bicycles, and pedestrian pathways. The data helps in creating seamless and sustainable urban mobility solutions. Additionally, geospatial data supports the development of Mobility-as-a-Service (MaaS) platforms. These platforms offer commuters personalized travel options and real-time information (Arias-Molinares & García-Palomares, 2020).

Infrastructure Planning and Maintenance activities can benefit from geospatial technologies. The technologies provide precise mapping and monitoring of transportation assets. LiDAR-based mapping technologies and satellite imagery allow the planning of new transportation infrastructure. They provide accurate, high-resolution 3D models of urban environments. These models are essential for evaluating the feasibility of constructing new roads, bridges, or tunnels and planning dedicated lanes for autonomous vehicles. Geospatial technologies enable better management of shared mobility services. Ride-hailing or bike-sharing programs analyze usage patterns. They optimize the distribution of vehicles or bikes and give vital inputs to the transport network company (TNC) or the platform (Dean & Kockelman, 2021). They help plan sharing frequency and fleet size (Zhu et al., 2020) and even improve urban air quality levels (Huang et al., 2022). Several cities are already using geospatial technologies to optimize traffic and transportation systems, including:

1. In Los Angeles, geospatial technologies helped implement a new traffic management system. The system uses sensors to collect data on traffic conditions in real-time. This data then adjusts traffic signals and gives drivers real-time information on traffic conditions.

2. In London, geospatial technologies helped plan the development of a new high-speed rail line. The city uses geospatial data to identify the best route for the line and assess the line's potential impact on the environment and surrounding communities.
3. In Singapore, geospatial technologies helped promote sustainable transportation. The city uses geospatial data to develop a network of bike lanes and pedestrian paths. The city also uses geospatial technologies to track public transit use and identify opportunities for improving public transit service.

3.4. Citizen Engagement and Public Safety Improvements

In the SC context, geospatial technologies are excellent citizen engagement and public safety tools. Traditional technologies complement these efforts using sensors, Internet of Things (IoT) devices, and AI to create interconnected urban environments. These technologies enable continuous observation of vital infrastructure, such as traffic flow, power grids, and surveillance systems. For example, IoT-enabled cameras provide live footage to law enforcement, helping them detect and prevent crime. Meanwhile, Smart traffic systems use geospatial data to optimize traffic patterns, reducing congestion and minimizing accident risks.

Mobile apps and platforms create two-way communication. They empower citizens to participate in public safety efforts. They provide authorities with real-time, crowdsourced data, improving decision-making. Interactive mapping platforms allow residents to report issues like potholes or graffiti. Location services can help pinpoint exact locations for swift municipal response. Real-time crime mapping helps law enforcement allocate resources more effectively and informs citizens of local safety concerns. Smart streetlights with integrated sensors can detect gunshots, monitor air quality, and adjust lighting for improved safety and energy efficiency. Mobile apps enable citizens to receive emergency alerts, access city services, and participate in community decision-making. Cities can use these technologies to promote a more connected, responsive, and secure urban environment. These actions contribute to a better quality of life for the community.

3.5. Geospatial technologies for Social Good in Smart Cities

Rapid urbanization should not come at the cost of sacrificing archeological and heritage sites. Geospatial technologies in Smart cities offer significant potential in balancing urban expansion with the conservation of archaeological and heritage sites. Rapid urbanization generally conflicts with the preservation of cultural landmarks. High-resolution multitemporal satellite data can help identify urban growth areas without surrendering archaeological heritage conservation (Alders, 2024).

Remote sensing technologies combined with AI-driven pattern recognition can give a potential solution. The duo can monitor urban encroachments near historical sites, ensuring that urban planning does not lead to irreversible damage to archaeological resources. Moreover, integrating geospatial data in heritage management can optimize conservation efforts by mapping urban pressures and assessing risks to cultural sites.

4. FUTURE RESEARCH DIRECTIONS IN GEOSPATIAL TECHNOLOGIES FOR SMART CITIES

Geospatial technologies are rapidly evolving, and new research directions are constantly emerging. While there are multiple directions for future research, there is a need to focus on advanced methodologies and innovative applications. As cities grow and get increasingly interconnected, the need for innovative geospatial tools to manage resources, optimize infrastructure, and promote sustainability has never been greater.

4.1. Advanced AI Techniques and Models for Geospatial Data Processing

1. **Responsible GeoAI**: Responsible Urban Geospatial AI practices will improve AI usage while minimizing potential negative consequences, adding to urban sustainability and equity (Marasinghe et al., 2024).
2. **Geospatial XAI**: This is the branch of AI systems that apply Explainable Artificial Intelligence (XAI) to geospatial data. It makes AI communication to the end users understandable (Roussel & Böhm, 2023). Similarly, a graph convolutional network (GCN) and a graph-based explainable AI (XAI) method can be combined. It results in building an explainable spatially explicit GeoAI-based analytical method (P. Liu et al., 2024).
3. **AI Integration with GIS Platforms**: GIS platforms allow the integration of AI models directly into them. Geospatial analysis workflow will become smooth. ML now powers several GeoAI platforms, such as ESRI's ArcGIS. Users can perform advanced analyses without extensive programming. Similarly, several open-source libraries are available. Libraries allow the development of custom AI solutions for geospatial data processing. The libraries allow data to be fed to an ML/DL to provide further insights. Two examples of this are:
    a. GeoPandas (famous for dealing with vector data)
    b. Rasterio (famous for dealing with raster data)

4. **Spatial Regression Models**: These models account for the spatial autocorrelation in geospatial data. Spatial autocorrelation is where data points close to each other are more likely to be similar. Hence, the models help in tasks such as predicting the value of a property based on its location and the surrounding environment.
5. **Hierarchical models**: These models capture the hierarchical structure of geospatial data. So, they can perform tasks such as modeling the relationships between different levels of a geographic hierarchy. For example, the models can explain the relationship between states, counties, and cities.
6. **Prioritization models**: Hyperspectral sensors placed on satellites, drones, or aircraft scan large geographic areas. This hybrid process combines spectroscopy and imaging and generates feature-rich images (Bhargava et al., 2024). They collect detailed information in each pixel across many wavelengths. Data volumes can easily reach terabytes depending on the resolution, the number of bands, and the size of the area being imaged. Managing and analyzing these large datasets often requires advanced storage solutions and powerful computational tools. Big Data and Cloud solutions address the storage aspects of this challenge. GCNs and DL tools can help schedule or prioritize which data segment needs to be processed first in image classification (Hong et al., 2021).

4.2. Smart City Modelling with 3D and 4D Data

Incorporating 3D spatial data (e.g., LiDAR point clouds) and 4D data (3D plus time) are becoming increasingly available. They present ample research opportunities to develop methods for processing and analyzing this data.

1. **3D CNNs and PointNet Architectures:** These models process 3D data directly and address the challenges of analyzing three-dimensional structures in urban environments. Unlike traditional 2D models, 3D CNNs can learn spatial hierarchies from volumetric data. Hence, they are more accurate with 3D object recognition. This feature fits for applications such as identifying buildings, roads, and other urban infrastructure. These models are influential in urban morphology analysis. Urban morphology studies physical shapes, forms, and spatial arrangement of urban elements. It can be buildings, roads, and open spaces. The architectures allow automated extraction of patterns related to city layout, building shapes, and spatial distribution.

PointNet is a neural network architecture. It operates directly on point clouds such as those generated by LiDAR sensors (Bai et al., 2020). By using PointNet, there is no need to turn the points into a regular grid structure like in 3D CNNs.

It enhances this by efficiently handling unordered point sets without voxelization. Two key benefits of doing so are:

a. **Surface classification**: *Identifying the types of surfaces present in the urban environment (e.g., roads, buildings, vegetation).*

b. **Volumetric change detection**: *Detecting structure changes, such as construction progress or structural degradation over time.*

These capabilities enable monitoring construction progress, assessing environmental impacts, and detecting structural changes in real-time, contributing to more efficient urban planning and management.

2. **Dynamic Modeling:** Dynamic modeling using AI techniques is critical in capturing and analyzing temporal changes in 3D structures, offering robust solutions for various Smart city applications. These models depend on spatio-temporal neural networks and recurrent architectures like LSTMs (Long Short-Term Memory). They allow for the continuous monitoring and prediction of structural transformations over time. In the context of construction monitoring, AI-powered dynamic models enable real-time tracking of building progress, ensuring adherence to design specifications and timelines. They can also detect deviations, delays, or irregularities that could lead to inefficiencies or safety hazards.

Additional research opportunities in Smart city modeling with 3D and 4D data include:

1. **3D Generative Adversarial Networks (3DGANs)**: GANs can generate realistic synthetic 3D data to train AI models and test new algorithms.
2. **3D Scene Understanding**: AI techniques can develop a deeper understanding of 3D scenes, including the relationships between objects and the spatial layout of the environment.
3. **4D City Modelling**: 4D city models capture the temporal dimension of urban environments, enabling the analysis of changes over time. AI techniques can develop new methods for creating, updating, and analyzing 4D city models.
4. **Urban Digital Twins (UDTs)**: UDTs provide a dynamic, real-time representation of urban environments. UDTs facilitate continuously updating 3D city models with temporal data from IoT sensors, enabling real-time monitoring and predictive analytics.
5. **Spatiotemporal data infrastructure**: They consolidate diverse data sources, leading to the accuracy and entirety of urban simulations.

6. **Game engines:** Game engines such as Unity and Unreal Engine allow working in interactive 3D and 4D (3D plus time) environments (Buyukdemircioglu & Kocaman, 2022).
7. **Web-based visualization platforms:** Web-browser-based visualization platforms support advanced urban planning and decision-making processes with collaboration (Murshed et al., 2018).
8. **Voxel-based representations and octree structures**: Researchers can consider advanced voxel-based representations and octree structures for efficient storage and processing of massive point clouds.
9. **Other considerations**: Adaptive algorithms are needed to update 3D city models in real-time, incorporating dynamic elements like traffic flow and pedestrian movement. Research should address the challenges of data uncertainty and incompleteness in 4D modeling, developing robust methods for interpolation and extrapolation of temporal urban dynamics.

4.3. Data Security and Privacy

Cybersecurity is vital for safe living, especially in digitally intensive Smart cities. As Smart cities increasingly rely on geospatial data, developing new methods for data security and privacy are required because:

a. Application developers should ensure that citizen data is always protected. Data includes location, transportation patterns, energy usage, and social interactions.
b. Promote trust and adoption.
c. Compliance with Data Protection Regulations
d. Security from cybercrimes, including security breaches and privacy violations
e. Geospatial Data Sensitivity to mask individual movements, routines, or private locations
f. Enabling Safe Data Sharing and Collaboration amongst stakeholders
g. Mitigating the Risk of Surveillance Overreach
h. Enhancing Data Utility While Preserving Privacy
i. Preventing De-anonymization

Research can focus on developing new encryption techniques, access control mechanisms, and data auditing methods. AI-driven cybersecurity safeguards IoT and smart city infrastructures from advanced cyber threats. The tools continuously adapt and evolve, ensuring a secure and resilient digital environment for urban areas (Sarker, 2024). Geospatial Blockchain provides some future research possibilities. Here are a few considerations:

1. One key area is the development of scalable and efficient consensus mechanisms tailored for large-scale geospatial datasets. Addressing the computational overhead and transaction speed becomes essential as geospatial data grows in volume.
2. Blockchain Smart contracts can automate spatial data transactions and securely handle them. They enable decentralized data-sharing frameworks.
3. Research should explore how blockchain-based geospatial data infrastructure can optimize for storage, retrieval, and interoperability with existing GIS platforms.
4. Privacy-preserving techniques such as zero-knowledge proofs and homomorphic encryption are crucial for smart cities to protect sensitive location data.
5. Standardized protocols for geospatial blockchain implementations will support wider adoption in Smart cities, fostering collaboration across multiple sectors while ensuring transparency, security, and data integrity. This research will lay the groundwork for more secure, privacy-aware geospatial systems in Smart urban environments.

4.4. Real-Time Geospatial Analytics and Decision-Making Systems

New and further research is called for on real-time geospatial analytics at various stages - data processing, visualization, and decision-making. ML/DL-based geospatial AI algorithms need optimization for high-velocity spatiotemporal data from IoT, mobile, and satellite sources. Timely availability and accurate decision-making are crucial to making an impact. Edge computing architectures can reduce latency and enable real-time responses. Its applications will be in dynamic urban events such as traffic congestion, natural disasters, and public safety issues. Another promising area is integrating augmented reality to enhance the visualization of real-time geospatial insights. City planners and decision-makers can interact with spatial data in new ways. Improving spatial data fusion techniques will allow for the smooth real-time combination of heterogeneous data sources and efficient and proactive city management.

5. CONCLUSION

Integrating geospatial technologies with computing technologies has led to several significant transformative opportunities for Smart cities to offer improved services to their citizens. As can be seen, the development of SCs coincided with the development of urbanization and the availability of technologies. Fortunately, city administrators and SC planners were able to catch up with the trend. Cities like

London, Singapore, Los Angeles, and New York are already using the technologies and reaping the benefits. Five possible areas where SCs can benefit have been highlighted. These include enhanced urban planning and land use management, environmental monitoring and sustainability, traffic and transportation, citizen engagement, public safety, and social good. Future research can significantly improve Geospatial AI. Researchers can be on advances in AI techniques, 3D/4D modeling, and data privacy-related aspects. The convergence of these technologies suggests supporting long-term sustainability and resilience, thereby improving the quality of life in urban areas. Practitioners should ensure these technologies are implemented inclusively, ethically, and aligned.

REFERENCES

Al-qudah, R., Khamayseh, Y., Aldwairi, M., & Khan, S. (2022). The Smart in Smart Cities: A Framework for Image Classification Using Deep Learning. *Sensors (Basel)*, 22(12), 4390. DOI: 10.3390/s22124390 PMID: 35746171

Albahri, A. S., Khaleel, Y. L., Habeeb, M. A., Ismael, R. D., Hameed, Q. A., Deveci, M., Homod, R. Z., Albahri, O. S., Alamoodi, A. H., & Alzubaidi, L. (2024). A systematic review of trustworthy artificial intelligence applications in natural disasters. *Computers & Electrical Engineering*, 118, 109409. DOI: 10.1016/j.compeleceng.2024.109409

Alders, W. (2024). Rural Settlement Dynamics in a Rapidly Urbanizing Landscape: Insights from Satellite Remote Sensing and Archaeological Field Surveys in Zanzibar, Tanzania. *Journal of Field Archaeology*, •••, 1–19. DOI: 10.1080/00934690.2024.2402962

Ang, K. L.-M., Seng, J. K. P., Ngharamike, E., & Ijemaru, G. K. (2022). Emerging Technologies for Smart Cities' Transportation: Geo-Information, Data Analytics and Machine Learning Approaches. *ISPRS International Journal of Geo-Information*, 11(2), 85. DOI: 10.3390/ijgi11020085

Arias-Molinares, D., & García-Palomares, J. C. (2020). The Ws of MaaS: Understanding mobility as a service from a literature review. *IATSS Research*, 44(3), 253–263. DOI: 10.1016/j.iatssr.2020.02.001

Bai, L., Lyu, Y., Xu, X., & Huang, X. (2020). PointNet on FPGA for Real-Time LiDAR Point Cloud Processing. *2020 IEEE International Symposium on Circuits and Systems (ISCAS)*, 1–5. DOI: 10.1109/ISCAS45731.2020.9180841

Balk, D., McPhearson, T., Cook, E. M., Knowlton, K., Maher, N., Marcotullio, P., Matte, T., Moss, R., Ortiz, L., Towers, J., Ventrella, J., & Wagner, G. (2024). NPCC4: Concepts and tools for envisioning New York City's futures. *Annals of the New York Academy of Sciences*, 1539(1), 277–322. DOI: 10.1111/nyas.15121 PMID: 38924595

Bhargava, A., Sachdeva, A., Sharma, K., Alsharif, M. H., Uthansakul, P., & Uthansakul, M. (2024). Hyperspectral imaging and its applications: A review. *Heliyon*, 10(12), e33208. DOI: 10.1016/j.heliyon.2024.e33208 PMID: 39021975

Buyukdemircioglu, M., & Kocaman, S. (2020). Reconstruction and Efficient Visualization of Heterogeneous 3D City Models. *Remote Sensing (Basel)*, 12(13), 2128. DOI: 10.3390/rs12132128

Buyukdemircioglu, M., & Kocaman, S. (2022). Development of a Smart City Concept in Virtual Reality Environment [Application/pdf]. https://doi.org/DOI: 10.3929/ETHZ-B-000557135

Cai, G., Zhang, J., Du, M., Li, C., & Peng, S. (2020). Identification of urban land use efficiency by indicator-SDG 11.3.1. *PLoS One*, 15(12), e0244318. DOI: 10.1371/journal.pone.0244318 PMID: 33370312

Cai, Z., Cvetkovic, V., & Page, J. (2020). How Does ICT Expansion Drive "Smart" Urban Growth? A Case Study of Nanjing, China. *Urban Planning*, 5(1), 129–139. DOI: 10.17645/up.v5i1.2561

Campbell, S. (1999). Planning: Green Cities, Growing Cities, Just Cities? Urban Planning and the Contradictions of Sustainable Development. In *The Earthscan Reader in Sustainable Cities* (1st ed., p. 23). Routledge.

Ceruzzi, P. E. (2021). Satellite Navigation and the Military-Civilian Dilemma: The Geopolitics of GPS and Its Rivals. In *Palgrave Studies in the History of Science and Technology* (pp. 343–367). Palgrave Macmillan UK. DOI: 10.1057/978-1-349-95851-1_13

Channi, H. K., & Kumar, R. (2022). The Role of Smart Sensors in Smart City. In Singh, U., Abraham, A., Kaklauskas, A., & Hong, T.-P. (Eds.), *Smart Sensor Networks* (Vol. 92, pp. 27–48). Springer International Publishing., DOI: 10.1007/978-3-030-77214-7_2

Cong, Y., & Inazumi, S. (2024). Integration of Smart City Technologies with Advanced Predictive Analytics for Geotechnical Investigations. *Smart Cities*, 7(3), 1089–1108. DOI: 10.3390/smartcities7030046

Cressie, N., & Moores, M. T. (2023). Spatial Statistics. In Daya Sagar, B. S., Cheng, Q., McKinley, J., & Agterberg, F. (Eds.), *Encyclopedia of Mathematical Geosciences* (pp. 1362–1373). Springer International Publishing., DOI: 10.1007/978-3-030-85040-1_31

d'Andrimont, R., Yordanov, M., Martinez-Sanchez, L., Eiselt, B., Palmieri, A., Dominici, P., Gallego, J., Reuter, H. I., Joebges, C., Lemoine, G., & Van Der Velde, M. (2020). Harmonised LUCAS in-situ land cover and use database for field surveys from 2006 to 2018 in the European Union. *Scientific Data*, 7(1), 352. DOI: 10.1038/s41597-020-00675-z PMID: 33067440

Danel, T., Spurek, P., Tabor, J., Śmieja, M., Struski, Ł., Słowik, A., & Maziarka, Ł. (2020). Spatial Graph Convolutional Networks. In Yang, H., Pasupa, K., Leung, A. C.-S., Kwok, J. T., Chan, J. H., & King, I. (Eds.), *Neural Information Processing* (Vol. 1333, pp. 668–675). Springer International Publishing., DOI: 10.1007/978-3-030-63823-8_76

Dean, M. D., & Kockelman, K. M. (2021). Spatial variation in shared ride-hail trip demand and factors contributing to sharing: Lessons from Chicago. *Journal of Transport Geography*, 91, 102944. DOI: 10.1016/j.jtrangeo.2020.102944

Ding, Y., Zhao, X., Zhang, Z., Cai, W., & Yang, N. (2022). Graph Sample and Aggregate-Attention Network for Hyperspectral Image Classification. *IEEE Geoscience and Remote Sensing Letters*, 19, 1–5. DOI: 10.1109/LGRS.2021.3062944

El Alaoui, D., Riffi, J., Sabri, A., Aghoutane, B., Yahyaouy, A., & Tairi, H. (2022). Deep GraphSAGE-based recommendation system: Jumping knowledge connections with ordinal aggregation network. *Neural Computing & Applications*, 34(14), 11679–11690. DOI: 10.1007/s00521-022-07059-x

Feng, Y., Ding, L., & Xiao, G. (2023). GeoQAMap—Geographic Question Answering with Maps Leveraging LLM and Open Knowledge Base [Application/pdf]. *LIPIcs, Volume 277, GIScience 2023*, 277, 28:1-28:7. DOI: 10.4230/LIPICS.GISCIENCE.2023.28

Gkontzis, A. F., Kontsiantis, S., Feretzakis, G., & Verykios, V. S. (2024). Enhancing Urban Resilience: Smart City Data Analyses, Forecasts, and Digital Twin Techniques at the Neighborhood Level. *Future Internet*, 16(2), 47. DOI: 10.3390/fi16020047

Gramacki, P., Leśniara, K., Raczycki, K., Woźniak, S., Przymus, M., & Szymański, P. (2023). SRAI: Towards Standardization of Geospatial AI. *Proceedings of the 6th ACM SIGSPATIAL International Workshop on AI for Geographic Knowledge Discovery*, 43–52. DOI: 10.1145/3615886.3627740

Hill, C., Young, M., Blainey, S., Cavazzi, S., Emberson, C., & Sadler, J. (2024). An integrated geospatial data model for active travel infrastructure. *Journal of Transport Geography*, 117, 103889. DOI: 10.1016/j.jtrangeo.2024.103889

Hong, D., Gao, L., Yao, J., Zhang, B., Plaza, A., & Chanussot, J. (2021). Graph Convolutional Networks for Hyperspectral Image Classification. *IEEE Transactions on Geoscience and Remote Sensing*, 59(7), 5966–5978. DOI: 10.1109/TGRS.2020.3015157

Huang, G., Zhang, W., & Xu, D. (2022). How do technology-enabled bike-sharing services improve urban air pollution? Empirical evidence from China. *Journal of Cleaner Production*, 379, 134771. DOI: 10.1016/j.jclepro.2022.134771

Javed, A. R., Shahzad, F., Rehman, S. U., Zikria, Y. B., Razzak, I., Jalil, Z., & Xu, G. (2022). Future smart cities: Requirements, emerging technologies, applications, challenges, and future aspects. *Cities (London, England)*, 129, 103794. DOI: 10.1016/j.cities.2022.103794

Jin, G., Liang, Y., Fang, Y., Shao, Z., Huang, J., Zhang, J., & Zheng, Y. (2024). Spatio-Temporal Graph Neural Networks for Predictive Learning in Urban Computing: A Survey. *IEEE Transactions on Knowledge and Data Engineering*, 36(10), 5388–5408. DOI: 10.1109/TKDE.2023.3333824

Kaginalkar, A., Kumar, S., Gargava, P., Kharkar, N., & Niyogi, D. (2022). SmartAirQ: A Big Data Governance Framework for Urban Air Quality Management in Smart Cities. *Frontiers in Environmental Science*, 10, 785129. DOI: 10.3389/fenvs.2022.785129

Kaginalkar, A., Kumar, S., Gargava, P., & Niyogi, D. (2021). Review of urban computing in air quality management as smart city service: An integrated IoT, AI, and cloud technology perspective. *Urban Climate*, 39, 100972. DOI: 10.1016/j.uclim.2021.100972

Klumbytė, G., & Athanasiadou, L. (2022). Algorithmic Governmentality and Managerial Fascism: The Case of Smart Cities. In *Deleuze and Guattari and Fascism* (pp. 84–104). Edinburgh University Press/Cambridge University Press. https://www.cambridge.org/core/books/abs/deleuze-and-guattari-and-fascism/algorithmic-governmentality-and-managerial-fascism-the-case-of-smart-cities/0B3DF569EEB012786A73B7703A16C458

Kowe, P., Mutanga, O., & Dube, T. (2021). Advancements in the remote sensing of landscape pattern of urban green spaces and vegetation fragmentation. *International Journal of Remote Sensing*, 42(10), 3797–3832. DOI: 10.1080/01431161.2021.1881185

Lee, I.-S. (2021). A Study on Geospatial Information Role in Digital Twin. *Journal of the Korea Academia-Industrial Cooperation Society*, 22(3), 268–278. DOI: 10.5762/KAIS.2021.22.10.268

Liu, C., Fan, C., & Mostafavi, A. (2024). Graph attention networks unveil determinants of intra- and inter-city health disparity. *Urban Informatics*, 3(1), 18. DOI: 10.1007/s44212-024-00049-5

Liu, P., Zhang, Y., & Biljecki, F. (2024). Explainable spatially explicit geospatial artificial intelligence in urban analytics. *Environment and Planning. B, Urban Analytics and City Science*, 51(5), 1104–1123. DOI: 10.1177/23998083231204689

Lv, Z., Shang, W.-L., & Guizani, M. (2022). Impact of Digital Twins and Metaverse on Cities: History, Current Situation, and Application Perspectives. *Applied Sciences (Basel, Switzerland)*, 12(24), 12820. DOI: 10.3390/app122412820

Marasinghe, R., Yigitcanlar, T., Mayere, S., Washington, T., & Limb, M. (2024). Towards Responsible Urban Geospatial AI: Insights From the White and Grey Literatures. *Journal of Geovisualization and Spatial Analysis*, 8(2), 24. DOI: 10.1007/s41651-024-00184-2

Masik, G., Sagan, I., & Scott, J. W. (2021). Smart City strategies and new urban development policies in the Polish context. *Cities (London, England)*, 108, 102970. DOI: 10.1016/j.cities.2020.102970

Mishra, P., & Singh, G. (2023). Enabling Technologies for Sustainable Smart City. In P. Mishra & G. Singh, *Sustainable Smart Cities* (pp. 59–73). Springer International Publishing. DOI: 10.1007/978-3-031-33354-5_3

Murshed, S. M., Al-Hyari, A. M., Wendel, J., & Ansart, L. (2018). Design and Implementation of a 4D Web Application for Analytical Visualization of Smart City Applications. *ISPRS International Journal of Geo-Information*, 7(7), 276. DOI: 10.3390/ijgi7070276

Ogryzek, M., Adamska-Kmieć, D., & Klimach, A. (2020). Sustainable Transport: An Efficient Transportation Network—Case Study. *Sustainability (Basel)*, 12(19), 8274. DOI: 10.3390/su12198274

Ozcelik, A. E. (2024). Blockchain-oriented geospatial architecture model for real-time land registration. *Survey Review*, 56(394), 1–17. DOI: 10.1080/00396265.2022.2156755

Pandya, S., Srivastava, G., Jhaveri, R., Babu, M. R., Bhattacharya, S., Maddikunta, P. K. R., Mastorakis, S., Piran, M., & Gadekallu, T. R. (2023). Federated learning for smart cities: A comprehensive survey. *Sustainable Energy Technologies and Assessments*, 55, 102987. DOI: 10.1016/j.seta.2022.102987

Pang, J., Huang, Y., Xie, Z., Li, J., & Cai, Z. (2021). Collaborative city digital twin for the COVID-19 pandemic: A federated learning solution. *Tsinghua Science and Technology*, 26(5), 759–771. DOI: 10.26599/TST.2021.9010026

Park, J., & Yoo, S. (2023). Evolution of the smart city: Three extensions to governance, sustainability, and decent urbanisation from an ICT-based urban solution. *International Journal of Urban Sciences, 27*(sup1), 10–28. DOI: 10.1080/12265934.2022.2110143

Pérez Del Hoyo, R., Visvizi, A., & Mora, H. (2021). Inclusiveness, safety, resilience, and sustainability in the smart city context. In *Smart Cities and the un SDGs* (pp. 15–28). Elsevier., DOI: 10.1016/B978-0-323-85151-0.00002-6

Postert, P., Wolf, A. E. M., & Schiewe, J. (2022). Integrating Visualization and Interaction Tools for Enhancing Collaboration in Different Public Participation Settings. *ISPRS International Journal of Geo-Information*, 11(3), 156. DOI: 10.3390/ijgi11030156

Reutov, V., Mottaeva, A., Varzin, V., Jallal, M. A. K., Burkaltseva, D., Shepelin, G., Blazhevich, O., Faskhutdinov, A., Trofimova, A., Niyazbekova, S., & Babin, M. (2023). Smart city development in the context of sustainable development and environmental solutions. *E3S Web of Conferences, 402*, 09020. DOI: 10.1051/e3sconf/202340209020

Rosa, L., Silva, F., & Analide, C. (2021). Mobile Networks and Internet of Things Infrastructures to Characterize Smart Human Mobility. *Smart Cities*, 4(2), 894–918. DOI: 10.3390/smartcities4020046

Roussel, C., & Böhm, K. (2023). Geospatial XAI: A Review. *ISPRS International Journal of Geo-Information*, 12(9), 355. DOI: 10.3390/ijgi12090355

Saha, R., Misra, S., & Deb, P. K. (2021). FogFL: Fog-Assisted Federated Learning for Resource-Constrained IoT Devices. *IEEE Internet of Things Journal*, 8(10), 8456–8463. DOI: 10.1109/JIOT.2020.3046509

Sarker, I. H. (2024). AI-Enabled Cybersecurity for IoT and Smart City Applications. In I. H. Sarker, *AI-Driven Cybersecurity and Threat Intelligence* (pp. 121–136). Springer Nature Switzerland. DOI: 10.1007/978-3-031-54497-2_7

Shamsuzzoha, A., Nieminen, J., Piya, S., & Rutledge, K. (2021). Smart city for sustainable environment: A comparison of participatory strategies from Helsinki, Singapore and London. *Cities (London, England)*, 114, 103194. DOI: 10.1016/j.cities.2021.103194

Smékalová, L., & Kučera, F. (2020). Smart City Projects in the Small-Sized Municipalities: Contribution of the Cohesion Policy. *Scientific Papers of the University of Pardubice, Series D. Faculty of Economics and Administration*, 28(2). Advance online publication. DOI: 10.46585/sp28021067

Tannous, H. O., Major, M. D., & Furlan, R. (2021). Accessibility of green spaces in a metropolitan network using space syntax to objectively evaluate the spatial locations of parks and promenades in Doha, State of Qatar. *Urban Forestry & Urban Greening*, 58, 126892. DOI: 10.1016/j.ufug.2020.126892

Tian, S., Kang, L., Xing, X., Tian, J., Fan, C., & Zhang, Y. (2022). A Relation-Augmented Embedded Graph Attention Network for Remote Sensing Object Detection. *IEEE Transactions on Geoscience and Remote Sensing*, 60, 1–18. DOI: 10.1109/TGRS.2021.3073269

Verma, J. P., Sharma, N., Krishnan, S., Gautam, S., & Balas, V. E. (Eds.). (2024). *Green Computing for Sustainable Smart Cities: A Data Analytics Applications Perspective*. CRC PRESS.

Wheeler, S. (2013). *Planning for Sustainability: Creating Livable, Equitable and Ecological Communities* (2nd ed.). Routledge., DOI: 10.4324/9780203134559

Wolniak, . (2023). Wolniak. (2023). Smart mobility in smart city – Copenhagen and Barcelona comparision. *Scientific Papers of Silesian University of Technology Organization and Management Series*, 2023(172). Advance online publication. DOI: 10.29119/1641-3466.2023.172.41

Yin, J., Dong, J., Hamm, N. A. S., Li, Z., Wang, J., Xing, H., & Fu, P. (2021). Integrating remote sensing and geospatial big data for urban land use mapping: A review. *International Journal of Applied Earth Observation and Geoinformation*, 103, 102514. DOI: 10.1016/j.jag.2021.102514

Zhang, P., Yi, W., Song, Y., Thomson, G., Wu, P., & Aghamohammadi, N. (2024). Geospatial learning for large-scale transport infrastructure depth prediction. *International Journal of Applied Earth Observation and Geoinformation*, 132, 103986. DOI: 10.1016/j.jag.2024.103986

Zhang, Y., & Raubal, M. (2022). Street-level traffic flow and context sensing analysis through semantic integration of multi-source geospatial data. *Transactions in GIS*, 26(8), 3330–3348. DOI: 10.1111/tgis.13005

Zhang, Y., Wei, C., He, Z., & Yu, W. (2024). GeoGPT: An assistant for understanding and processing geospatial tasks. *International Journal of Applied Earth Observation and Geoinformation*, 131, 103976. DOI: 10.1016/j.jag.2024.103976

Zhang, Z., Qian, Z., Zhong, T., Chen, M., Zhang, K., Yang, Y., Zhu, R., Zhang, F., Zhang, H., Zhou, F., Yu, J., Zhang, B., Lü, G., & Yan, J. (2022). Vectorized rooftop area data for 90 cities in China. *Scientific Data*, 9(1), 66. DOI: 10.1038/s41597-022-01168-x PMID: 35236863

Zheng, Z., Shen, W., Li, Y., Qin, Y., & Wang, L. (2020). Spatial equity of park green space using KD2SFCA and web map API: A case study of zhengzhou, China. *Applied Geography (Sevenoaks, England)*, 123, 102310. DOI: 10.1016/j.apgeog.2020.102310

Zhu, R., Zhang, X., Kondor, D., Santi, P., & Ratti, C. (2020). Understanding spatio-temporal heterogeneity of bike-sharing and scooter-sharing mobility. *Computers, Environment and Urban Systems*, 81, 101483. DOI: 10.1016/j.compenvurbsys.2020.101483

Zou, X., Yan, Y., Hao, X., Hu, Y., Wen, H., Liu, E., Zhang, J., Li, Y., Li, T., Zheng, Y., & Liang, Y. (2025). Deep learning for cross-domain data fusion in urban computing: Taxonomy, advances, and outlook. *Information Fusion*, 113, 102606. DOI: 10.1016/j.inffus.2024.102606

KEY TERMS AND DEFINITIONS

3D and 4D Modeling: Techniques for creating three-dimensional (spatial) and four-dimensional (spatial + temporal) representations of urban environments. These models are used for advanced urban planning, simulation, and analysis.

Digital Twins: Digital twins are virtual representations of physical objects, processes, environments, or systems of the real world. They are created using data collected from sensors, IoT devices, and other sources. Digital twins allow for simulation, visualization, and analysis of complex systems, providing insights for decision-making, optimization, and predictive maintenance. In the context of smart cities, it often refers to a detailed digital replica of urban environments used for simulation, analysis, and planning.

Edge Computing: A distributed computing system that brings computation and data storage closer to the data sources, i.e., to the location where it is needed to improve response times and save bandwidth. In Smart Cities, Edge computing processes geospatial data from IoT devices more efficiently, reducing latency and bandwidth usage. In Smart Cities, edge computing is used for processing data from IoT devices locally, enabling faster decision-making and reducing network congestion.

Geoinformation and Communication Technology (GeoICT): GeoICT is the use of information and communication technologies (ICT) in the context of geographic information science (GIS) and systems. It involves the integration of GIS with sensors, networks, mobile devices, and other technologies. The technologies support data collection, visualization, analysis, and decision-making in spatial applications.

Geospatial Blockchain: Integrating blockchain technology (a distributed ledger technology) with geospatial data and systems. It enhances data security, privacy, and integrity in Smart city applications. It is used in smart cities to create decentralized and transparent systems for data sharing, asset management, and secure voting.

Real-time Analytics: Analyzing data as soon as it becomes available allows immediate insights and decision-making. Real-time analytics in Smart cities is used to monitor traffic, public safety, and urban services, enabling rapid responses to dynamic urban events.

Spatial Analysis: Spatial analysis involves analyzing data with a geographic component. It is a method used to examine geographic patterns to understand spatial relationships and the arrangement of objects across locations. It uses both statistical and GIS techniques. It explores data's spatial distribution, identifies patterns and relationships, and makes inferences about the underlying processes that shape spatial phenomena. Its applications in Smart Cities include urban planning, transportation management, and environmental monitoring.

Spatial Statistics: Spatial statistics combines spatial analysis and statistical methods to analyze data that has a geographic component. It provides techniques for modeling spatial autocorrelation, analyzing spatial patterns, and making statistical inferences about spatial data.

Sustainability: The practice of managing urban growth and resources to meet present needs without compromising the ability of future generations to meet theirs. Smart Cities achieve sustainability through energy efficiency, waste reduction, and green technologies.

Urban Simulations: Urban simulations are computational models that emulate the complex dynamics of cities. They incorporate data on land use, transportation, demographics, and other factors. They create virtual representations of urban environments. Urban simulations can assess the impact of urban planning decisions, explore alternative scenarios, and support decision-making processes.

Chapter 10
Benefits and Opportunities of Geospatial AI Adoption

Sucheta Yambal
Dr. Babasaheb Ambedkar Marathwada University, India

Yashwant Arjunrao Waykar
https://orcid.org/0000-0002-9693-9738
Dr. Babasaheb Ambedkar Marathwada University, India

ABSTRACT

The convergence of artificial intelligence (AI) and geospatial intelligence (GEOINT) is transforming industries such as agriculture, disaster management, urban planning, environmental conservation, and defense. AI's data processing power, combined with GEOINT's geographic insights, is enhancing decision-making and predictive models. Urban planners are using this technology to create smarter cities and optimize infrastructure amid growing urbanization. In agriculture, AI-powered precision farming improves crop yields and food security while reducing environmental impact. In disaster management, the integration enables faster evaluations and better coordination of relief efforts. Despite its potential, challenges like AI biases, ethical concerns, and data privacy issues must be addressed for responsible deployment. Together, AI and GEOINT offer innovative solutions for a resilient, sustainable future.

DOI: 10.4018/979-8-3693-8054-3.ch010

1. PROLOGUE ON GEOGRAPHICAL INTELLIGENCE

With the use of satellite images, geographic information systems (GIS), and data analytics, GEOINT is a rapidly growing subject that offers in-depth understanding of the real world. GEOINT was first created for military and defense applications, but it has now spread to other industries, including agriculture, urban planning, environmental management, and disaster response. The integration of cutting-edge technology such as machine learning and artificial intelligence (AI) with geospatial intelligence has opened up new avenues for situational awareness and data-driven decision making.

1.1 The Geospatial Intelligence Foundations

The gathering and examination of geographic and geographical data is the foundation of geospatial intelligence. Usually, satellite images, aerial photography, and remote sensing technologies are the sources of these data points. To produce intricate maps, models, and visual representations of physical surroundings, they are processed using GIS. In the past, GEOINT has been essential to military operations, giving strategic advantages by monitoring enemy movements, locating vital infrastructure, and mapping terrain Dunbar, 2021 . Nevertheless, GEOINT's uses have grown considerably over time.

1.2 Increasing the Use of GEOINT

While tackling a broad variety of civilian concerns remains a major application, GEOINT has emerged as a critical instrument. Geospatial intelligence, for instance, is utilized in urban planning to control urban expansion and maximize infrastructure development. Geospatial data in real time plays a crucial role in the development of smart cities, better traffic control, and enhanced public services. Using data to increase efficiency and sustainability, cities like Singapore and Barcelona have already included GEOINT into their urban planning strategies Tao et al., 2022 .

Precision farming in agriculture is greatly enhanced by the use of geospatial information. Using satellite imaging and GIS data, farmers can assess crop conditions, keep an eye on the health of their soil, and manage irrigation. This increases farming's sustainability by lowering resource waste and increasing output. According to Jensen and Bolstad (2021), GEOINT assists farmers in anticipating crop yields and tracking environmental factors, enabling them to adapt to weather fluctuations and avert agricultural losses.

The capabilities of geospatial intelligence have been enhanced by the integration of AI and machine learning, becoming it more potent and effective. Artificial intelligence (AI) systems are able to automatically analyze satellite pictures, identifying patterns and abnormalities that are hard for humans to see in real time. According to Zhang et al. (2020), machine learning models demonstrate exceptional efficacy when applied to tasks including environmental change tracking, deforestation monitoring, and land-use categorization.

For example, GEOINT systems driven by artificial intelligence are being used to track deforestation in the Amazon jungle. Machine learning techniques are used to evaluate satellite data and identify illicit forestry activity almost instantly. This makes it possible for authorities and environmental groups to respond quickly to save threatened ecosystems Silva et al., 2021 . Similar to this, AI-enhanced GEOINT is used in disaster management to anticipate and track natural catastrophes like hurricanes, floods, and wildfires, assisting governments and relief agencies in better preparedness and response to catastrophic occurrences Hogan et al., 2023 .

1.3 GEOINT for Disaster Relief

Disaster response is one of the most revolutionary uses of geospatial information. Real-time surveillance of disaster-affected regions is made possible by GEOINT, which provides crucial information on survivor location, damage assessment, and resource allocation. Early warning systems for natural catastrophes may be provided by AI-driven prediction models based on geospatial data, allowing emergency responders time to gather resources and evacuate affected areas Heinzelman & Mitnick, 2022 .

For example, GEOINT supplied vital information for organizing rescue and relief activities following the 2010 Haiti earthquake. Aid agencies were able to determine the extent of the damage and pinpoint locations that need urgent help because to satellite imagery Hogan et al., 2023 . In recent times, geospatial intelligence has been used to track the evolution of wildfires and forecast their spread, aiding firefighters in more efficient containment of the flames in California and Australia Tao et al., 2022 .

1.4 Difficulties and Ethical Issues

Although geospatial intelligence has great potential, it is not without difficulties. Privacy issues are a big worry, especially as more and more GEOINT is being used for monitoring. The capacity to gather precise location information and high-resolution photos might potentially result in misuse, especially if the technology is used without appropriate control or regulation Dunbar, 2021 . Moreover, algorithmic bias is a concern that comes with integrating AI into GEOINT systems. The quality of machine learning models is contingent upon the quality of the training data. The analysis that results from training these models on skewed data may result in biased decision-making. According to Zhang et al. (2020), it is essential that GEOINT specialists create frameworks that guarantee the moral use of these technologies, including openness about the gathering and utilization of data.

2. GEOSPATIAL INTELLIGENCE (GEOINT) COMPONENTS

The field of Geospatial Intelligence (GEOINT) integrates photography, geospatial data, and intelligence information to provide a thorough picture of the physical characteristics of the Earth and the actions that occur there. The intricate area of geospatial intelligence (GEOINT) encompasses the gathering, manipulation, evaluation, and distribution of geographic information, which is subsequently used in diverse domains including emergency preparedness, urban development, ecological preservation, and military affairs. These are the main elements that make up GEOINT.

2.1. Imagery

One of the core elements of GEOINT is imagery. It consists of images or other visual depictions of the Earth's surface taken by various devices, including airplanes, drones, and satellites. One may gather imagery in a variety of ways:

- Electro-optical imagery: Conventional digital camera images that record visible light.
- Infrared imagery: Identifies heat-emitting things, such as cars or power plants, and records the heat signatures of those items. It is especially helpful for monitoring throughout the night.
- Radar imaging: Obscured by clouds or darkness, radar imagery is a detailed representation of the Earth's surface that is obtained via the use of synthetic aperture radar (SAR) devices (Smith et al., 2021).

Different information is provided by each kind of imaging, which adds to our comprehension of the investigated region on several levels. For instance, infrared photography may reveal hidden structures or temperature anomalies, but electro-optical imagery is useful for recognizing things that are apparent.

2.2. Geographic Information

Data that is associated with particular spots on the surface of the Earth is referred to as geospatial data. Numerous pieces of data, including height, coordinates (latitude and longitude), and other location-related information that depicts physical and cultural aspects, might be included in this. Geographic Information Systems (GIS) are often where geospatial data is kept and used to generate intricate maps and models of certain regions. There are several methods for gathering geospatial data:

- Remote sensing: gathering data about the Earth's surface via satellites or airborne platforms such as drones. In order to research regions of interest, remote sensing involves gathering several types of electromagnetic data, such as radar and infrared data.
- GPS (Global Positioning System): Offers accurate geolocation information that is useful for a range of tasks, including tracking, navigation, and mapping.
- Geographic Information Systems (GIS): An organizing principle for collecting, organizing, and interpreting geographical data. To provide a visual depiction of a place, GIS combines many data sources (maps, images, and sensor data) Nguyen & Brown, 2021 .

2.3. Geographical Humanities

By fusing physical geography with social, cultural, political, and economic data, human geography concentrates on the human element of GEOINT. Data on political borders, infrastructure, cultural sites, infrastructure, and other man-made elements are all included. Understanding human geography is crucial for comprehending the context of geographic data because it sheds light on how people interact with and alter their surroundings Garcia & Singh, 2021 . For instance, while examining military movements, GEOINT analysts may also consider population density and cultural trends in addition to topography and infrastructure, since these factors may provide light on possible areas of cooperation or conflict. For homeland security, emergency response, and urban planning, human geography data is crucial.

2.4. Analyzing and Interpreting Data

The actual usefulness of GEOINT is found in the analysis and interpretation of the geographical data that has been gathered. In order to evaluate the data, extract pertinent information, and provide actionable intelligence, GEOINT analysts use a variety of analytical approaches. Typical techniques include the following:

- Pattern recognition: Recognizing patterns or irregularities in geographical data that are repeated, since this might indicate hidden trends or dangers.
- Temporal analysis: tracking changes in geographic characteristics over time is helpful for keeping an eye on military operations, urban growth, and environmental changes.
- Predictive modeling: using geographic data to forecast future occurrences, such as military movements, weather patterns, or traffic jams Silva et al., 2021 . Artificial intelligence (AI) and machine learning (ML) techniques are often used in data analysis to swiftly and effectively manage massive amounts of geographical data. These algorithms are capable of automating processes like object recognition in satellite images and spatial trend prediction based on historical data.

2.5. Data Management and Metadata

Another essential part of GEOINT is metadata, or the information about the actual geographic data. Analysts may confirm the legitimacy, correctness, and applicability of the geographical data by using metadata, which contains details about the location, time, and method of data collection. Metadata may provide important information for assessing the usefulness of a study, such as the resolution of a satellite picture or the duration of a drone flight over a certain area Johnson & Lee, 2020 . Large amounts of geographic data must be stored, arranged, and retrieved using effective data management techniques. In order to guarantee that data is readily available and can be shared across stakeholders across various locations, cloud computing systems are being employed more and more in GEOINT.

2.6. Automation and Artificial Intelligence

Artificial intelligence (AI) and machine learning (ML) are becoming essential elements of current GEOINT due to the growing amount of geospatial data. By automating the examination of geographic data, artificial intelligence (AI) techniques may improve accuracy and shorten the processing time of enormous datasets. AI systems, for instance, may be taught to identify certain items in satellite data, such as cars or buildings, and then identify regions of interest for more research (Nguyen & Brown, 2021).

Furthermore, the relocation of natural hazards or urban expansion are two examples of how changes in geographic characteristics are predicted using AI-based predictive models. AI automation boosts productivity and frees up analysts to work on more difficult jobs like strategic decision-making and result interpretation.

2.7. Reporting and Distribution

The transmission of information and insights to decision-makers, policymakers, or military commanders constitutes the last element of GEOINT. Reports, maps, charts, and visualizations are often used to convey GEOINT results so that stakeholders may comprehend the data and make defensible judgments. A military commander could want a comprehensive terrain map, but a city planner might require a 3D model of an urban region. The structure and level of complexity of the reports vary depending on the target audience (Smith et al., 2021).

Technological developments have enhanced the availability of geospatial information; teams can now cooperate more easily and obtain GEOINT data from any location thanks to interactive maps, real-time updates, and cloud-based sharing platforms.

Figure 1. Showing Components of GEOINT

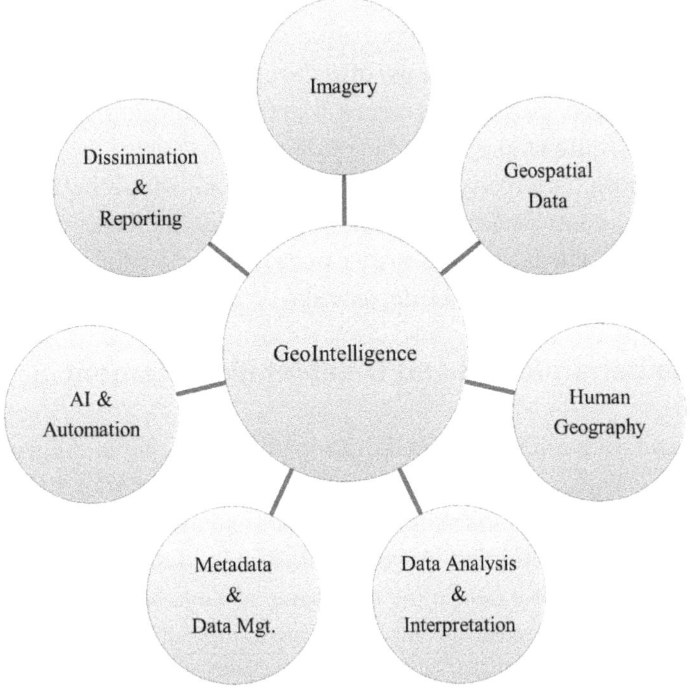

3. GEOSPATIAL INTELLIGENCE APPLICATIONS (GEOINT)

GEOINT, or geospatial intelligence, is the process of gathering, evaluating, and interpreting information about the surface of the Earth. GEOINT, which was once only connected to the military and the defense industry, has spread to a number of industries, including agriculture, transportation, urban planning, disaster relief, and environmental conservation. Geographic information systems (GIS), satellite imaging, and remote sensors are all used by GEOINT to deliver crucial insights that assist decision-making in a variety of businesses. We examine a few of the most important uses of GEOINT across a range of industries below.

3. 1. National Security and Defense

GEOINT has its roots in military and defense activities, where it is still a vital instrument. Military forces can follow troop movements, monitor hotspots throughout the world, and evaluate adversary capabilities with the use of GEOINT. Defense organizations may get real-time information about military sites, weaponry, and topographical characteristics via the use of high-resolution satellite images and remote sensing. Furthermore, GEOINT is essential to the monitoring and reconnaissance of hostile areas (Smith et al., 2021).

By spotting possible terrorist training grounds, monitoring rebel movements, and charting territory controlled by extremist groups, GEOINT aids counterterrorism operations as well. The nation's security can be improved, military actions may be planned, and troop hazards can be decreased.

3.2. Humanitarian Relief and Disaster Management

For humanitarian relief and catastrophe management, geointelligence is essential. Real-time geospatial data gives authorities situational awareness during natural disasters like hurricanes, floods, and earthquakes and aids in assessing the damage and prioritizing relief activities. According to Tao et al. (2022), satellite imaging has the ability to promptly determine the degree of devastation, identify impacted regions, and assess the condition of vital infrastructure such as hospitals, bridges, and highways.

GEOINT can help with disaster planning and risk reduction in addition to supporting immediate disaster response. Governments and organizations may simulate the possible effect of future catastrophes and make plans appropriately by using predictive models that are based on geographical data. As an example, the U.S., in order to map flood zones and help with the creation of evacuation routes, the Federal Emergency Management Agency (FEMA) employs GEOINT Johnson & Lee, 2020 .

3.3. Smart Cities and Urban Planning

The creation of smart cities and contemporary urban planning both depend on geospatial information. To make judgments concerning the expansion of cities, the development of transportation networks, and zoning laws, urban planners employ GEOINT to examine land use patterns, infrastructure, and population density. Planners may visualize possible developments and evaluate their effects on traffic, the environment, and community resources by adding 3D representations of metropolitan regions using GEOINT (Jensen & Bolstad, 2021).

Real-time geospatial data is utilized in smart cities to monitor and control urban processes including trash management, energy use, and public transit. By giving municipal authorities real-time insights into patterns of congestion, GEOINT improves traffic management by allowing them to optimize traffic flow and lower carbon emissions. According to Silva et al. (2021), two of the best-known cities that have included GEOINT into their smart infrastructure projects are Singapore and Dubai.

3.4. Protection of the Environment and Climate Monitoring

Monitoring and protecting the environment is one of the most significant uses of GEOINT. Scientists and environmentalists may monitor changes in land use, deforestation, animal habitats, and biodiversity with the use of GEOINT, which uses satellite images and GIS data. According to Garcia and Singh (2021), this data is essential for overseeing protected areas, keeping an eye on illicit activities like poaching and logging, and assessing how human activity affects ecosystems.

Additionally, GEOINT is essential for tracking and reducing the consequences of climate change. Policymakers may better evaluate the dangers associated with global warming by using geospatial intelligence to analyze data on temperature changes, sea-level rise, and polar ice melt. Furthermore, models powered by GEOINT can forecast the effects of climate change in the future, directing efforts to create mitigation and adaptation plans Nguyen et al., 2022 .

3.5. Food Security and Agriculture

Precision farming methods that increase crop yields while reducing resource consumption are made possible by geospatial information, which is revolutionizing the agricultural industry. Farmers use GEOINT to track crop health, weather trends, and soil conditions in real time. Farmers may optimize their use of water, fertilizers, and pesticides by using satellite data, which offers precise insights regarding soil moisture, nutrient levels, and insect infestations Zhang et al., 2020 .

By offering early warning systems for environmental hazards including droughts, floods, and other situations that might endanger agricultural output, GEOINT also aids in the management of food security. This knowledge is essential for planning agricultural operations and reducing the consequences of unfavorable weather conditions in areas where there is food insecurity. For instance, the World Food Programme (WFP) forecasts food shortages in critical regions and tracks agricultural output using geospatial information (Hogan et al., 2021).

3.6. Logistics and Transportation

GEOINT facilitates supply chain management, infrastructure development, and route optimization in the transportation and logistics industries. Planning road networks, airports, and seaports requires the use of geospatial data to make sure that the transportation infrastructure is built to accommodate future demand. Furthermore, GEOINT analyzes traffic patterns, meteorological data, and real-time road condition information to assist logistics organizations in optimizing delivery routes Smith et al., 2021 .

Geospatial intelligence is also necessary for self-driving cars to traverse highways, identify impediments, and make judgments in real time. These vehicles can operate safely and effectively thanks to AI-powered GEOINT systems, which provide them access to comprehensive maps and environmental data Nguyen et al., 2022 . GEOINT is also used by the aviation and marine sectors for air traffic control, flight route optimization, and shipping channel surveillance.

3.7. Management of Natural Resources and Energy

When it comes to managing natural resources like water, minerals, oil, and gas, geospatial information is crucial. Energy firms use geointelligence (GEOINT) to identify prospective drilling locations, evaluate the ecological consequences of extraction operations, and keep an eye on pipelines and other infrastructure (Johnson & Lee, 2020). Energy firms may prevent environmental harm and make educated choices by using GEOINT's precise maps and geospatial data.

Through data analysis on sunlight exposure, wind patterns, and land availability, GEOINT assists in determining the best sites for energy production facilities in the renewable energy sectors, such as solar and wind power. As a result, resources may be used more effectively, and the production of renewable energy projects can be maximized (Silva et al., 2021).

3.8. Public Health

Additionally useful in monitoring the spread of infectious illnesses and detecting health concerns, GEOINT has shown promise in the field of public health. Geospatial data was utilized during the COVID-19 pandemic to map infection rates, pinpoint hotspots, and forecast the virus's path of propagation. In order to manage the epidemic, public health authorities were able to better allocate resources, enforce quarantine regulations, and make data-driven decisions Garcia & Singh, 2021.

Environmental influences on public health may also be studied with the use of geospatial information. GEOINT, for instance, is used to map air pollution levels and evaluate how they affect respiratory conditions. According to Tao et al. (2022), this data is essential for formulating plans to enhance public health outcomes and reduce the dangers brought on by environmental hazards.

4. TOGETHER, AI AND GEOINT

The revolution in the collection, analysis, and use of geographical data is being brought about by the merger of artificial intelligence (AI) and geospatial intelligence (GEOINT), which is reshaping whole sectors. This dynamic interplay is opening up new opportunities in a variety of fields, including agriculture, urban planning, military, disaster relief, and environmental protection. Although GEOINT has historically been linked to national security and military, the development of AI has broadened its use, allowing for better decision-making, quicker data processing, and predictive analytics across a range of industries. This essay examines how data analysis and operational efficiency may change in the future as a result of AI and GEOINT working together.

4.1 Rudiments of Geographical Intelligence

The gathering and analysis of geographic data using satellites, sensors, and geographic information systems (GIS) constitutes geospatial intelligence. It includes the development of intricate visual and analytical models of real-world settings via the use of remote sensing technologies like radar and aerial photography. GEOINT is an effective tool for many industries since it enables a thorough knowledge and visualization of both natural and man-made settings. Since its initial development for military use, it has been used to a variety of fields, including disaster assistance, agriculture, urban planning, and more Dunbar, 2022.

4.2 AI's Place in Increasing Geographical Intelligence

Through the automation of data processing and analysis, artificial intelligence significantly contributes to the enhancement of GEOINT capabilities. In order to glean insights from raw data, traditional geographic data analysis needed a substantial investment of time and human effort. Artificial intelligence (AI), in particular machine learning (ML) and deep learning algorithms, can now handle enormous volumes of geospatial data and satellite images at previously unheard-of rates. With the use of historical data, AI is now able to identify trends, categorize land use, track environmental changes, and forecast future events Zhang et al., 2021 .

AI-powered systems, for example, may analyze satellite photos in real-time to automatically identify unlawful forestry operations. Artificial intelligence has been used in the Amazon rainforest to detect and monitor deforestation-affected regions and notify authorities to take prompt action Silva et al., 2021 . AI-driven automation has greatly expedited this procedure, which would have been labor-intensive and time-consuming if done by hand.

4.2.1 Calamity Management using AI and GEOINT

The discipline of catastrophe management has benefited greatly from the merging of AI and GEOINT. Conventional tragedy response mostly depended on human evaluations and historical data. Artificial intelligence (AI)-driven GEOINT systems are now making it possible to make decisions during natural catastrophes like hurricanes, floods, wildfires, and earthquakes more quickly and accurately.

Artificial intelligence (AI) models examine real-time geographic data to forecast the course of catastrophes, evaluate the degree of damage, and suggest the best course of action. For instance, AI-based geospatial intelligence systems aided firefighters in tracking fire movement and assessing danger regions during the 2017 California wildfires, allowing for speedier and more efficient responses Tao et al., 2022 . Furthermore, by giving emergency services real-time situational information, GEOINT helps them allocate resources more wisely, minimizing deaths and infrastructure damage.

4.2.2 IT-Based Cities and it's Planning

Additionally, AI-powered geospatial intelligence is essential to the growth of smart cities. Real-time data is essential for smart city resource management, transportation optimization, and public safety. While AI examines traffic patterns, population density, and energy use to influence choices on infrastructure development and urban design, GEOINT offers the geographical foundation for comprehending urban environments Jensen & Bolstad, 2022 .

Urban planners may evaluate urban sprawl, predict future population increase, and maximize land usage with the use of AI-enabled GEOINT technologies. By anticipating infrastructure requirements, these technologies allow for proactive urban planning. AI-driven GEOINT technologies are already being used by cities like Singapore and Amsterdam to improve public services, lower carbon emissions, and manage traffic better Hogan et al., 2023 .

4.2.3 Sustainability and Environmental Conservation

Because AI and GEOINT make it possible to monitor ecosystems and species on a wide scale, they are proving to be very useful in environmental conservation efforts. Real-time geospatial data is becoming crucial for monitoring environmental changes due to the rising danger of habitat loss and climate change. Artificial intelligence systems are capable of analyzing satellite footage to detect illicit mining or logging operations, follow animal movement patterns, and monitor deforestation Silva et al., 2021 .

For instance, in Africa, where poaching of endangered animals like elephants is still a serious problem, AI-powered GEOINT has been used to monitor these species. Conservationists may take immediate action to save fragile species by monitoring migration patterns and identifying hotspots for poaching Zhang et al., 2021 . Furthermore, the repercussions of climate change are predicted by GEOINT-driven AI systems, giving policymakers the information they need to put effective mitigation plans in place.

4.2.4 Innovations in Agriculture and Food Security

Precision farming in agriculture is being revolutionized by the combination of AI and GEOINT. AI-powered geospatial information is now available to farmers, enabling them to monitor soil health, improve crop yields, and use less water. Artificial intelligence (AI) systems can monitor insect infestations, forecast weather, and recommend the best times to plant by evaluating satellite photos and real-time geospatial data Jensen & Bolstad, 2022 .

This invention is especially important in areas where there is a problem with food security. Farmers may adjust their operations and avoid crop loss by using GEOINT-enabled AI systems to foresee environmental hazards such as floods, droughts, and other natural disasters. Furthermore, more environmentally friendly farming methods have been adopted as a result of the capacity to precisely monitor vast agricultural regions, improving food security and lowering the agricultural sector's environmental impact Silva et al., 2021.

4.2.5 Difficulties and Ethical Issues

The combination of artificial intelligence with geographic intelligence has many advantages, but there are drawbacks as well. Privacy issues are becoming more important, especially with the increasing use of high-resolution satellite photography and data collecting. There is increasing worry about the possibility of these technologies being abused for surveillance purposes and to violate people's right to privacy Dunbar, 2022. Legislators must enact rules to guarantee that geospatial data gathering and usage follow moral guidelines.

Algorithmic prejudice is another problem. The quality of AI models, particularly machine learning algorithms, depends on the quality of the training data. The results of AI-driven geographic analysis may be faulty if biased data is employed, which might result in unjust or inaccurate decision-making. To reduce these dangers, it is essential to guarantee AI systems' accountability and transparency Zhang et al., 2021.

5. PROSPECTS FOR THE FUTURE: AI AND GEOINT TOGETHER

The continuing integration of geospatial intelligence with artificial intelligence (AI) and cutting-edge technologies like edge computing, quantum computing, and the Internet of Things (IoT) is key to the field's future. While edge computing will provide quicker analysis by processing data closer to the source, quantum computing promises to greatly boost the speed and accuracy of geographical data processing. The increasing number of IoT devices will provide AI-driven GEOINT systems access to increasingly more real-time data for analysis, hence improving their capacity for making decisions Hogan et al., 2023.

We are only now starting to comprehend the potential of AI and GEOINT to completely transform businesses. Technology's uses will grow as it develops, providing fresh approaches to pressing global issues like urbanization, climate change, and security. Our comprehension of the real world will be improved by the combination of AI and GEOINT, as will our capacity to anticipate and address new threats.

GEOINT's impact on international decision-making is expected to grow as it develops further. It is anticipated that cutting-edge technologies like Internet of Things (IoT), quantum computing, and sophisticated artificial intelligence will improve geospatial capabilities even further. For example, quantum computing may significantly speed up the processing of geographical data, producing even more precise and timely insights Jensen & Bolstad, 2021.

GEOINT will be more crucial than ever in the field of environmental sustainability, helping to manage natural resources, keep an eye on climate change, and preserve biodiversity. Policymakers and environmentalists will need access to vital data from the application of geospatial intelligence in monitoring environmental transformations, such as the melting of polar ice caps or the expansion of deserts Silva et al., 2021 .

Moreover, the need for sustainable agriculture, smart cities, and communities that can withstand natural disasters will propel the ongoing advancement and use of GEOINT. Global geopolitical conflicts, population expansion, and other complex issues will continue to be major global concerns, and geospatial intelligence will play a critical role in driving future solutions.

5.1 Advantages of Combining Geospatial Intelligence (GEOINT) with AI

There are several benefits for a variety of sectors when Artificial Intelligence (AI) and Geospatial Intelligence (GEOINT) are combined. Large datasets, sophisticated computing capacity, and data analysis methods are all used by GEOINT and AI to improve situational awareness and prediction capabilities. Together, they provide a synergy that significantly improves operational effectiveness, prediction accuracy, and decision-making. The main advantages of combining AI with GEOINT are listed below.

5.1.1. Enhanced Efficiency in Data Processing

Efficient processing and analysis of vast amounts of data is one of the main advantages of integrating AI with GEOINT. Massive volumes of data from many sources, such as satellite images, sensor data, maps, and knowledge of human geography, are often used in geospatial intelligence. This data might be laborious and error-prone to handle manually.

Numerous geographic data processing tasks, including anomaly detection, data categorization, and picture identification, may be automated using AI. Much more quickly than human analysts, machine learning algorithms may be taught to interpret satellite data, identify things of interest (such as cars, buildings, or natural features), and detect changes in terrain Smith et al., 2021 . Organizations can handle real-time data streams and swiftly provide meaningful insights thanks to automation, which enables them to make choices faster.

5.1.2. Improved Precision and Accuracy

AI technologies improve the precision and accuracy of GEOINT data analysis, particularly machine learning and deep learning models. With the help of these models, which are made to learn from data patterns and gradually enhance their capabilities, geographical data may be more accurately used for item identification, categorization, and change detection.

Artificial intelligence (AI) algorithms, for instance, are capable of seeing minute patterns in satellite data that human analysts could miss, including the presence of illicit deforestation or military bases that are hidden. AI helps decrease human error and improves the dependability of GEOINT outputs by streamlining the analysis process Nguyen & Brown, 2021 .

Furthermore, by using historical and current geographical data, AI-based predictive models enable analysts to provide very precise predictions. One example of this is the ability to anticipate the possibility of natural catastrophes like earthquakes or floods.

5.1.3. Analyzing and Making Decisions in Real Time

Time is of the importance in vital domains including emergency management, military operations, and catastrophe relief. Real-time data analysis is made possible by the combination of AI with GEOINT, which facilitates quicker decision-making. AI systems are able to handle real-time data from drones, satellite feeds, or distant sensors and provide almost immediate insights into circumstances that are changing quickly.

For example, AI-driven GEOINT systems in disaster management are able to evaluate damage, evaluate incoming data from impacted locations, and provide the best possible response plans in real-time Tao et al., 2022 . First responders are able to prioritize rescue operations and devote resources more efficiently as a result. In a similar vein, military leaders may swiftly prepare strategic reactions and follow enemy movements by using real-time GEOINT.

5.1.4. Predictive Forecasting and Analytics

Artificial intelligence (AI) is a potent tool for predictive analytics because of its capacity to examine past data and identify trends. AI may assist in making very accurate predictions about future events or patterns when used with GEOINT. To forecast the effects of climate change on particular areas, AI models, for instance, might use topographical data, historical weather patterns, and environmental variables. For governments and organizations preparing for potential threats, this may be quite helpful.

AI and GEOINT may be used in agriculture to anticipate drought conditions, identify possible insect outbreaks, and predict crop yields. According to Zhang et al. (2020), these insights assist farmers and agricultural organizations in minimizing risks, improving resource allocation, and scheduling plantings optimally.

Predictive models in urban planning may foresee infrastructure requirements, population increase, and traffic congestion, which enable planners to make well-informed choices that result in more sustainable and productive communities.

5.1.5. Minimizing Expenses and Optimizing Resources

AI-enabled GEOINT systems may decrease operating expenses and the need for human involvement by automating the data collecting, processing, and analysis procedures. Combining AI with GEOINT results in considerable cost reductions in sectors like military, agriculture, and energy where extensive monitoring and surveillance are necessary. Since artificial intelligence (AI) systems can independently evaluate data from distant sources like satellites and drones, organizations no longer need to deploy as many workers or physical resources to acquire information.

AI also makes resource use more effective. AI models, for instance, may identify locations that need conservation efforts in environmental monitoring by analyzing geospatial data; this enables organizations to concentrate their resources on high-priority areas Garcia & Singh, 2021 . AI-powered GEOINT in logistics may improve delivery routes, saving gasoline and enhancing supply chain effectiveness Smith et al., 2021 .

5.1.6. Advanced Pattern and Object Identification

AI is very good at finding patterns and objects in big datasets, which is helpful for GEOINT applications. Artificial intelligence (AI) may be used in defense and security operations to follow the movement of military assets, discover concealed or camouflaged facilities, and identify suspicious activity by analyzing satellite or drone footage.

The ability of AI to recognize images is especially important for border security, since it may be used to detect unauthorized activity or unlawful border crossings over lengthy, difficult-to-monitor borders Nguyen et al., 2022 . Artificial intelligence (AI) may be used in environmental applications to monitor animal populations, identify illicit activities like poaching and logging, and identify changes in forest cover.

5.1.7. Flexibility and Scalability

Scaling across many industries and geographies is one of the advantages of AI-powered GEOINT systems. Artificial Intelligence systems provide adaptability and may be taught to handle geographical data from many businesses and geographies. This facilitates the use of GEOINT solutions in a variety of industries, including public health, energy, transportation, and military.

With very little modifications to the underlying algorithms, AI-based GEOINT technologies employed in military operations, for instance, may be repurposed to monitor urban growth or natural catastrophes. Without having to invest in completely new systems, firms may modify their GEOINT tactics to match changing demands thanks to AI's scalability and flexibility Silva et al., 2021 .

5.1.8. Improved Contextual Awareness

In both static and dynamic situations, the integration of AI with GEOINT improves situational awareness. Effective decision-making in complicated situations, such military operations, border control, or disaster management, requires a thorough awareness of the environment, human activity, and infrastructure. Artificial intelligence (AI) systems continually evaluate geographical data in real-time and provide current insights that assist decision-makers in keeping a clear picture of the situation Smith et al., 2021 .

AI-powered GEOINT systems, for example, can monitor traffic patterns and urban infrastructure in real-time in smart cities, providing insights that enhance safety and city administration. Being attentive of the surroundings is essential for foreseeing issues and averting mishaps or disturbances.

5.2 Cons of using GEOINT and AI together

In the modern world, geospatial intelligence, or GEOINT, has become increasingly significant, influencing everything from urban planning and disaster management to environmental preservation and national security. Fundamentally, GEOINT uses data analytics, satellite imaging, and geographic information to deliver comprehensive geographical insights that help enterprises make quick and accurate decisions. GEOINT is becoming more and more important in tackling complicated issues as the amount of data increases and as global issues like climate change and fast urbanization become more urgent (Harris, 2020).

5.2.1. Strengthening Defense and National Security

GEOINT is used extensively in national security, where it is essential to intelligence and defense activities. GEOINT allows defense agencies to track movements, monitor threats, and defend borders in real time by combining artificial intelligence, satellite imaging, and GPS data. This capacity lowers risks and improves operational success by enabling military operations to be planned with exact spatial information (Garcia & Li, 2021). For example, GEOINT is used in military intelligence to help locate enemies, forecast potential routes, and protect soldiers. Traditional defense mechanisms have been altered by GEOINT's real-time nature, which has made reactions more effective and nimble (Miller et al., 2022).

5.2.2 Urban Planning and Smart Cities

Cities struggle to manage resources, infrastructure, and population expansion as a result of fast urbanization. By giving city planners data-driven insights into resource allocation, transportation patterns, and land use, GEOINT helps to address these problems (Smith & Kline, 2020). For instance, by evaluating traffic flow data, locating areas of high congestion, and facilitating the creation of effective routes, GEOINT tools aid in the optimization of transportation systems.

GEOINT makes it possible to incorporate Internet of Things (IoT) devices that track conditions in real time, improving urban sustainability and efficiency in the creation of smart cities (Silva et al., 2021). To make sure cities are prepared to manage future expansion, encourage energy efficiency, and enhance general quality of life, urban planners use GEOINT (Al-Turjman, F., & Malekloo, A. (2020)).

5.2.3 Agriculture and Food Security

GEOINT's ability to facilitate resource management and precision farming is extremely advantageous to the agriculture industry. By combining satellite imagery with soil and weather data, GEOINT provides farmers with insights into soil health, crop status, and yield potential, helping them make data-driven decisions that enhance productivity and reduce waste (Zhang, 2021). This application of GEOINT is especially important for ensuring food security as global demand for food rises. Precision agriculture, which relies heavily on GEOINT, not only optimizes crop management but also minimizes environmental impacts by reducing the overuse of fertilizers and water resources (Al-Turjman & Malekloo, 2020).

5.2.4 Disaster Management and Resilience Building

Because it gives governments and organizations the ability to anticipate, prepare for, and respond to both natural and man-made disasters, GEOINT has proven crucial in disaster management (Ahmed, R., & Chowdhury, S. (2021)). By tracking environmental conditions and examining past data, GEOINT facilitates the creation of early warning systems in areas vulnerable to storms, earthquakes, and flooding. For instance, authorities were able to predict storm courses and organize evacuations thanks to real-time GEOINT applications during recent storms (Cova & Goodchild, 2021). GEOINT makes response activities more effective and efficient by assisting in damage assessment, safe path determination, and rescue operation coordination both during and after a disaster (Anderson et al., 2020).

5.2.5 Environmental Monitoring and Conservation

GEOINT has emerged as a crucial instrument for environmental monitoring and conservation in the wake of the global climate crisis. GEOINT facilitates monitoring of changes in biodiversity, water quality, land use, and forest cover through remote sensing and spatial data processing (Nguyen & Brown, 2021). GEOINT, for example, is used by conservationists to track deforestation, identify illicit activities such as poaching, and comprehend the dynamics of endangered species' habitats. GEOINT helps establish policies and initiatives that safeguard natural resources and advance sustainable development by facilitating accurate tracking of environmental changes (White, 2020).

5.2.6 Public Health and Epidemiology

Because geographical data enabled authorities to follow the virus's progress, pinpoint infection hotspots, and allocate resources, the COVID-19 pandemic demonstrated the promise of GEOINT in public health (Tao et al., 2022). GEOINT improves response times and permits focused interventions by enabling health agencies to track and forecast disease outbreaks. GEOINT systems, for instance, were utilized to develop spatial models of infection rates and resource requirements during the pandemic, which aided public health organizations in their response (Garcia & Singh, 2021). GEOINT is a potent instrument for handling present and upcoming public health emergencies because of its capacity to evaluate spatial health data.

5.2.7 Ethical Considerations and Challenges

Although GEOINT has many advantages, there are also moral dilemmas, especially with regard to data security and privacy. Concerns regarding data privacy and surveillance are raised by the usage of extremely precise geographic data, which frequently contains sensitive information about people and communities (Nguyen & Brown, 2021). Furthermore, biases in the AI algorithms employed for GEOINT can produce unfair or erroneous results, particularly in areas like public health and law enforcement. For GEOINT to be used properly and to benefit society overall, these ethical concerns must be addressed (Smith et al., 2021).

6. CONCLUSION

Numerous advantages are made possible by the combination of artificial intelligence and geospatial intelligence, including increased accuracy, scalability, real-time decision-making, predictive analytics, cost savings, advanced pattern recognition, and situational awareness. When combined, AI and GEOINT may provide greater insights, quicker reactions, and more efficient solutions for a range of industries, including urban planning, agriculture, and disaster relief in addition to military. The potential for innovation in geospatial intelligence will only grow as AI technologies advance, opening the door to future generations of intelligent and effective decision-makers.

Although AI and GEOINT have enormous potential to solve global issues, their drawbacks must be recognized and resolved. Algorithmic biases, data security threats, privacy and ethical issues, and possible environmental effects are all serious problems that require prompt resolution. Furthermore, the substantial operating costs and dependence on technology underscore the necessity of deploying GEOINT and AI responsibly and under strict regulations. Stakeholders can endeavor to maximize the advantages of these technologies while reducing their disadvantages by tackling these problems, guaranteeing that they have a favorable impact on society.

References:

Ahmed, R., & Chowdhury, S. (2021). Leveraging GEOINT for disaster management and resilience building: A case study of hurricane forecasting. *Disaster Management Journal*, 14(3), 45–58.

Al-Turjman, F., & Malekloo, A. (2020). The role of artificial intelligence in smart cities: A review of geospatial applications. *Journal of Smart City Planning*, 10(1), 15–30.

Anderson, M., Brown, J., & Stevens, L. (2020). The role of geospatial intelligence in disaster response: Lessons learned from recent flood and earthquake events. *Journal of Emergency Response and Planning*, 18(1), 13–28.

Cova, T., & Goodchild, M. (2021). Real-time geospatial intelligence applications in natural disaster management: Insights from the 2020 hurricane season. *Journal of Natural Disaster and Geospatial Solutions*, 12(2), 90–104.

Dunbar, R. (2022). Geospatial intelligence and AI: Revolutionizing data analysis for multiple sectors. *Journal of Geospatial Technology*, 14(3), 45–58.

Garcia, L., & Li, X. (2021). The role of geospatial intelligence in national defense and security: Enhancing border protection and military strategy. *Geospatial Intelligence Review*, 16(4), 132–146.

Garcia, L., & Singh, P. (2021). Human geography in geospatial intelligence: Understanding human-environment interactions. *Journal of Geospatial Humanities*, 4(2), 85–99.

Garcia, L., & Singh, R. (2021). AI-enabled GEOINT in public health: Tracking pandemics and managing healthcare resources. *Public Health Technology Review*, 7(2), 101–115.

Garcia, M., & Singh, R. (2021). Geospatial intelligence in environmental conservation. *Journal of Earth Science*, 12(4), 44–58.

Harris, J. (2020). Geospatial intelligence: The growing importance of spatial data in solving global challenges. *International Journal of Geospatial Research*, 8(3), 119–134.

Heinzelman, J., & Mitnick, L. (2022). AI and GEOINT: Transforming disaster preparedness and response. *Journal of Emergency Management (Weston, Mass.)*, 18(3), 29–40.

Hogan, M., Williams, T., & Patel, A. (2021). Geospatial data in food security: Using GEOINT for agricultural resilience and resource management. *Journal of Agricultural Geospatial Analysis*, 8(2), 58–70.

Hogan, M., Williams, T., & Patel, A. (2023). AI-powered GEOINT for urban planning and smart cities: Innovations in public service and resource management. *Urban Development and Technology Review*, 12(2), 123–136.

Hogan, R., Anderson, J., & Patel, M. (2023). AI-driven GEOINT in environmental and disaster management. *Environmental Data Science Review*, 7(1), 102–118.

Jensen, J., & Bolstad, P. (2021). The role of geospatial intelligence in urban planning and smart city infrastructure. *Journal of Urban Studies and Geospatial Sciences*, 19(4), 112–126.

Jensen, J., & Bolstad, P. (2022). AI and geospatial intelligence in urban planning: Optimizing smart city growth and sustainability. *International Journal of Urban Studies and Smart Technologies*, 8(4), 77–91.

Jensen, M., & Bolstad, P. (2021). Precision agriculture through GEOINT applications: Enhancing crop management. *Agriculture and Technology Innovations*, 5(2), 53–68.

Johnson, K., & Lee, H. (2020). Geospatial intelligence for resource management and defense applications. *Journal of Geo-Intelligence Research*, 10(3), 34–49.

Johnson, K., & Lee, H. (2020). Metadata management in geospatial intelligence: Ensuring data reliability and accuracy. *Journal of Data Science and Intelligence*, 8(3), 47–58.

Miller, P., Johnson, R., & Hart, D. (2022). GEOINT and artificial intelligence in defense: The integration of AI for real-time national security operations. *Military Technology and Intelligence*, 10(1), 47–59.

Nguyen, T., & Brown, M. (2021). Geospatial intelligence and artificial intelligence: The future of predictive analysis. *Journal of Geospatial Technologies*, 15(2), 22–38.

Nguyen, T., & Brown, R. (2021). Geospatial intelligence in environmental conservation: Monitoring ecosystems with advanced geospatial data and AI. *Environmental Monitoring & Management*, 22(4), 75–89.

Nguyen, T., & Brown, R. (2021). GIS and geospatial data systems: Foundations and modern applications. *Geospatial Technology Journal*, 15(1), 32–49.

Nguyen, T., & Brown, R. (2021). Improving precision in geospatial data analysis through AI and deep learning models. *International Journal of Geospatial Intelligence*, 16(1), 48–62.

Nguyen, T., Zhang, Q., & Lee, M. (2022). AI-powered object and pattern recognition in geospatial intelligence for security applications. *Geospatial Security Review*, 18(3), 123–137.

Silva, A., Zhang, Y., & Lee, M. (2021). AI and geospatial intelligence for environmental protection and sustainability. *Environmental Monitoring and AI Research*, 19(1), 102–115.

Silva, A., Zhang, Y., & Lee, M. (2021). AI and GEOINT in environmental conservation: Monitoring ecosystems and biodiversity. *Environmental Monitoring and AI Research*, 19(2), 98–110.

Silva, A., Zhang, Y., & Lee, M. (2021). The role of geospatial intelligence in smart cities: Integrating IoT for urban sustainability. *Journal of Smart City Innovations*, 9(1), 24–38.

Silva, C., Zhang, Y., & Hogan, M. (2021). Geospatial intelligence for smart city development. *Journal of Urban Technology*, 10(3), 27–42.

Smith, A., Patel, D., & Williams, H. (2021). Geospatial intelligence in military operations and national security. *Military Intelligence Journal*, 13(1), 5–20.

Smith, J., & Kline, G. (2020). Urban planning and smart cities: The role of geospatial intelligence in modernizing infrastructure. *Journal of Urban Development*, 13(2), 56–68.

Smith, J., Tan, R., & Chen, H. (2021). Ethical considerations in geospatial intelligence: Data privacy, surveillance, and biases in AI algorithms. *Geospatial Ethics Review*, 11(3), 101–113.

Smith, R., Patel, K., & Williams, S. (2021). Advances in satellite imaging: Radar and infrared technology in geospatial intelligence. *Remote Sensing and Geospatial Intelligence Review*, 6(3), 60–78.

Tao, H., Liu, Y., & Chen, S. (2022). Smart cities and GEOINT integration: Case studies from Singapore and Barcelona. *Urban Systems & Planning Journal*, 15(4), 211–227.

Tao, P., Zhang, Y., & Silva, C. (2022). The impact of geospatial intelligence on disaster response and recovery. *International Journal of Emergency Management*, 11(1), 48–62.

Tao, Q., Garcia, L., & Chen, H. (2022). AI-driven geospatial intelligence for disaster management: Case studies from natural catastrophes. *Journal of Disaster Response and AI Solutions*, 15(3), 53–65.

Tao, Q., Garcia, L., & Chen, H. (2022). Using GEOINT for disaster management and public health: Case studies in real-time geospatial applications. *International Journal of Disaster Response and Public Health*, 20(3), 65–83.

Tao, Q., Garcia, L., & Chen, H. (2022). AI-driven geospatial intelligence in public health: Real-time tracking of disease outbreaks during the COVID-19 pandemic. *Journal of Public Health Informatics*, 18(4), 55–69.

White, R. (2020). Geospatial intelligence in environmental monitoring: Tracking biodiversity and combating climate change. *Environmental Science and GeoData*, 19(2), 102–115.

Zhang, Q. (2021). Geospatial intelligence in precision agriculture: Supporting food security through satellite imagery and AI. *Agricultural Geospatial Intelligence*, 12(3), 67–80.

Zhang, Q., Li, J., & Xu, W. (2020). Machine learning applications in GEOINT: Tracking environmental changes and land use. *Geospatial Technology Journal*, 8(5), 73–89.

Zhang, Q., Nguyen, T., & Brown, R. (2020). Geospatial intelligence in precision agriculture: Enhancing food security and agricultural efficiency. *Agricultural Geointelligence Review*, 6(4), 149–163.

Zhang, Q., Nguyen, T., & Brown, R. (2021). Enhancing agriculture through AI and GEOINT: Precision farming for

Chapter 11
Geospatial AI Future Perspectives

Dina Darwish
Ahram Canadian University, Egypt

ABSTRACT

Geocomputation and geospatial artificial intelligence (GeoAI) play crucial roles in propelling geographic information science (GIS) and Earth observation into a new era. GeoAI has transformed conventional geospatial analysis and mapping, changing the approaches for comprehending and overseeing intricate human–natural systems. Nonetheless, challenges persist in multiple facets of geospatial applications concerning natural, built, and social environments, as well as in the integration of distinctive geospatial features into GeoAI models. At the same time, geospatial and Earth data play essential roles in geocomputation and GeoAI studies, as they can efficiently uncover geospatial patterns, factors, relationships, and decision-making processes. This chapter focuses on several topics related to geospatial AI, including advancements in this field and future perspectives of Geospatial AI.

INTRODUCTION

The fundamental concept of spatial prediction involves estimating the values of a geographic variable at locations that are not known, by utilizing values from known locations or through multivariate data analysis (for further details, refer to Zhu et al. 2018, mentioned under Spatially Explicit AI Models). Spatial interpolation represents a specific form of spatial prediction capability within GIS. Common approaches to spatial interpolation consist of Inverse Distance Weighting (IDW) and Triangulated Irregular Networks (TIN). The innovative application of machine learning and deep learning in spatial prediction encompasses various advancements. These include

DOI: 10.4018/979-8-3693-8054-3.ch011

the development of spatial interpolation methods utilizing conditional generative adversarial neural networks, the interpolation and prediction of activity locations derived from sparsely sampled mobile phone location data, the classification of GPS noise levels through convolutional neural networks for precise distance estimation, the recognition of traffic signs to facilitate traffic rule updates, and the enhancement of trip distribution prediction. Furthermore, numerous human activities take place along road networks. Consequently, the prediction of traffic flows, urban mobility patterns, and crime occurrences over time and space has garnered significant interest (for further details, see Zhang et al. 2019; Zhao et al. 2019; Ren et al. 2020; and Zhang and Cheng 2020).

GeoAI, also known as geospatial artificial intelligence, represents a dynamic field of research that combines advanced AI technologies to address geospatial challenges (Li 2020). Over the last ten years, remarkable advancements have been achieved in the realm of AI, especially in the areas of machine learning and deep learning. The convolutional neural network (CNN) framework represents a significant advancement (Reichstein et al. 2019). The CNN framework utilizes the innovative idea of artificial neural networks (ANN) to create a computer model that simulates the biological neural network of the human brain, while also introducing transformative changes with the implementation of convolution modules (Li et al. 2012; Li 2021; Fukushima 2007; Zhang 1988). These modules are capable of performing information extraction, often referred to as feature extraction, where each feature is considered as the independent variable X in a regression process, directly from the raw data. CNN-based techniques can effectively engage with raw data, revealing concealed patterns through extensive mining and continuous learning processes. This type of analysis, grounded in data, alleviates the limitations found in conventional spatial analytics that rely on established rules or relationships between the input data and the desired outcome. It facilitates the direct discovery and recognition of patterns from the data itself. This is referred to as data-driven discovery (Yuan et al. 2004; Miller and Goodchild 2015). A significant advancement in CNN design is that each convolution layer (Albawi et al. 2017) executes local operations on the data, enabling the possibility of parallel computation. This design alleviates the computational limitations found in conventional artificial neural networks, which heavily rely on the interdependencies among artificial neurons within fully connected layers. The recent advancement of high-speed GPUs, featuring a few hundred to several thousand micro-processing units, enables the efficient training of CNNs, even those with intricate structures, utilizing their computing units in parallel. This also enables a deep learning model to handle large datasets, enhancing its capacity to identify new patterns, gather valuable information, and generate high-quality foundational datasets to support the clarification of significant scientific enquiries (Arundel et al. 2020). Additionally, deep learning models are often more effective at managing noise in training labels compared to conventional statistical methods (Rolnick et al. 2017). Many of these models are crafted to grasp intricate relationships, which often leads them to overfit the training data. Overfitting happens when a model perfectly matches the training data.

In such cases, the model's performance on new data will be subpar. A potential approach involves introducing noise into the training data, which can lead to a less perfect fit for the model. This adjustment helps to lower the chances of overfitting and enhances predictive accuracy. Moreover, approaches like enhancing the batch size to present the model with a greater number of samples for parameter updates throughout the iterative learning process, reducing learning rates to facilitate a more comprehensive search for optimal solutions, and ensuring an adequate supply of accurately labelled samples will empower a deep learning model to effectively address even highly noisy data (Rolnick et al. 2017). While noise in large datasets is unavoidable, the design of deep learning and its data handling capabilities render it more resilient to noise compared to conventional spatial analytical methods. Conversely, deep learning necessitates a vast number of training examples, ranging from thousands to billions, to create abstractions that the human brain can effortlessly grasp through clear, verbal definitions (Marcus 2018). The results' interpretability and the ability to extend beyond the training data's scope present limitations for deep learning systems (Reichstein et al. 2019) that need to be addressed.

BACKGROUND

Recent research highlights the significant potential of AI techniques, particularly deep learning, in the realms of cartographic design and map style transfer. Studies have shown advancements in the detection and extraction of map features, symbols, and texts, as well as in the area of cartographic generalization. Initially, the application of generative adversarial networks (GAN) can be broadened to various mapping scenarios, such as assisting cartographers in analyzing the most prominent stylistic features that define the distinctive appearance and essence of current designs, and leveraging this knowledge to enhance cartographic creations. Additionally, it is essential to maintain the topology of geographic features effectively, and the map symbols and texts might necessitate distinct pattern recognition models from styling to achieve improved results. Ultimately, the combination of AI and cartographic design has the potential to completely or partially streamline the map generalization process.

Recent advancements in deep learning have transformed various fields in both scientific and practical aspects (see LeCun, et al. 2015, cited under Historical Roots and General Overviews). Researchers have discovered that combining spatiotemporal features derived from remote sensing big data with deep learning models enhances our understanding of both data-driven and physical process-based Earth system science, as highlighted by Reichstein et al. (2019) and Zhu et al. (2017). This integration supports a range of Earth observation applications, including land cover and land use classification (Scott et al. 2017; Huang et al. 2018), air quality monitoring (Li et al. 2016).

Geospatial Artificial Intelligence (GeoAI) has shown remarkable potential. When we look at remote sensing data, which has effectively captured the physical attributes of the Earth's surface, we can see how social sensing data enhances this by providing insights into human dynamics and the socioeconomic factors at play. This is achieved through a variety of data sources, including mobile phone data, taxi GPS trajectories, location-based social networks, and social media, as outlined in research (Liu et al. 2015). The process of social sensing entails generating complex semantic data signatures, which include spatial, temporal, and thematic features as outlined by research (Janowicz et al. in 2020), derived from location-based digital traces. A significant portion of social sensing research is based on the idea of place, which encompasses the analysis of place characteristics within geographic contexts (Zhu, et al. 2020) and the extraction of human emotions in various locations through facial expressions (Kang, et al. 2019). As drive-by sensors, computer vision, and deep learning techniques continue to evolve, street-level images emerge as a valuable data source for gaining insights into both physical and social environments. This includes estimating the demographic composition of neighborhoods (Gebru, et al. 2017), exploring human perception of locations through semantically segmented scene elements (Zhang, et al. 2018), and investigating the relationship between street green and blue spaces and geriatric depression (Helbich, et al. 2019). Additionally, the rise of different forms of geospatial big data opens up fresh avenues for social sensing. The integration of various geospatial data through deep learning is an emerging area of research. This includes the merging of remote sensing and social sensing data to identify urban functional regions (Cao, et al. 2020), the combination of street view images with social media check-ins to reveal hidden locations (Zhang, et al. 2020), and the unification of street view images with OpenStreetMap data for the classification of street frontages (Law, et al. 2020).

MAIN FOCUS OF THE CHAPTER

Current Developments and Advancements in GIS Technology

GIS technologies are increasingly becoming prominent in various aspects of our lives. Let's explore the new trends and innovations in GIS. In recent years, geographic information systems (GIS) have advanced significantly, transforming the methods we use to collect, analyze, and visualize data, making them more effective.

Innovations such as drone-based data analysis and machine learning algorithms are creating new opportunities for GIS applications across various industries and sectors. Several advancements are enhancing the accessibility of GIS technology, including:

- Cloud technology
- Mobile Geographic Information Systems
- Artificial Intelligence
- Geographic Information Systems utilizing drones
- Three-dimensional representations and virtual replicas
- Automating processes
- Autonomous vehicles
- Immediate and comprehensive data examination
- Reduction in size of sensors
- AI
- Augmented reality technology

Miniaturization of Sensors

The miniaturization of sensors is propelling the market growth to unprecedented levels. Their compact dimensions have demonstrated remarkable efficiency, affordability, and ease of use. Utilizing these offers a cost-effective and efficient method for gathering real-time data.

AI and ML

Handling disorganized and chaotic data can feel overwhelming, yet the integration of artificial intelligence (AI) and machine learning (ML) allows for the revelation of concealed patterns that might have previously escaped attention. These technologies help to arrange and format data, ensuring it is understandable for users. GIS seeks to improve the present landscape while also working towards a future that is more resilient and sustainable. Consequently, AI and ML are essential for GIS professionals as they enhance data management and understanding, going beyond just the collection of information.

Analysis of Real-Time and Large Data Sets

Real-time data analysis in GIS projects brings together a multitude of data sources, including sensors and satellites. It handles and examines the data, displays it on maps, and offers insights for prompt decision-making. Automation, predictive analytics, and integration with IoT and AI significantly improve its functionalities for superior results.

Autonomous Cars

The trend of driverless vehicles is growing every day. The industry is advancing swiftly, and GIS is a crucial component of this evolution. LiDAR and Radar play a crucial role in their operation, while geospatial details are essential for the navigation of these vehicles.

GIS plays a crucial role in self-driving vehicles by generating detailed high-definition maps, incorporating real-time data from GPS, sensors, and cameras, identifying the best routes, ensuring accurate localization, establishing geofencing boundaries, detecting hazards, supporting remote fleet management, and performing simulation tests for refining algorithms. These GIS applications significantly improve navigation, decision-making, and overall safety in autonomous driving.

Geospatial Artificial Intelligence

Geospatial AI fundamentally employs AI algorithms to examine and make sense of extensive geospatial data, encompassing geographic information like maps, satellite imagery, GPS data, and various other spatial data sources. Utilizing AI techniques like machine learning and deep learning, geospatial AI has the capability to autonomously recognize patterns, observe changes, and forecast future events or outcomes associated with particular locations.

The uses of geospatial AI are varied and extensive. It can be utilized in urban planning to enhance infrastructure development, in environmental monitoring to observe changes in ecosystems, and in disaster response to evaluate the effects of natural disasters and organize relief efforts efficiently. Furthermore, geospatial AI significantly impacts agriculture, transportation, logistics, and marketing by empowering businesses to make informed location-based decisions and improve their operations.

Geographic Information Systems in the Cloud

The main application of GIS cloud computing usually focusses on storage, such as the remote access and analysis of spatial data through desktop GIS software, but the trend of adopting GIS as a service is increasing swiftly.

One significant advantage of adopting cloud technology is the capacity to utilize the economy of scale. By utilizing virtualization, service providers can deliver GIS features to a wide array of users, all leveraging the same hardware while enjoying private instances of the cloud environment. This effective method facilitates the economical provision of GIS services to a wide audience. Figure 1 illustrates GIS data collection, processing, modelling and analysis.

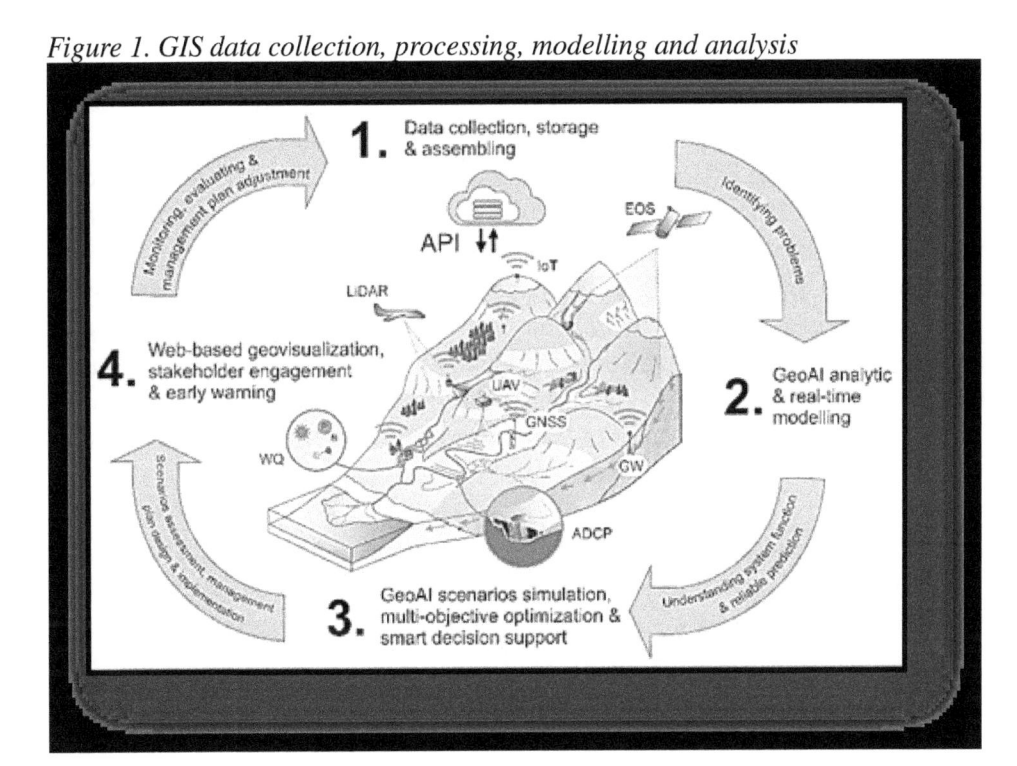

Figure 1. GIS data collection, processing, modelling and analysis

Geospatial Augmented Reality Technology

Geospatial AR technology is a system that merges real-world geographic data with computer-generated information, enriching the user's understanding of their environment. It superimposes digital components, including images, graphics, or data, onto the actual environment using a device such as a smartphone or AR headset. Geospatial AR makes use of GNSS and various location-based technologies to precisely position and align virtual elements with the physical environment.

This technology enables users to engage with and discover their environments in innovative ways. For instance, individuals have the ability to access more details regarding structures or notable sites by directing their device towards them, observe virtual navigation cues superimposed on actual roads while travelling, or visualize subsurface utilities or concealed infrastructure lying beneath the ground. Geospatial AR offers a wide range of applications in fields such as navigation, urban planning, architecture, tourism, gaming, and beyond, serving as an essential tool for improving spatial awareness and informed decision-making.

What Lies Ahead for GIS?

GIS technology significantly impacts our lives in various ways. The outlook for GIS appears to be very promising. A greater number of industries and sectors will become aware of the extensive capabilities of GIS technology and the significant value found in geospatial data. With data analysis, AR, automated devices, and various trends in GIS reaching their pinnacle worldwide, significant transformation is on the horizon for the sectors utilizing these technologies. GIS is significantly contributing to the management of population dynamics, disaster response, emergency situations, and the health sector. With the ongoing evolution of GIS technology, innovative methods for analyzing, visualizing, and managing spatial data will consistently arise.

Exploring the exciting future of GIS technology, there are several noteworthy prospects to consider:

> The integration of Virtual Reality (VR): is anticipated to accelerate alongside GIS, allowing users to engage with spatial data through a range of innovative methods. It is utilized in the realms of urban management and environmental monitoring.
>
> ***The growth of Cloud GIS:*** Cloud GIS, which has already gained significant traction for its broad capabilities, is expected to experience continued growth owing to its rising popularity in the GIS technology landscape.
>
> ***The Growing Importance of Machine Learning:*** With machine learning enhancing convenience and effectiveness for users, its significance is expected to rise in the coming years. GIS seeks to promote the widespread adoption of these technologies and trends to strengthen the world's resilience.

The Reach of GIS is Substantial Across Multiple Sectors

Environmental Management: GIS plays a crucial role in tracking weather patterns, managing natural resources, monitoring wildlife, and assessing biodiversity.

Retail Market: GIS plays a crucial role in examining foot traffic, enhancing store placements, and investigating sales trends within the retail sector.
HR/Recruitment: GIS transforms HR and recruiting through focused talent searches, improved commutes, strategic growth initiatives, streamlined facilities management, customized HR policies, and valuable insights into employee well-being.
Urban Planning: GIS plays a crucial role in examining land use, population demographics, growth trends, utility mapping, and forecasting areas susceptible to disasters.

The ongoing advancement and incorporation of GIS within these sectors emphasize its essential role in fostering progress and ensuring sustainable management.

Reflections on Trends and Innovations in Geospatial Technology

With an understanding of the transformative impact of GIS technologies, it becomes clear that they are propelling advancement and improvement in our world. The emergence of numerous trends and innovations, including Machine Learning, Artificial Intelligence, Cloud GIS, and Automation, aims to enhance our lives and offer greater convenience.

These advancements create exciting possibilities for us, and the market value of GIS is poised to rise significantly in the future. A wide array of industries and sectors has undergone a transformation thanks to the swift embrace of GIS.

As we move forward, the future of GIS appears bright, with its importance set to gain greater recognition across diverse industries and educational institutions. This growing acknowledgement will enhance awareness of its crucial role and significance in the marketplace.

Geospatial Analytics Enhanced by AI: Revolutionizing the Sector

The combination of AI and geospatial data has led to significant advancements in various fields, such as logistics, agriculture, urban planning, and disaster management. Geospatial analytics driven by AI enables businesses to uncover significant patterns from extensive datasets that conventional methods struggle to manage efficiently.

The capacity to analyze and interpret intricate spatial information has transformed decision-making, offering precise predictions and insights that were once out of reach. Heliware offers innovative AI-driven solutions like Heli Mapper and Heli AI, aimed at empowering industries to leverage AI in geospatial applications. Utilizing sophisticated machine learning (ML) algorithms, these tools are capable of handling vast quantities of geospatial data in real time, providing predictive insights that enable businesses to remain nimble and adaptable to evolving circumstances. In the realm of agriculture, asset management in supply chains, and urban development forecasting, geospatial technology driven by AI is emerging as a crucial resource for contemporary businesses.

Significant Developments in AI-Driven Geospatial Technology

1. Processing Data in Real-Time and Analyzing Predictions

A significant trend influencing the future of geospatial technology is the capability to process real-time data. AI algorithms enable the analysis and interpretation of spatial data in real time, resulting in quicker and more informed decision-making. In the logistics sector, AI-driven geospatial tools have the capability to monitor shipments and enhance routes instantly by considering traffic patterns, weather conditions, and various other elements.

Predictive analytics, a remarkable capability facilitated by AI, empowers organizations to foresee future occurrences by examining historical geospatial data. Through the identification of trends and correlations, AI has the capability to predict outcomes like traffic congestion, urban growth, or crop health, allowing for proactive interventions. Heliware's solutions provide predictive analytics that assist businesses in planning for the future, minimizing risks and enhancing operational efficiency.

2. Enhancing the Process of Gathering and Analyzing Geospatial Data Through Automation

The automation of data collection and analysis is a significant trend that is set to gain momentum in the years ahead. Historically, the process of collecting geospatial data has involved significant effort, necessitating field surveys, manual data entry, and intricate processing methods. Nonetheless, tools powered by AI have made this process easier, enabling the automated gathering and examination of spatial data. Unmanned Aerial Vehicles (UAVs) and satellites equipped with AI-powered sensors have the ability to autonomously capture high-resolution geospatial data. These AI systems have the capability to analyze data instantaneously, delivering real-time insights autonomously. This automation greatly minimizes the time and expenses linked to geospatial data collection, while also creating new opportunities for sectors such as agriculture. In this field, AI-enabled drones can autonomously oversee crop health, soil conditions, and water levels.

3. Improved Data Representation and Engaging Technologies

Geospatial technology driven by AI is revolutionizing the visualization and interpretation of data. Conventional 2D maps and charts are being replaced by more sophisticated methods of data visualization, such as 3D models, augmented reality (AR), and virtual reality (VR) environments. These immersive technologies enable users to engage with geospatial data in ways that were once beyond imagination.

Urban planners have the opportunity to utilize AR for layering real-time traffic information onto city maps, whereas construction firms can leverage VR to visualize building designs in relation to their geographical surroundings. Heliware's Heli 3D platform allows users to visualize intricate geospatial data in three dimensions, improving the comprehension of spatial relationships and facilitating more intuitive data-driven decisions.

4. Geospatial AI for Climate and Environmental Oversight

With the rising worries about climate change and the need for environmental sustainability, AI-driven geospatial technology is becoming more crucial in observing and addressing environmental effects. Cutting-edge AI algorithms have the capability to scrutinize satellite imagery, weather data, and environmental sensors to observe deforestation, assess air and water quality, and forecast natural disasters like floods and wildfires.

Heliware's geospatial platforms provide immediate environmental monitoring solutions that enable governments and organizations to take proactive measures in safeguarding the environment. Utilizing AI-driven geospatial data allows decision-makers to more effectively evaluate environmental risks and establish policies that foster sustainability and resilience.

5. Intelligent Urban Areas and Spatial Artificial Intelligence

The idea of smart cities is increasingly becoming prominent, with geospatial technology playing a crucial role in this evolution. Geospatial tools driven by AI play a vital role in the advancement and oversight of smart cities, facilitating real-time monitoring, predictive analytics, and effective resource management. AI-powered geospatial data plays a crucial role in enhancing public transport systems, managing energy usage, and overseeing the health of infrastructure within the smart city framework.

A notable trend is the implementation of digital twins—virtual representations of physical assets and environments—enhanced by AI technology. These digital twins have the capability to replicate city operations, enabling urban planners and city officials to enhance infrastructure, minimize energy usage, and elevate the living standards for residents. Heliware's Heli AI platform is crafted to facilitate the creation of digital twins, equipping cities with essential tools to efficiently manage and enhance their resources.

6. Geospatial AI and Issues Related to Data Privacy

With the increasing prevalence of AI-driven geospatial tools, there is a rising apprehension regarding data privacy and security. Geospatial data frequently includes sensitive information, like the locations of essential infrastructure or individual movement trends. The rise of AI and the collection of real-time data have made the safeguarding of this information a crucial focus.

To address these concerns, new data protection regulations are being put into place, including the Digital Personal Data Protection Bill in India, aimed at ensuring the safety of geospatial data. Organizations need to guarantee that their AI-driven geospatial tools adhere to these regulations to safeguard privacy and uphold public trust.

Envisioning the Path Forward: The Evolution of Geospatial Technology

As we progress, AI-driven geospatial technology will keep advancing, revealing new possibilities and reshaping sectors. Processing real-time data, automating geospatial tasks, and delivering predictive analytics will play crucial roles in this transformation. Moreover, progress in immersive data visualization, environmental monitoring, and smart city development will enhance the significance of geospatial AI in influencing the future.

For businesses aiming to maintain a competitive edge, embracing AI-driven geospatial tools such as those provided by Heliware will be essential. With the increasing accessibility and integration of these technologies across different sectors, the outlook for geospatial technology is bright, as AI takes on a crucial role in fostering innovation and enhancing operational efficiency. Figure 2 illustrates developments in AI-driven Geospatial Technology, and Figure 3 illustrates use of UAVs in collecting spatial data.

Figure 2. Developments in AI-driven Geospatial Technology

Figure 3. Use of UAVs in collecting spatial data

Challenges Facing GeoAI Implementation

1. Creating Models of Dynamic Spatial Settings

There are a number of opportunities and difficulties, many of which are of a geospatial character, that are brought about by the ever-increasing digitalization of ordinary life and the push of artificial intelligence systems out of research laboratories and into our everyday lives. The concept of digital twins, self-driving vehicles, robots in healthcare, retail, and our homes, as well as the fight against global pandemics, all act and occur in (geographic) space, require some understanding of space, and require the need to solve spatial problems. Smart cities and smart homes are examples of these types of technologies. They are all similar in that they function in conditions that are (very) dynamic, with changes that are frequently difficult to completely forecast, a high degree of uncertainty, and that they need interaction (in a broad sense) with individuals who are not experts on the inner workings of the respective systems. The use of spatial simulation approaches, such as cellular automata (Batty, 1997) or agent-based simulations, which may be utilized to research health treatments in a city (Sonnenschein, 2022), amongst other applications, is one way that is particularly effective for modelling such systems.

2. Having Interactions with GeoAI Frameworks

As an alternative to utilizing geoAI to engage with geographic data sources in innovative ways or to model complicated spatial systems, "intelligence" is also required on another level, namely to assist users in interacting with geoAI systems on their own. This is necessary because, despite the fact that geoAI systems have a tendency to replace human abilities, eliminating people totally from the loop has shown to be impossible or is not desired. Moreover, because geoAI models frequently continue to be opaque, it becomes difficult for people to communicate with them. Therefore, human-computer interaction is playing an increasingly important role in the field of geoAI research.

As a result of the proliferation of contemporary machine learning techniques, in particular deep learning, the well-known problems that these techniques bring with them have also been imported. These problems include problems with transparency, explainability, fairness, and other similar difficulties. Due to the fact that many of the concerns that these approaches are used to have far-reaching repercussions, such as in urban planning (Liu and Biljecki, 2022), demography (Liu, 2015), or environmental conservation (Tuia et al., 2022), the issues become extremely important and pertinent. When it comes to geoAI, some writers suggest that explainable artificial intelligence (XAI) approaches cannot be implemented "out of the box" because of the unique characteristics of geo-spatial events and data. Instead, they argue that spatially explicit XAI is necessary (Xing and Sieber, 2021).

Explicitly modelling the methods and the type of data they work on in terms of geo-analytic goals and the accompanying data transformations is one technique that may be taken to make geoAI more explainable. In the present day, human intelligence continues to play a significant role in the collection of information on the origin and quality of data products, as well as the selection of data and processes that are geared towards certain objectives. With that being said, it is necessary to address the goals of geographic information in order to scale up the intelligent use of data across a wide variety of geoinformation sources (Couclelis, 2009).

The current geoAI approaches are not even close to being able to incorporate the goals and processes necessary for automating geo calculation, which continues to be a significant bottleneck. It is possible that geoAI will need to draw on previous research concerning workflow synthesis, service description and composition, and cyberinfrastructures in order to make progress in addressing this difficulty. In addition, geoAI necessitates the possession of pragmatic knowledge in order to effectively manage the information possibilities that are presented in geodata. This is true whether the objective is to retrieve geo-information, automate Geo computational workflows, or answer geo-analytic (indirect) questions.

Case study: Geo-explicit Artificial Intelligence for the Purpose of Social-Economic Distinction in Urban Cities

The rapid urban development process not only changes the scale of the urban spatial organization but also leads to an alteration in the social organization of people's activities, which consequently results in the evident clustering of groups of people sharing similar socio-economic characteristics into particular parts of the cities. These spatial groupings have been recognized for a long time in a variety of urban and economically active cities (Van Liempt, 2011; Walks, 2020). It is for this reason that the social differentiation of places, such as population demographics and segregation, is frequently a result of the urban development and expansion, also known as urbanization.

The examination of such differentiation systems at the urban or national size may frequently be accomplished through the use of geodemographic categorization, socio-economic characterization, and quantitative learning segregation. Those methods are often created to summarize indicators for small areas' (e.g. pre-defined neighborhoods) socio-economic, demographic and built environment characteristics, and have been heavily relied on the clustering methods (e.g. k-means clustering, or shadow (one-layer) neural network (e.g. self-organizing map)). Although there has been some research that has investigated the possibility of using computer vision techniques in conjunction with street view images in order to analyze the social differentiation of urban areas (Gebru et al., 2017), those methods are considered to be a direct import from artificial intelligence to geography. There has been a limited amount of research that has utilized or created spatially explicit methods based on deep learning approaches to such concerns.

Privacy and Ethical Considerations when Dealing with GeoAI

The purpose of this section is to examine the ways in which privacy and ethical considerations are evolving within GeoAI and to highlight some of the most pressing problems that the community ought to solve. Despite the fact that there are several approaches that may be taken to further address privacy concerns within GeoAI, there are three areas that are shown below as beginning points.

- *Privacy that is built in.* Privacy considerations are often a secondary factor in the development of artificial intelligence, despite the fact that there is a considerable body of work on privacy that has been produced by legal experts, policy makers, and ethical AI researchers.

This holds true not only for GeoAI but also for the field of artificial intelligence and other technologies that are connected to it. The concepts of data privacy should be incorporated into the future orientations of GeoAI research from the very beginning, rather than being regarded an afterthought.

In addition, the protection of personal information must to be taken into consideration throughout the entirety of the development process, beginning with the conception stage and continuing through the delivery phase. When it comes to the creation and evaluation of new algorithms that will have an effect on the privacy of people or certain demographic groups, it is important to conduct consultations with experts who have knowledge in privacy and ethics. An example of a potential remedy may be the implementation of privacy effect evaluations (Clarke, 2009) or audits, which are analogous to ethics-based audits (Mokander and Floridi, 2021).

- *The concept of spatial privacy is unique.* Continuing on from the phrase "Spatial is Special," which is frequently used by those who work in the field of geographic information science, there is still a need for a wider recognition within the community of artificial intelligence as to the fact that geographic data are distinct due to the relationship between entity similarity and spatiotemporal proximity. In situations when the privacy of an individual is at risk, this is especially an important consideration. According to research (Griffith, 2018), forgetting about the geographical features of a dataset might have a significant influence on an individual's privacy. When working with geographic data, it is necessary to have a fundamental grasp of geographic concepts such as spatial heterogeneity, autocorrelation, and inference, as well as an awareness of how these ideas may be utilized to either conceal or reveal private information.
- *Incorporating new rules*. Given that data serve as the basis upon which almost all artificial intelligence systems are constructed, it is imperative that access to such data for the development of AI be carefully examined. At the moment, there is a lack of control or transparency regarding the sorts of data that are gathered, the methods that are used to obtain them, and the purposes that are being served by the data.

It is necessary for us to have independent evaluations and rules that are intergovernmental in nature regarding the gathering, storage, and use of data. Although it is a welcome first step, the General Data Protection Regulation (GDPR) that was passed by the European Union has certain flaws. By way of illustration, it is the responsibility of every European nation to conduct investigations on the businesses that are registered inside its borders. This means that a country like Ireland is responsible for regulating a massive percentage of big tech. The amount of fines that have been imposed on those who have violated the General Data Protection Regulation (GDPR) is far fewer than what was anticipated five years ago (Burgess, 2022). In order to guarantee that users of digital platforms have the right to govern the manner in which their data is gathered, kept, and analyzed, more efforts need to be taken. The requirement for such openness is of the utmost importance.

New Trends in GeoAI Field

Direct techniques of exposure assessment, and indirect approaches, such as exposure modelling, are utilized by environmental epidemiologists in order to ascertain the elements to which we may be exposed and, consequently, may have an impact on our health. Exposure modelling is the process of developing a model to reflect a specific environmental variable by making use of a variety of data inputs (such as environmental measurements) and statistical methods (such as land use regression and generalized additive mixed models for example) (Nieuwenhuijsen, 2015). When compared to the use of direct approaches, exposure modelling is a more cost-effective tool for determining the distribution of exposures in research populations that are particularly big (Nieuwenhuijsen, 2015). Basic proximity-based measurements, such as buffers and measured distance, are included in exposure models. More complex modelling techniques, such as kriging, are also included in these models. Environmental epidemiologists have been able to use geographic information system (GIS) technologies to create and link exposure models to health outcome data using geographic variables (for example, geocoded addresses) in order to investigate the effects of factors such as air pollution on the risk of developing diseases such as cardiovascular disease (Nuckols et al., 2004; Art et al., 2015).

This has been made possible by the importance of spatial science in the field of exposure modelling for epidemiologic studies It is possible to use geoAI methods and big data infrastructures, with applications such as Spark and Hadoop, to address the challenges that are associated with exposure modelling in environmental epidemiology. These challenges include inefficiency in computational processing and time (especially when big data are compounded with large geographic study areas), as well as data-related constraints that affect spatial and/or temporal resolution. For instance, previous efforts to model exposure have frequently been associated with coarse spatial resolutions, which has an effect on the degree to which the exposure model is able to accurately estimate individual-level exposure (also known as exposure measurement error) (Nieuwenhuijsen, 2015). This is especially true in terms of high-performance computing to handle big data (big in space and time; spatiotemporal), as well as the development and application of machine learning and deep learning algorithms, as well as big data infrastructures, in order to extract the most meaningful and relevant pieces of input information in order to, for instance, predict the amount of an environmental factor at a particular time and location. GeoAI advancements have made it possible to make accurate, high-resolution exposure modelling for environmental epidemiologic studies.

Future Research Directions in GeoAI Field

GeoAI has not yet been able to tackle a great number of global issues. Because of the ever-increasing amounts of geospatial data, the development of new technical advancements is predicted in the field of geoinformation. To achieve near-human performance with artificial intelligence, however, it appears that the availability of ground-truth data for training is essential. Taking advantage of the huge volume of unlabeled data for the purpose of training unsupervised AI-based systems might result in the opening of further opportunities. In addition, it is necessary to have a comprehensive understanding of the domain challenges, and it is necessary to involve solution owners as well as AI specialists from both academia and industry in the process of designing the GeoAI solutions. In the following, potential research paths that might be pursued by GeoAI are categorized into different groups:

- *Creating models that are spatially and temporally clear.* It is essential to make use of spatiotemporally explicit models and assess the outcomes by integrating both human intelligence and machine intelligence evaluations (Yan et al., 2017). This is necessary in order to achieve a better understanding of the intricate geospatial settings and geographical processes that occur on the ground.

The enhancement of the generalizability of the model within the framework of geography." When it comes to geographic data sets, some regions are always the ones that are gathered. In order to guarantee that the GeoAI models that have been trained with data from a certain geographic region can generalize to other regions, what measures can we take?

- *Taking into account the presence of uncertainty in geographical information and challenges*. Despite the fact that uncertainty is a key notion in geography, deep learning approaches have traditionally not been constructed with data uncertainty in mind. To what extent is it possible for geography to make a contribution to deep learning in a more general sense by inventing ways to infuse models with uncertainty analysis?
- *The creation of new geospatial data infrastructures* that are available to the public. ImageNet (Deng at al., 2009) was a significant contributor to the revolution that took place in the field of computer vision. The investigation

of open and indexable rich geospatial data archives might allow for the development of future GeoAI applications that could benefit from comparable platforms. The BigEarthNet platform (Sumbul et al., 2019) is a wonderful example since it has been shown to be substantially bigger than the archives that are currently available in the field of remote sensing. Additionally, it has been utilized as a diversified training source within the framework of deep learning.

- ***Combining geographical data from several sources in order to uncover new information.*** Innovative geographic knowledge may be made possible by the combination of several geographical datasets at varying spatiotemporal resolutions, which can be accomplished through feature engineering and deep learning.

CONCLUSION

In summary, the trends discussed—real-time data processing, automation, immersive visualization, environmental monitoring, smart cities, and data privacy—illustrate the promising future of AI-driven geospatial technology. By remaining knowledgeable and embracing these advancements, companies can utilize geospatial AI to secure a competitive advantage in a world that is becoming more and more driven by data.

Environmental sustainability involves the careful conservation and management of natural resources to fulfil the needs of the present while ensuring that future generations can also meet their own needs. The objective is to harmonize ecological, social, and economic objectives while ensuring fair access to resources. In recent decades, sustainability has focused on the human-environment interface, the intricate boundary where bio-physical and socio-cultural systems converge. It gathered extensive knowledge about the earth's surface features and natural phenomena. It is essential to explore the advancements and innovations in the ever-evolving field of earth and environmental monitoring. It is crucial to examine and utilize advanced technologies to attain environmental sustainability across various observational scales. Geospatial big data and artificial intelligence (AI) present significant and promising opportunities to address the challenges related to environmental sustainability through sophisticated analytics. It has the capability to facilitate immediate spatial analysis and display of data through maps.

Recent advancements in remote sensing technologies have brought us to an era characterized by spatially explicit big Earth data and Artificial Intelligence. The deployment of multiple satellites equipped with sensors has ushered in a groundbreaking phase in earth observation methods for environmental research. A range of spaceborne and airborne multispectral sensors, UAV-based sensors, along with IoT and social media data, play a significant role in the collection of geospatial big data. In recent decades, there has been a significant increase in the volume, collection, and intricacy of spatial environmental data. Data from various sources can be combined to create a more detailed picture and to tackle real-world environmental and sustainability challenges. Additionally, cloud-based computing platforms that incorporate AI are now accessible for addressing large and intricate computational challenges. In addition to the extensive use of machines, there is a growing fascination with utilizing advanced methods to manage the vast amounts of geospatial big data available. Advanced modelling tools combined with Geospatial information fulfil the need for precise and prompt analysis in environmental monitoring, risk assessment, and strategic decision-making for sustainable development. It facilitates research across various disciplines, offering fresh perspectives on the connections between air, water, soil, food, and energy for a robust society and a sustainable future.

The integration of extensive Earth data with remote sensing and geospatial technologies has gained significant importance for environmental studies. Advanced geospatial data analytics possess the capability to create and implement analytical and precise computer-based data science techniques for assessing and managing the Earth's natural resources, aiming to achieve a sustainable society. This topical collection invites contributions that delve into the realm of geospatial data analytics, particularly in the areas of Earth observation, environmental monitoring, and management. It seeks to enhance our overall grasp of current environmental events and steer us towards sustainable development methods.

REFERENCES

ALBAWI, S., MOHAMMED, T. A., AND AL-ZAWI, S. 2017. UNDERSTANDING OF A CONVOLUTIONAL NEURAL NETWORK. IN 2017 INTERNATIONAL CONFERENCE ON ENGINEERING AND TECHNOLOGY (ICET) (PP. 1-6).

Batty, M., Couclelis, H., & Eichen, M. (1997). Urban systems as cellular automata. *Environment and Planning. B, Planning & Design*, 24(2), 159–164. DOI: 10.1068/b240159

Burgess, M. (2022). *How gdpr is failing*. Wired Magazine.

Cao, R., Tu, W., Yang, C., Li, Q., Liu, J., Zhu, J., Zhang, Q., Li, Q., & Qiu, G. (2020). Deep Learning-Based Remote and Social Sensing Data Fusion for Urban Region Function Recognition. *ISPRS Journal of Photogrammetry and Remote Sensing*, 163, 82–97. DOI: 10.1016/j.isprsjprs.2020.02.014

Clarke, R. (2009). Privacy impact assessment: Its origins and development. *Computer Law & Security Report*, 25(2), 123–135. DOI: 10.1016/j.clsr.2009.02.002

Couclelis, H. (2009). The abduction of geographic information science: Transporting spatial reasoning to the realm of purpose and design. In Hornsby, K. S., Claramunt, C., Denis, M., & Ligozat, G. (Eds.), Lecture Notes in Computer Science: Vol. 5756. *Spatial Information Theory. COSIT 2009* (pp. 342–356). Springer. DOI: 10.1007/978-3-642-03832-7_21

Deng, J., Dong, W., Socher, R., Li, L.-J., Li, K., & Fei-Fei, L. (2009). *Imagenet: A large-scale hierarchical image database. In 2009 IEEE conference on computer vision and pattern recognition*. IEEE.

Fukushima, K. 2007. Neocognitron. *Scholarpedia*, 2(1): 1717. GAO, (2021). 2020 Census: Innovations helped with implementation, but Bureau can do more to realize future benefits. United States Government Accountability Office (GAO). https://www.gao.gov/assets/gao-21-478.pdf

Gebru, T., Krause, J., Wang, Y., Chen, D., Deng, J., Aiden, E. L., & Fei-Fei, L. (2017). Using Deep Learning and Google Street View to Estimate the Demographic Makeup of Neighborhoods across the United States. *Proceedings of the National Academy of Sciences of the United States of America*, 114(50), 13108–13113. DOI: 10.1073/pnas.1700035114 PMID: 29183967

Gebru, T., Krause, J., Wang, Y., Chen, D., Deng, J., Aiden, E. L., & Fei-Fei, L. (2017). Using deep learning and google street view to estimate the demographic makeup of neighborhoods across the United States. *Proceedings of the National Academy of Sciences of the United States of America*, 114(50), 13108–13113. DOI: 10.1073/pnas.1700035114 PMID: 29183967

Griffith, D. A. (2018). Uncertainty and context in geography and giscience: Reflections on spatial autocorrelation, spatial sampling, and health data. *Annals of the American Association of Geographers*, 108(6), 1499–1505. DOI: 10.1080/24694452.2017.1416282

Hart, J. E., Puett, R. C., Rexrode, K. M., Albert, C. M., & Laden, F. (2015). Effect modification of long-term air pollution exposures and the risk of incident cardiovascular disease in US women. *Journal of the American Heart Association*, 4(12), e002301. DOI: 10.1161/JAHA.115.002301 PMID: 26607712

Helbich, M., Yao, Y., Liu, Y., Zhang, J., Liu, P., & Wang, R. (2019). Using Deep Learning to Examine Street View Green and Blue Spaces and Their Associations with Geriatric Depression in Beijing, China. *Environment International*, 126, 107–117. DOI: 10.1016/j.envint.2019.02.013 PMID: 30797100

Huang, B., Zhao, B., & Song, Y. (2018). Urban Land-Use Mapping Using a Deep Convolutional Neural Network with High Spatial Resolution Multispectral Remote Sensing Imagery. *Remote Sensing of Environment*, 214, 73–86. DOI: 10.1016/j.rse.2018.04.050

Janowicz, K., Gao, S., McKenzie, G., Hu, Y., & Bhaduri, B. (2020). GeoAI: Spatially Explicit Artificial Intelligence Techniques for Geographic Knowledge Discovery and Beyond. *International Journal of Geographical Information Science*, 34(4), 625–636. DOI: 10.1080/13658816.2019.1684500

Kang, Y., Jia, Q., Gao, S., Zeng, X., Wang, Y., Angsuesser, S., Liu, Y., Ye, X., & Fei, T. (2019). Extracting Human Emotions at Different Places Based on Facial Expressions and Spatial Clustering Analysis. *Transactions in GIS*, 23(3), 450–480. DOI: 10.1111/tgis.12552

Law, S., Seresinhe, C. I., Shen, Y., & Gutierrez-Roig, M. (2020). Street-Frontage-Net: Urban Image Classification Using Deep Convolutional Neural Networks. *International Journal of Geographical Information Science*, 34(4), 681–707. DOI: 10.1080/13658816.2018.1555832

LeCun, Y., Bengio, Y., & Hinton, G. (2015). Deep Learning. *Nature*, 521(7553), 436–444. DOI: 10.1038/nature14539 PMID: 26017442

Li, W. (2020). GeoAI: Where machine learning and big data converge in GIScience. *Journal of Spatial Information Science*, (20), 71–77. DOI: 10.5311/JOSIS.2020.20.658

Li, W. (2021). GeoAI and Deep Learning. International Encyclopedia of Geography: People, the Earth. *Environmental Technology*, •••, 1–6. DOI: 10.1002/9781118786352.wbieg2083 PMID: 34223810

Li, W., Raskin, R., & Goodchild, M. F. (2012). Semantic similarity measurement based on knowledge mining: An artificial neural net approach. *International Journal of Geographical Information Science*, 26(8), 1415–1435. DOI: 10.1080/13658816.2011.635595

Li, W., Shao, H., Wang, S., Zhou, X., & Wu, S. (2016). A2CI: A cloud-based, service-oriented geospatial cyberinfrastructure to support atmospheric research. In *Cloud Computing in Ocean and Atmospheric Sciences* (pp. 137–161). Academic Press. DOI: 10.1016/B978-0-12-803192-6.00009-8

Liu P, Biljecki F (2022) A review of spatially-explicit GeoAI applications in urban geography. Int J Appl Earth Obs Geoinf. . jag. 2022. 102936DOI: 10. 1016/j

Liu, Y., Liu, X., Gao, S., Gong, L., Kang, C., Zhi, Y., Chi, G., & Shi, L. (2015). Social Sensing: A New Approach to Understanding Our Socioeconomic Environments. *Annals of the Association of American Geographers*, 105(3), 512–530. DOI: 10.1080/00045608.2015.1018773

Liu Y, Liu X, Gao S, Gong L, Kang C, Zhi Y, Chi G, Shi L (2015) Social sensing: A new approach to understanding our socioeconomic environments. Ann Assoc Am Geogr 105(3):512–530. . 2015. 10187 73DOI: 10. 1080/ 00045 608

Marcus, G. (2018). Deep learning: A critical appraisal. 1–27. arXiv preprint arXiv:1801.00631.

Miller, H. J., & Goodchild, M. F. (2015). Data-driven geography. *GeoJournal*, 80(4), 449–461. DOI: 10.1007/s10708-014-9602-6

Mokander, J., & Floridi, L. (2021). Ethics-based auditing to develop trustworthy ai. *Minds and Machines*, 31(2), 323–327. DOI: 10.1007/s11023-021-09557-8

Nieuwenhuijsen, M. J. (2015). *Exposure assessment in environmental epidemiology* (2nd ed.). Oxford University Press. DOI: 10.1093/med/9780199378784.001.0001

Nuckols, J. R., Ward, M. H., & Jarup, L. (2004). Using geographic information systems for exposure assessment in environmental epidemiology studies. *Environmental Health Perspectives*, 112(9), 1007–1015. DOI: 10.1289/ehp.6738 PMID: 15198921

Reichstein, M., Camps-Valls, G., Stevens, B., Jung, M., Denzler, J., Carvalhais, N., & Prabhat, . (2019). Deep learning and process understanding for data-driven Earth system science. *Nature*, 566(7743), 195–204. DOI: 10.1038/s41586-019-0912-1 PMID: 30760912

Ren, Y., Chen, H., Han, Y., Cheng, T., Zhang, Y., & Chen, G. (2020). A Hybrid Integrated Deep Learning Model for the Prediction of Citywide Spatio-Temporal Flow Volumes. *International Journal of Geographical Information Science*, 34(4), 802–823. DOI: 10.1080/13658816.2019.1652303

Rolnick, D., Veit, A., Belongie, S., & Shavit, N. (2017). Deep learning is robust to massive label noise. arXiv preprint arXiv:1705.10694.

Scott, G. J., England, M. R., Starms, W. A., Marcum, R. A., & Davis, C. H. (2017). Training Deep Convolutional Neural Networks for Land-Cover Classification of High-Resolution Imagery. *IEEE Geoscience and Remote Sensing Letters*, 14(4), 549–553. DOI: 10.1109/LGRS.2017.2657778

Sonnenschein, T., Scheider, S., de Wit, G. A., Tonne, C. C., & Vermeulen, R. (2022). Agent-based modeling of urban exposome interventions: Prospects, model architectures, and methodological challenges. *Exposome*, 2(1), osac009. DOI: 10.1093/exposome/osac009 PMID: 37811475

G. Sumbul, M. Charfuelan, B. Demir, and V. Markl. (2019). Bigearthnet: A large-scale benchmark archive for remote sensing image understanding. CoRR, abs/1902.06148.

Tuia, D., Kellenberger, B., Beery, S., Costelloe, B. R., Zuffi, S., Risse, B., Mathis, A., Mathis, M. W., van Langevelde, F., Burghardt, T., Kays, R., Klinck, H., Wikelski, M., Couzin, I. D., van Horn, G., Crofoot, M. C., Stewart, C. V., & Berger-Wolf, T. (2022). Perspectives in machine learning for wildlife conservation. *Nature Communications*, 13(1), 792. Advance online publication. DOI: 10.1038/s41467-022-27980-y PMID: 35140206

Van Liempt, I. (2011). From dutch dispersal to ethnic enclaves in the UK: The relationship between segregation and integration examined through the eyes of somalis. *Urban Studies (Edinburgh, Scotland)*, 48(16), 3385–3398. DOI: 10.1177/0042098010397401

Walks, A. (2020). On the meaning and measurement of the ghetto as a form of segregation. In *Handbook of Urban Segregation*. Edward Elgar Publishing. DOI: 10.4337/9781788115605.00032

Xing, J., & Sieber, R. (2021) Integrating XAI and GeoAI. In: GIScience 2021 Short Paper Proceedings, UC Santa Barbara: Center for Spatial Studies. DOI: 10.25436/ E2301473

Yan, B., Janowicz, K., Mai, G., & Gao, S. (2017). From itdl to place2vec: Reasoning about place type similarity and relatedness by learning embeddings from augmented spatial contexts. In *Proceedings of the 25th ACM SIGSPATIAL International Conference on Advances in Geographic Information Systems*, page 35. ACM. DOI: 10.1145/3139958.3140054

Yuan, M., Buttenfield, B. P., Gahegan, M. N., & Miller, H. (2004). Geospatial data mining and knowledge discovery. In *A Research Agenda for Geographic Information Science* (p. 24). CRC Press. DOI: 10.1201/9781420038330-14

Zhang, F., Wu, L., Zhu, D., & Liu, Y. (2019). Social Sensing from Street-Level Imagery: A Case Study in Learning Spatio-Temporal Urban Mobility Patterns. *ISPRS Journal of Photogrammetry and Remote Sensing*, 153, 48–58. DOI: 10.1016/j.isprsjprs.2019.04.017

Zhang, F., Zhou, B., Liu, L., Liu, Y., Fung, H. H., Lin, H., & Ratti, C. (2018). Measuring Human Perceptions of a Large-Scale Urban Region Using Machine Learning. *Landscape and Urban Planning*, 180, 148–160. DOI: 10.1016/j.landurbplan.2018.08.020

Zhang, W. (1988). Shift-invariant pattern recognition neural network and its optical architecture. *Proceedings of Annual Conference of* the Japan Society of Applied Physics.

Zhang, Y., & Cheng, T. (2020). Graph Deep Learning Model for Network-Based Predictive Hotspot Mapping of Sparse Spatio-Temporal Events. *Computers, Environment and Urban Systems*, 79, 101403. DOI: 10.1016/j.compenvurbsys.2019.101403

Zhang, Y., & Cheng, T. (2020). Graph Deep Learning Model for Network-Based Predictive Hotspot Mapping of Sparse Spatio-Temporal Events. *Computers, Environment and Urban Systems*, 79, 101403. DOI: 10.1016/j.compenvurbsys.2019.101403

Zhao, L., Song, Y., Zhang, C., Liu, Y., Wang, P., Lin, T., Deng, M., & Li, H. (2019). T-GCN: A Temporal Graph Convolutional Network for Traffic Prediction. *IEEE Transactions on Intelligent Transportation Systems*, 21(9), 3848–3858. DOI: 10.1109/TITS.2019.2935152

Zhu, A. X., Lu, G., Liu, J., Qin, C. Z., & Zhou, C. (2018). Spatial Prediction Based on Third Law of Geography. *Annals of GIS*, 24(4), 225–240. DOI: 10.1080/19475683.2018.1534890

Zhu, D., Zhang, F., Wang, S., Wang, Y., Cheng, X., Huang, Z., & Liu, Y. (2020). Understanding Place Characteristics in Geographic Contexts through Graph Convolutional Neural Networks. *Annals of the American Association of Geographers*, 110(2), 408–420. DOI: 10.1080/24694452.2019.1694403

Zhu, X. X., Tuia, D., Mou, L., Xia, G.-S., Zhang, L., Xu, F., & Fraundorfer, F. (2017). Deep Learning in Remote Sensing: A Comprehensive Review and List of Resources. *IEEE Geoscience and Remote Sensing Magazine*, 5(4), 8–36. DOI: 10.1109/MGRS.2017.2762307

KEY TERMS AND DEFINITIONS

Augmented Reality (AR): is the integration of digital information with the user's environment in real time.

Generative adversarial networks (GAN): comprises two deep neural networks—the generator network and the discriminator network.

Global navigation satellite system (GNSS): is a general term describing any satellite constellation that provides positioning, navigation, and timing (PNT) services on a global or regional basis.

Inverse Distance Weighting (IDW): is also known as inverse distance-based weighted interpolation. It is the estimation of the value z at location x by a weighted mean of nearby observations.

Light Detection and Ranging (LiDAR): is a remote sensing method used to examine the surface of the Earth.

Triangulated Irregular Networks (TIN): is a commonly-used data structure in GIS software. It is a standard implementation techniques for digital terrain models, but it can also be used to represent any continuous field.

Virtual Reality (VR): is a simulated experience that employs 3D near-eye displays and pose tracking to give the user an immersive feel of a virtual world.

Compilation of References

Abelha, M., Fernandes, S., Mesquita, D., Seabra, F., & Ferreira-Oliveira, A. T. (2020). Graduate employability and competence development in higher education—A systematic literature review using PRISMA. *Sustainability (Basel)*, 12(15), 5900. DOI: 10.3390/su12155900

Abramov, N., Lankegowda, H., Liu, S., Barazzetti, L., Beltracchi, C., & Ruttico, P. (2024). Implementing Immersive Worlds for Metaverse-Based Participatory Design through Photogrammetry and Blockchain. *ISPRS International Journal of Geo-Information*, 13(6), 211. DOI: 10.3390/ijgi13060211

Ahmad, L., & Nabi, F. (2021). IoT (Internet of things) based agricultural systems. *Agriculture 5.0: Artificial Intelligence, IoT, and Machine Learning*, 69-121. DOI: 10.1201/9781003125433-4

Ahmed, R., & Chowdhury, S. (2021). Leveraging GEOINT for disaster management and resilience building: A case study of hurricane forecasting. *Disaster Management Journal*, 14(3), 45–58.

Ahnaf, M. M., Rafizul, I. M., & Shuvo, M. B. (2023). Development of water quality indices for the assessment of groundwater quality: A case study of tubewells adjacent to the waste landfill in Khulna city. *AIP Conference Proceedings*, 2713, 060002. DOI: 10.1063/5.0129962

Al Jawarneh, I. M., Bellavista, P., Corradi, A., Foschini, L., & Montanari, R. (2022). Efficient Geospatial Analytics on Time Series Big Data. *ICC 2022 - IEEE International Conference on Communications*, 3002–3008. DOI: 10.1109/ICC45855.2022.9839005

Alastal, A. I., & Shaqfa, A. H. (2022). Geoai technologies and their application areas in urban planning and development: Concepts, opportunities and challenges in the smart city (Kuwait, study case). *Journal of Data Analysis and Information Processing*, 10(2), 110–126. DOI: 10.4236/jdaip.2022.102007

Albahri, A. S., Khaleel, Y. L., Habeeb, M. A., Ismael, R. D., Hameed, Q. A., Deveci, M., Homod, R. Z., Albahri, O. S., Alamoodi, A. H., & Alzubaidi, L. (2024). A systematic review of trustworthy artificial intelligence applications in natural disasters. *Computers & Electrical Engineering*, 118, 109409. DOI: 10.1016/j.compeleceng.2024.109409

ALBAWI, S., MOHAMMED, T. A., AND AL-ZAWI, S. 2017. UNDERSTANDING OF A CONVOLUTIONAL NEURAL NETWORK. IN 2017 INTERNATIONAL CONFERENCE ON ENGINEERING AND TECHNOLOGY (ICET) (PP. 1-6).

Alders, W. (2024). Rural Settlement Dynamics in a Rapidly Urbanizing Landscape: Insights from Satellite Remote Sensing and Archaeological Field Surveys in Zanzibar, Tanzania. *Journal of Field Archaeology*, •••, 1–19. DOI: 10.1080/00934690.2024.2402962

Alfieri, L., Bisselink, B., Dottori, F., Naumann, G., Roo, A. D., Salamon, P., & Feyen, L. (2018). Global projections of river flood risk in a warmer world. *Earth's Future*, 6(2), 704–717.

Al-qudah, R., Khamayseh, Y., Aldwairi, M., & Khan, S. (2022). The Smart in Smart Cities: A Framework for Image Classification Using Deep Learning. *Sensors (Basel)*, 22(12), 4390. DOI: 10.3390/s22124390 PMID: 35746171

Al-Turjman, F., & Malekloo, A. (2020). The role of artificial intelligence in smart cities: A review of geospatial applications. *Journal of Smart City Planning*, 10(1), 15–30.

Anand, A., Batra, G., & Uitto, J. I. (2024). Harnessing Geospatial approaches to strengthen evaluative evidence. *Artificial Intelligence and Evaluation*, 196-218. DOI: 10.4324/9781003512493-10

Anand, T., Sinha, S., Mandal, M., Chamola, V., & Yu, F. R. (2021). AgriSegNet: Deep aerial semantic segmentation framework for IoT-assisted precision agriculture. *IEEE Sensors Journal*, 21(16), 17581–17590. DOI: 10.1109/JSEN.2021.3071290

Anderson, M., Brown, J., & Stevens, L. (2020). The role of geospatial intelligence in disaster response: Lessons learned from recent flood and earthquake events. *Journal of Emergency Response and Planning*, 18(1), 13–28.

Ang, K. L.-M., Seng, J. K. P., Ngharamike, E., & Ijemaru, G. K. (2022). Emerging Technologies for Smart Cities' Transportation: Geo-Information, Data Analytics and Machine Learning Approaches. *ISPRS International Journal of Geo-Information*, 11(2), 85. DOI: 10.3390/ijgi11020085

Arellano, L., Alcubilla, P., & Leguízamo, L. (2023). *Ethical considerations in informed consent*. Ethics - Scientific Research, Ethical Issues, Artificial Intelligence and Education. [Working Title], DOI: 10.5772/intechopen.1001319

Arias-Molinares, D., & García-Palomares, J. C. (2020). The Ws of MaaS: Understanding mobility as a service from a literature review. *IATSS Research*, 44(3), 253–263. DOI: 10.1016/j.iatssr.2020.02.001

Arundel, S. T., Li, W., & Wang, S. (2020). GeoNat v1. 0: A dataset for natural feature mapping with artificial intelligence and supervised learning. *Transactions in GIS*, 24(3), 556–572. DOI: 10.1111/tgis.12633

Badidi, E., Mahrez, Z., & Sabir, E. (2020). Fog Computing for Smart Cities' Big Data Management and Analytics: A Review. *Future Internet*, 12(11), 190. DOI: 10.3390/fi12110190

Bai, L., Lyu, Y., Xu, X., & Huang, X. (2020). PointNet on FPGA for Real-Time LiDAR Point Cloud Processing. *2020 IEEE International Symposium on Circuits and Systems (ISCAS)*, 1–5. DOI: 10.1109/ISCAS45731.2020.9180841

Balasubramanian, S. (2024). AI-driven Geospatial Decision Support Systems for Sustainable Development and Natural Resource Management. *International Journal of Artificial Intelligence In Geosciences*, 2(1), 1–11.

Balk, D., McPhearson, T., Cook, E. M., Knowlton, K., Maher, N., Marcotullio, P., Matte, T., Moss, R., Ortiz, L., Towers, J., Ventrella, J., & Wagner, G. (2024). NPCC4: Concepts and tools for envisioning New York City's futures. *Annals of the New York Academy of Sciences*, 1539(1), 277–322. DOI: 10.1111/nyas.15121 PMID: 38924595

Batty, M., Couclelis, H., & Eichen, M. (1997). Urban systems as cellular automata. *Environment and Planning. B, Planning & Design*, 24(2), 159–164. DOI: 10.1068/b240159

Bernini, G., Piscione, P., & Seder, E. (2023). AI-driven service and Slice orchestration. *Shaping the Future of IoT with Edge Intelligence*, 15-36. DOI: 10.1201/9781032632407-3

Bhaduri, B., & Waddell, P. (2018). Advances in spatiotemporal big data analytics for urban planning. *Computers, Environment and Urban Systems*, 69, 1–11. DOI: 10.1016/j.compenvurbsys.2018.03.003

Bhambri, P., & Bajdor, P. (Eds.). (2024a). *Handbook of Technological Sustainability: Innovation and Environmental Awareness* (1st ed., p. 412). CRC Press., DOI: 10.1201/9781003475989

Bhargava, A., Sachdeva, A., Sharma, K., Alsharif, M. H., Uthansakul, P., & Uthansakul, M. (2024). Hyperspectral imaging and its applications: A review. *Heliyon*, 10(12), e33208. DOI: 10.1016/j.heliyon.2024.e33208 PMID: 39021975

Bianchini, D., De Antonellis, V., & Melchiori, M. (2019). AI-based bus scheduling and routing optimization in smart cities: The case of Rome. *IEEE Transactions on Smart Cities*, 1(1), 56–68.

Biljecki, F., Chow, Y. S., & Lee, K. (2023). Quality of crowdsourced geospatial building information: A global assessment of OpenStreetMap attributes. *Building and Environment*, 237, 110295. DOI: 10.1016/j.buildenv.2023.110295

Bisht, S., Bhardwaj, R., & Roy, D. (2024). Optimizing role assignment for scaling innovations through AI in agricultural frameworks: An effective approach. DOI: 10.1016/j.aac.2024.07.004

Blaga, L., Ilie, D. C., Wendt, J. A., Rus, I., Zhu, K., & Dávid, L. D. (2023). Monitoring forest cover dynamics using orthophotos and satellite imagery. *Remote Sensing (Basel)*, 15(12), 3168. DOI: 10.3390/rs15123168

Borisova, B., Semerdzhieva, L., Dimitrov, S., Valchev, S., Iliev, M., & Georgiev, K. (2024). Geospatial Prioritization of Terrains for "Greening" Urban Infrastructure. *Land (Basel)*, 13(9), 1487. DOI: 10.3390/land13091487

Burgess, M. (2022). *How gdpr is failing*. Wired Magazine.

Buyukdemircioglu, M., & Kocaman, S. (2020). Reconstruction and Efficient Visualization of Heterogeneous 3D City Models. *Remote Sensing (Basel)*, 12(13), 2128. DOI: 10.3390/rs12132128

Buyukdemircioglu, M., & Kocaman, S. (2022). Development of a Smart City Concept in Virtual Reality Environment [Application/pdf]. https://doi.org/DOI: 10.3929/ETHZ-B-000557135

Cai, G., Zhang, J., Du, M., Li, C., & Peng, S. (2020). Identification of urban land use efficiency by indicator-SDG 11.3.1. *PLoS One*, 15(12), e0244318. DOI: 10.1371/journal.pone.0244318 PMID: 33370312

Cai, Z., Cvetkovic, V., & Page, J. (2020). How Does ICT Expansion Drive "Smart" Urban Growth? A Case Study of Nanjing, China. *Urban Planning*, 5(1), 129–139. DOI: 10.17645/up.v5i1.2561

Campbell, S. (1999). Planning: Green Cities, Growing Cities, Just Cities? Urban Planning and the Contradictions of Sustainable Development. In *The Earthscan Reader in Sustainable Cities* (1st ed., p. 23). Routledge.

Cao, R., Tu, W., Yang, C., Li, Q., Liu, J., Zhu, J., Zhang, Q., Li, Q., & Qiu, G. (2020). Deep Learning-Based Remote and Social Sensing Data Fusion for Urban Region Function Recognition. *ISPRS Journal of Photogrammetry and Remote Sensing*, 163, 82–97. DOI: 10.1016/j.isprsjprs.2020.02.014

Ceruzzi, P. E. (2021). Satellite Navigation and the Military-Civilian Dilemma: The Geopolitics of GPS and Its Rivals. In *Palgrave Studies in the History of Science and Technology* (pp. 343–367). Palgrave Macmillan UK. DOI: 10.1057/978-1-349-95851-1_13

Chadzynski, A., Krdzavac, N., Farazi, F., Lim, M. Q., Li, S., Grisiute, A., Herthogs, P., von Richthofen, A., Cairns, S., & Kraft, M. (2021). Semantic 3D City Database—An enabler for a dynamic geospatial knowledge graph. *Energy and AI*, 6, 100106. DOI: 10.1016/j.egyai.2021.100106

Chafiq, T., Azmi, R., Fadil, A., & Mohammed, O. (2024). Investigating the Potential of Blockchain Technology for Geospatial Data Sharing: Opportunities, Challenges, and Solutions. *Geomatica*, 100026(2), 100026. Advance online publication. DOI: 10.1016/j.geomat.2024.100026

Chakravarthy, A. S., Sinha, S., Narang, P., Mandal, M., Chamola, V., & Yu, F. R. (2022). DroneSegNet: Robust aerial semantic segmentation for UAV-based IoT applications. *IEEE Transactions on Vehicular Technology*, 71(4), 4277–4286. DOI: 10.1109/TVT.2022.3144358

Chandrashekhar, B. N., Sanjay, H. A., & Geetha, V. (2024). Impact of hybrid [CPU+GPU] HPC infrastructure on AI/ML techniques in industry 4.0. *AI-Driven Digital Twin and Industry*, 4(0), 280–295. DOI: 10.1201/9781003395416-18

Channi, H. K., & Kumar, R. (2022). The Role of Smart Sensors in Smart City. In Singh, U., Abraham, A., Kaklauskas, A., & Hong, T.-P. (Eds.), *Smart Sensor Networks* (Vol. 92, pp. 27–48). Springer International Publishing., DOI: 10.1007/978-3-030-77214-7_2

Chauhan, P. S. Lokendra, & Shekhar, Shashi. (2021). GeoAI–Accelerating a Virtuous Cycle between AI and Geo. 2021 Thirteenth International Conference on Contemporary Computing (IC3-2021), 355–370.

Chen, C., Jiang, L., & Yuan, Z. (2020). Challenges and prospects of AI in smart city traffic management. *Sustainable Cities and Society*, 63, 102–114.

Cheng, Q., Zhao, J., & Wu, H. (2022). Ensemble learning for flood hazard modeling using multi-source remote sensing data. *Remote Sensing of Environment*, 276, 113043.

Chen, L., & Yang, L. (2019). Deep learning for geospatial data analysis: A review. *Remote Sensing*, 11(12), 1432. DOI: 10.3390/rs11121432

Chen, Z., & Sun, Y. (2022). Smart Cities and Traffic Management: Innovations and Future Prospects. *Urban Planning and Development*, 148(3), 123–136.

Chiang, K. (2021, November 24). *Combating Traffic-Related Pollution in the Bay Area*. https://storymaps.arcgis.com/stories/8c1fb9facc774ada87d92e900e421b45

Clarke, R. (2009). Privacy impact assessment: Its origins and development. *Computer Law & Security Report*, 25(2), 123–135. DOI: 10.1016/j.clsr.2009.02.002

Cong, Y., & Inazumi, S. (2024). Integration of Smart City Technologies with Advanced Predictive Analytics for Geotechnical Investigations. *Smart Cities*, 7(3), 1089–1108. DOI: 10.3390/smartcities7030046

Costa, D. G., Bittencourt, J. C. N., Oliveira, F., Peixoto, J. P. J., & Jesus, T. C. (2024). Achieving Sustainable Smart Cities through Geospatial Data-Driven Approaches. *Sustainability (Basel)*, 16(2), 640. DOI: 10.3390/su16020640

Couclelis, H. (2009). The abduction of geographic information science: Transporting spatial reasoning to the realm of purpose and design. In Hornsby, K. S., Claramunt, C., Denis, M., & Ligozat, G. (Eds.), Lecture Notes in Computer Science: Vol. 5756. *Spatial Information Theory. COSIT 2009* (pp. 342–356). Springer. DOI: 10.1007/978-3-642-03832-7_21

Cova, T., & Goodchild, M. (2021). Real-time geospatial intelligence applications in natural disaster management: Insights from the 2020 hurricane season. *Journal of Natural Disaster and Geospatial Solutions*, 12(2), 90–104.

Cressie, N., & Moores, M. T. (2023). Spatial Statistics. In Daya Sagar, B. S., Cheng, Q., McKinley, J., & Agterberg, F. (Eds.), *Encyclopedia of Mathematical Geosciences* (pp. 1362–1373). Springer International Publishing., DOI: 10.1007/978-3-030-85040-1_31

Cugurullo, F. (2020). Urban artificial intelligence: From automation to autonomy in the smart city. *Frontiers in Sustainable Cities*, 2, 38. DOI: 10.3389/frsc.2020.00038

d'Andrimont, R., Yordanov, M., Martinez-Sanchez, L., Eiselt, B., Palmieri, A., Dominici, P., Gallego, J., Reuter, H. I., Joebges, C., Lemoine, G., & Van Der Velde, M. (2020). Harmonised LUCAS in-situ land cover and use database for field surveys from 2006 to 2018 in the European Union. *Scientific Data*, 7(1), 352. DOI: 10.1038/s41597-020-00675-z PMID: 33067440

Danel, T., Spurek, P., Tabor, J., Śmieja, M., Struski, Ł., Słowik, A., & Maziarka, Ł. (2020). Spatial Graph Convolutional Networks. In Yang, H., Pasupa, K., Leung, A. C.-S., Kwok, J. T., Chan, J. H., & King, I. (Eds.), *Neural Information Processing* (Vol. 1333, pp. 668–675). Springer International Publishing., DOI: 10.1007/978-3-030-63823-8_76

Dean, M. D., & Kockelman, K. M. (2021). Spatial variation in shared ride-hail trip demand and factors contributing to sharing: Lessons from Chicago. *Journal of Transport Geography*, 91, 102944. DOI: 10.1016/j.jtrangeo.2020.102944

Deng, J., Dong, W., Socher, R., Li, L.-J., Li, K., & Fei-Fei, L. (2009). *Imagenet: A large-scale hierarchical image database. In 2009 IEEE conference on computer vision and pattern recognition.* IEEE.

Dhananjaya, A. S., Vinayak, B. K., Mahajan, S., & Agrawal, S. (2021). Geospatial Technology in Disaster Management: Harnessing Spatial Intelligence for Effective Preparedness, Response, and Recovery. *International Journal of Open Publication and Exploration*, 9, 36–41.

Dhibar, K., & Maji, P. (2023). Future outlier detection algorithm for smarter industry application using ML and AI. *Advances in Systems Analysis, Software Engineering, and High Performance Computing*, •••, 152–166. DOI: 10.4018/978-1-6684-8785-3.ch008

Dia, F., Bayar, N., & Abdellatif, T. (2024). From Functional Requirements to NoSQL Database Models: Application to IoT Geospatial Data. In Mosbah, M., Kechadi, T., Bellatreche, L., Gargouri, F., Guegan, C. G., Badir, H., Beheshti, A., & Gammoudi, M. M. (Eds.), *Advances in Model and Data Engineering in the Digitalization Era* (Vol. 2071, pp. 224–236). Springer Nature Switzerland., DOI: 10.1007/978-3-031-55729-3_18

Díaz-Díaz, R., Muñoz, L., & Pérez-González, D. (2017). Business model analysis of public services operating in the smart city ecosystem: The case of smart grid and smart mobility. *Future Internet*, 9(3), 24.

Ding, Y., Zhao, X., Zhang, Z., Cai, W., & Yang, N. (2022). Graph Sample and Aggregate-Attention Network for Hyperspectral Image Classification. *IEEE Geoscience and Remote Sensing Letters*, 19, 1–5. DOI: 10.1109/LGRS.2021.3062944

Döllner, J. (2020). Geospatial Artificial Intelligence: Potentials of Machine Learning for 3D Point Clouds and Geospatial Digital Twins. *PFG –. Journal of Photogrammetry, Remote Sensing and Geoinformation Science*, 88(1), 15–24. DOI: 10.1007/s41064-020-00102-3

Dunbar, R. (2022). Geospatial intelligence and AI: Revolutionizing data analysis for multiple sectors. *Journal of Geospatial Technology*, 14(3), 45–58.

El Alaoui, D., Riffi, J., Sabri, A., Aghoutane, B., Yahyaouy, A., & Tairi, H. (2022). Deep GraphSAGE-based recommendation system: Jumping knowledge connections with ordinal aggregation network. *Neural Computing & Applications*, 34(14), 11679–11690. DOI: 10.1007/s00521-022-07059-x

El Baba, M., Kayastha, P., Huysmans, M., & De Smedt, F. (2020). Groundwater vulnerability and nitrate contamination assessment and mapping using drastic and Geostatistical analysis. *Water (Basel)*, 12(7), 2022. DOI: 10.3390/w12072022

Elgarroussi, K., Wang, S., Banerjee, R., & Eick, C. F. (2018). *"Aconcagua: A Novel Spatiotemporal Emotion Change Analysis Framework," Proc. 2Nd ACM SIGSPATIAL Int. Work. AI Geogr. Kwl. Discov., no.* Ccdm. DOI: 10.1145/3281548.3281552

Eseosa Halima, I., & Hiroaki, S. (2022). Assessing the disparities of the population exposed to flood hazards in Nigeria. *IOP Conference Series. Earth and Environmental Science*, 1016(1), 012007. DOI: 10.1088/1755-1315/1016/1/012007

Ethical considerations in artificial intelligence development. (2024). Filosofiya Referativnyi Zhurnal, (1). DOI: 10.31249/rphil/2024.01.03

Ezhilarasan, K., & Jeevarekha, A. (2023). Powering the geothermal energy with AI, ML, and IoT. *Power Systems*, 271-286. DOI: 10.1007/978-3-031-15044-9_13

Feng, Y., Ding, L., & Xiao, G. (2023). GeoQAMap—Geographic Question Answering with Maps Leveraging LLM and Open Knowledge Base [Application/pdf]. *LIPIcs, Volume 277, GIScience 2023, 277*, 28:1-28:7. DOI: 10.4230/LIPICS.GISCIENCE.2023.28

FinancesOnline. (2024). *35 IoT Device Statistics You Must Read: 2024 Data on Market Size, Adoption & Usage*. https://financesonline.com/iot-device-statistics/

Flood prediction in Nigeria using ensemble machine learning techniques. (2023). *Ilorin Journal of Science, 10*(1). DOI: 10.54908/iljs.2023.10.01.004

Fukushima, K. 2007. Neocognitron. *Scholarpedia, 2*(1): 1717. GAO, (2021). 2020 Census: Innovations helped with implementation, but Bureau can do more to realize future benefits. United States Government Accountability Office (GAO). https://www.gao.gov/assets/gao-21-478.pdf

G. Sumbul, M. Charfuelan, B. Demir, and V. Markl. (2019). Bigearthnet: A large-scale benchmark archive for remote sensing image understanding. CoRR, abs/1902.06148.

G. Xi and S. Mei,(2009). "A Deep Residual Network Integrating Spatial-temporal Properties to Predict Influenza Trends at an Intra-urban Scale".

Ganz, S., Adler, P., & Kändler, G. (2020). Forest cover mapping based on a combination of aerial images and Sentinel-2 satellite data compared to National Forest Inventory data. *Forests*, 11(12), 1322. DOI: 10.3390/f11121322

Gao, J., & Li, X. (2023). The Role of Autonomous Vehicles and AI in Future Traffic Management. *Journal of Intelligent Transport Systems*, 27(4), 362–378.

Garcia, L., & Li, X. (2021). The role of geospatial intelligence in national defense and security: Enhancing border protection and military strategy. *Geospatial Intelligence Review*, 16(4), 132–146.

Garcia, L., & Singh, P. (2021). Human geography in geospatial intelligence: Understanding human-environment interactions. *Journal of Geospatial Humanities*, 4(2), 85–99.

Garcia, L., & Singh, R. (2021). AI-enabled GEOINT in public health: Tracking pandemics and managing healthcare resources. *Public Health Technology Review*, 7(2), 101–115.

Garcia, M., & Singh, R. (2021). Geospatial intelligence in environmental conservation. *Journal of Earth Science*, 12(4), 44–58.

Garg, P., Chakravarthy, A. S., Mandal, M., Narang, P., Chamola, V., & Guizani, M. (2021). ISDNet: AI-enabled instance segmentation of aerial scenes for smart cities. *ACM Transactions on Internet Technology*, 21(3), 1–18. DOI: 10.1145/3418205

Gavali, V., Jagtap, A., Jagzap, G., Garud, A., & Digraskar, V. (2024). Pothole Detection a Geospatial Approach to Prioritize Road Repairs. In Tripathi, S. L. (Ed.), *Emerging trends in IoT and Computing Technologies*. CRC Press. DOI: 10.1201/9781003535423-41

Gebru, T., Krause, J., Wang, Y., Chen, D., Deng, J., Aiden, E. L., & Fei-Fei, L. (2017). Using Deep Learning and Google Street View to Estimate the Demographic Makeup of Neighborhoods across the United States. *Proceedings of the National Academy of Sciences of the United States of America*, 114(50), 13108–13113. DOI: 10.1073/pnas.1700035114 PMID: 29183967

Gkontzis, A. F., Kontsiantis, S., Feretzakis, G., & Verykios, V. S. (2024). Enhancing Urban Resilience: Smart City Data Analyses, Forecasts, and Digital Twin Techniques at the Neighborhood Level. *Future Internet*, 16(2), 47. DOI: 10.3390/fi16020047

González, A., Kelly, C., & Rymszewicz, A. (2020). Advancements in web-mapping tools for land use and marine spatial planning. *Transactions in GIS*, 24(2), 253–267. DOI: 10.1111/tgis.12603

Goodall, N. J., Smith, B. L., & Park, B. B. (2020). Traffic signal control with connected and autonomous vehicles: A review of potential benefits and challenges. *Transportation Research Part C, Emerging Technologies*, 106, 1–18.

Gramacki, P., Leśniara, K., Raczycki, K., Woźniak, S., Przymus, M., & Szymański, P. (2023). SRAI: Towards Standardization of Geospatial AI. *Proceedings of the 6th ACM SIGSPATIAL International Workshop on AI for Geographic Knowledge Discovery*, 43–52. DOI: 10.1145/3615886.3627740

Gregory, D., Johnston, R., Pratt, G., Watts, M., & Whatmore, S. (2011). *The dictionary of human geography*. John Wiley & Sons.

Griffith, D. A. (2018). Uncertainty and context in geography and giscience: Reflections on spatial autocorrelation, spatial sampling, and health data. *Annals of the American Association of Geographers*, 108(6), 1499–1505. DOI: 10.1080/24694452.2017.1416282

Gupta, S., & Vyas, S. (2023). Contemporary role of edge-AI in IoT and IoE in healthcare and digital marketing. *Edge-AI in Healthcare*, 75-84. DOI: 10.1201/9781003244592-6

Gupta, M., & Pandya, S. D. (2024). Predictive modeling of recruitment and selection in campus placement using machine learning algorithms. DOI: 10.2139/ssrn.4862743

Gupta, S., Mittal, S., & Agarwal, V. (2022). Identification and Analysis of Challenges and Their Solution in Implementation of Decision Support System (DSS) in Smart Cities. In Gaur, L., Agarwal, V., & Chatterjee, P. (Eds.), *Decision Support Systems for Smart City Applications* (1st ed., pp. 99–118). Wiley., DOI: 10.1002/9781119896951.ch6

Han, Q., Nesi, P., Pantaleo, G., & Paoli, I. (2020). Smart City Dashboards: Design, Development, and Evaluation. *2020 IEEE International Conference on Human-Machine Systems (ICHMS)*, 1–4. DOI: 10.1109/ICHMS49158.2020.9209493

Harris, J. (2020). Geospatial intelligence: The growing importance of spatial data in solving global challenges. *International Journal of Geospatial Research*, 8(3), 119–134.

Hart, J. E., Puett, R. C., Rexrode, K. M., Albert, C. M., & Laden, F. (2015). Effect modification of long-term air pollution exposures and the risk of incident cardiovascular disease in US women. *Journal of the American Heart Association*, 4(12), e002301. DOI: 10.1161/JAHA.115.002301 PMID: 26607712

Hassani, H., Amiri Andi, P., Ghodsi, A., Norouzi, K., Komendantova, N., & Unger, S. (2021). Shaping the future of smart dentistry: From artificial intelligence (AI) to intelligence augmentation (IA). *IoT*, 2(3), 510–523. DOI: 10.3390/iot2030026

Heinzelman, J., & Mitnick, L. (2022). AI and GEOINT: Transforming disaster preparedness and response. *Journal of Emergency Management (Weston, Mass.)*, 18(3), 29–40.

Helbich, M., Yao, Y., Liu, Y., Zhang, J., Liu, P., & Wang, R. (2019). Using Deep Learning to Examine Street View Green and Blue Spaces and Their Associations with Geriatric Depression in Beijing, China. *Environment International*, 126, 107–117. DOI: 10.1016/j.envint.2019.02.013 PMID: 30797100

Heras, J., Marani, R., & Milella, A. (2021). 39. semi-supervised semantic segmentation for grape bunch identification in natural images. *Precision Agriculture*, 21(11211), 331–337. DOI: 10.3920/978-90-8686-916-9_39

Hill, C., Young, M., Blainey, S., Cavazzi, S., Emberson, C., & Sadler, J. (2024). An integrated geospatial data model for active travel infrastructure. *Journal of Transport Geography*, 117, 103889. DOI: 10.1016/j.jtrangeo.2024.103889

Hoch, J. M., Neal, J., Baart, F., & Winsemius, H. C. (2019). Data scarcity for hydrological impact studies in data-sparse regions: Challenges and potential solutions. *Frontiers of Earth Science*, 7, 257.

Hogan, M., Williams, T., & Patel, A. (2021). Geospatial data in food security: Using GEOINT for agricultural resilience and resource management. *Journal of Agricultural Geospatial Analysis*, 8(2), 58–70.

Hogan, M., Williams, T., & Patel, A. (2023). AI-powered GEOINT for urban planning and smart cities: Innovations in public service and resource management. *Urban Development and Technology Review*, 12(2), 123–136.

Hogan, R., Anderson, J., & Patel, M. (2023). AI-driven GEOINT in environmental and disaster management. *Environmental Data Science Review*, 7(1), 102–118.

Hoggart, K., (2002). Researching human geography.

Hohmeier, K. C., Turner, K., Harland, M., Frederick, K., Rein, L., Atchley, D., Woodyard, A., Wasem, V., & Desselle, S. (2024). Scaling the optimizing care model in community pharmacy using implementation mapping and COM-B theoretical frameworks. *JAPhA Practice Innovations*, 1(1), 100002. DOI: 10.1016/j.japhpi.2023.100002

Hong, D., Gao, L., Yao, J., Zhang, B., Plaza, A., & Chanussot, J. (2021). Graph Convolutional Networks for Hyperspectral Image Classification. *IEEE Transactions on Geoscience and Remote Sensing*, 59(7), 5966–5978. DOI: 10.1109/TGRS.2020.3015157

Hossain, M. S., Muhammad, G., & Amin, S. U. (2021). Smart cities with artificial intelligence, big data, and internet of things. *IEEE Communications Magazine*, 59(1), 40–46.

Hou, Y., & Biljecki, F. (2022). A comprehensive framework for evaluating the quality of street view imagery. *International Journal of Applied Earth Observation and Geoinformation*, 115, 103094. DOI: 10.1016/j.jag.2022.103094

Huang, B., Zhao, B., & Song, Y. (2018). Urban Land-Use Mapping Using a Deep Convolutional Neural Network with High Spatial Resolution Multispectral Remote Sensing Imagery. *Remote Sensing of Environment*, 214, 73–86. DOI: 10.1016/j.rse.2018.04.050

Huang, C., Chen, Y., & Wu, J. (2020). Mapping flood susceptibility using geospatial machine learning techniques. *Water (Basel)*, 12(4), 1026.

Huang, G., Zhang, W., & Xu, D. (2022). How do technology-enabled bike-sharing services improve urban air pollution? Empirical evidence from China. *Journal of Cleaner Production*, 379, 134771. DOI: 10.1016/j.jclepro.2022.134771

Huang, H., Yao, X. A., Krisp, J. M., & Jiang, B. (2021). Analytics of location-based big data for smart cities: Opportunities, challenges, and future directions. *Computers, Environment and Urban Systems*, 90, 101712. DOI: 10.1016/j.compenvurbsys.2021.101712

Huang, J., & He, Y. (2021). Spatial-temporal modeling for environmental hazard prediction using deep learning. In *Proceedings of the IEEE Conference on Computer Vision and Pattern Recognition* (pp. 6342–6351). IEEE. DOI: 10.1109/CVPR46437.2021.00634

Huang, J., Liang, S., & Wang, L. (2021). Flood forecasting with integrated hydrological models and deep learning techniques. *Journal of Hydrometeorology*, 22(3), 577–594.

Ighile, E. H., Shirakawa, H., & Tanikawa, H. (2022). Application of GIS and machine learning to predict flood areas in Nigeria. *Sustainability (Basel)*, 14(9), 5039. DOI: 10.3390/su14095039

Ilesanmi, K. S., & Timothy, O. I. (2024). Possibility of Land Ownership Transaction with Non-Fungible Token Technology: Minting Survey Plan. *African Journal on Land Policy and Geospatial Sciences*, 7(2), 488–497. DOI: 10.48346/IMIST.PRSM/AJLP-GS.V7I2.41704

Ioanid, A., & Andrei, N. (2024). *Artificial Intelligence and Geospatial Technologies for Sustainable Maritime Logistics. Case Study: Port Of Constanta, Romania*. 15. https://marlog.aast.edu/files/marlog13/MARLOG13_paper_112.pdf

J. Murphy, (2017). "Image-based Classification of GPS Noise Level using Convolutional Neural Networks for Accurate Distance Estimation."

Jafarzadeh, M., Shabani, M., & Alizadeh, M. (2023). Hybrid AI-based flood risk assessment using fuzzy logic and machine learning. *The Science of the Total Environment*, 868, 161542.

Janowicz, K., Gao, S., McKenzie, G., Hu, Y., & Bhaduri, B. (2020). GeoAI: Spatially explicit artificial intelligence techniques for geographic knowledge discovery and beyond. [Taylor & Francis.]. *International Journal of Geographical Information Science*, 34(4), 625–636. DOI: 10.1080/13658816.2019.1684500

Javed, A. R., Shahzad, F., Rehman, S. U., Zikria, Y. B., Razzak, I., Jalil, Z., & Xu, G. (2022). Future smart cities: Requirements, emerging technologies, applications, challenges, and future aspects. *Cities (London, England)*, 129, 103794. DOI: 10.1016/j.cities.2022.103794

Jensen, J., & Bolstad, P. (2021). The role of geospatial intelligence in urban planning and smart city infrastructure. *Journal of Urban Studies and Geospatial Sciences*, 19(4), 112–126.

Jensen, J., & Bolstad, P. (2022). AI and geospatial intelligence in urban planning: Optimizing smart city growth and sustainability. *International Journal of Urban Studies and Smart Technologies*, 8(4), 77–91.

Jensen, M., & Bolstad, P. (2021). Precision agriculture through GEOINT applications: Enhancing crop management. *Agriculture and Technology Innovations*, 5(2), 53–68.

Jha, D., & Singh, R. (2018). "Swimming pool detection and classification using deep learning." [Online]. Available: https://medium.com/geoai/swimming-pool-detection-and-classification-using-deep-learning-aaf4a3a5e652

Ji, Q. (2020). *Geospatial Inference and Management of Utility Infrastructure Networks* [Ph. D. Thesis, School of Engineering, New Castle University]. https://theses.ncl.ac.uk/jspui/handle/10443/4985

Jin, G., Liang, Y., Fang, Y., Shao, Z., Huang, J., Zhang, J., & Zheng, Y. (2024). Spatio-Temporal Graph Neural Networks for Predictive Learning in Urban Computing: A Survey. *IEEE Transactions on Knowledge and Data Engineering*, 36(10), 5388–5408. DOI: 10.1109/TKDE.2023.3333824

Johnson, K., & Lee, H. (2020). Geospatial intelligence for resource management and defense applications. *Journal of Geo-Intelligence Research*, 10(3), 34–49.

Johnson, K., & Lee, H. (2020). Metadata management in geospatial intelligence: Ensuring data reliability and accuracy. *Journal of Data Science and Intelligence*, 8(3), 47–58.

Jurišić, M., Radočaj, D., Plaščak, I., & Rapčan, I. (2022). A UAS and machine learning classification approach to suitability prediction of expanding natural habitats for endangered flora species. *Remote Sensing (Basel)*, 14(13), 3054. DOI: 10.3390/rs14133054

Kaginalkar, A., Kumar, S., Gargava, P., Kharkar, N., & Niyogi, D. (2022). SmartAirQ: A Big Data Governance Framework for Urban Air Quality Management in Smart Cities. *Frontiers in Environmental Science*, 10, 785129. DOI: 10.3389/fenvs.2022.785129

Kaginalkar, A., Kumar, S., Gargava, P., & Niyogi, D. (2021). Review of urban computing in air quality management as smart city service: An integrated IoT, AI, and cloud technology perspective. *Urban Climate*, 39, 100972. DOI: 10.1016/j.uclim.2021.100972

Kalpana, Y. B., Nirmaladevi, J., Sabitha, R., Ammal, S. G., Dhiyanesh, B., & Radha, R. (2024). Revolutionizing agriculture: Integrating IoT cloud, and machine learning for smart farm monitoring and precision agriculture. *Studies in Computational Intelligence*, 79-108. DOI: 10.1007/978-3-031-67450-1_4

Kaluarachchi, Y. (2022). Implementing Data-Driven Smart City Applications for Future Cities. *Smart Cities*, 5(2), 455–474. DOI: 10.3390/smartcities5020025

Kamel Boulos, M. N., Peng, G., & Vopham, T. (2019). An overview of GeoAI applications in health and healthcare. *International Journal of Health Geographics*, 18(1), 1–9. DOI: 10.1186/s12942-019-0171-2 PMID: 31043176

Kang, Y., Jia, Q., Gao, S., Zeng, X., Wang, Y., Angsuesser, S., Liu, Y., Ye, X., & Fei, T. (2019). Extracting Human Emotions at Different Places Based on Facial Expressions and Spatial Clustering Analysis. *Transactions in GIS*, 23(3), 450–480. DOI: 10.1111/tgis.12552

Keskin, M., & Sekerli, Y. E. (2024). Mitigation of the effects of climate change on agriculture through the adoption of precision agriculture technologies. *Climate-Smart and Resilient Food Systems and Security*, 435-458. DOI: 10.1007/978-3-031-65968-3_20

Khankhoje, R. (2024). Future trends and ethical challenges in transforming gender healthcare using AI and ML. *Transforming Gender-Based Healthcare with AI and Machine Learning*, 239-259. DOI: 10.1201/9781003473435-14

Kharad, V., & Thakur, N. V. (2023). Analysis of decision support system for crop health management in smart and precision agriculture based on Internet of things (IoT) and artificial intelligence (AI). *2023 1st DMIHER International Conference on Artificial Intelligence in Education and Industry 4.0 (IDICAIEI)*, *25*, 1-6. DOI: 10.1109/IDICAIEI58380.2023.10406812

Khardia, N., Meena, R. H., Jat, G., Sharma, S., Kumawat, H., Dhayal, S., Meena, A. K., & Sharma, K. (2022). Soil properties influenced by the foliar application of Nano fertilizers in maize (Zea mays L.) crop. *International Journal of Plant and Soil Science*, •••, 99–111. DOI: 10.9734/ijpss/2022/v34i1430996

Khasgiwala, Y., Castellino, D. T., & Deshmukh, S. (2022). A Decentralized Federated Learning Paradigm for Semantic Segmentation of Geospatial Data. In Vasant, P., Zelinka, I., & Weber, G.-W. (Eds.), *Intelligent Computing & Optimization* (Vol. 371, pp. 196–206). Springer International Publishing., DOI: 10.1007/978-3-030-93247-3_20

Khatri, M. (2023). Transforming Indian business landscapes: The impact of AI, IoT, Metaverse, and emerging technologies. [IJSR]. *International Journal of Science and Research (Raipur, India)*, *12*(10), 373–378. DOI: 10.21275/SR231003023311

Khosravi, K., Shahabi, H., & Chen, W. (2022). Transfer learning for flood susceptibility modeling: A step towards universal flood risk assessment. *Environmental Modelling & Software*, *153*, 105403.

Kibona, I., Mkoma, S., & Mjemah, I. (2011). Nitrate pollution of Neogene alluvium aquifer in Morogoro municipality, Tanzania. *International Journal of Biological and Chemical Sciences*, *5*(1). Advance online publication. DOI: 10.4314/ijbcs.v5i1.68095

Klumbytė, G., & Athanasiadou, L. (2022). Algorithmic Governmentality and Managerial Fascism: The Case of Smart Cities. In *Deleuze and Guattari and Fascism* (pp. 84–104). Edinburgh University Press/Cambridge University Press. https://www.cambridge.org/core/books/abs/deleuze-and-guattari-and-fascism/algorithmic-governmentality-and-managerial-fascism-the-case-of-smart-cities/0B3DF569EEB012786A73B7703A16C458

Knauer, U., von Rekowski, C. S., Stecklina, M., Krokotsch, T., Pham Minh, T., Hauffe, V., Kilias, D., Ehrhardt, I., Sagischewski, H., Chmara, S., & Seiffert, U. (2019). Tree species classification based on hybrid ensembles of a convolutional neural network (CNN) and random forest classifiers. *Remote Sensing (Basel)*, 11(23), 2788. DOI: 10.3390/rs11232788

Kost, G. J. (2020). Geospatial Hotspots Need Point-of-Care Strategies to Stop Highly Infectious Outbreaks. *Archives of Pathology & Laboratory Medicine*, 144(10), 1166–1190. DOI: 10.5858/arpa.2020-0172-RA PMID: 32298139

Kowe, P., Mutanga, O., & Dube, T. (2021). Advancements in the remote sensing of landscape pattern of urban green spaces and vegetation fragmentation. *International Journal of Remote Sensing*, 42(10), 3797–3832. DOI: 10.1080/01431161.2021.1881185

Kraak, M.-J., & Ormeling, F. (2020). *Cartography: Visualization of Geospatial Data* (4th ed.). CRC Press., DOI: 10.1201/9780429464195

Kramers, A., Höjer, M., Lövehagen, N., & Wangel, J. (2018). Smart cities and climate targets: An exploration of smart city implementation in Stockholm. *Journal of Cleaner Production*, 172, 4039–4046.

Krishna, E. S. P., Praveena, N., Manju, I., Malathi, N., Giri, R. K., & Preetha, M.E. S. Phalguna Krishna. (2024). IoT-Enabled Wireless Sensor Networks and Geospatial Technology for Urban Infrastructure Management. *Journal of Electrical Systems*, 20(4s), 2248–2256. DOI: 10.52783/jes.2395

Kudinov, D., Hedges, D., & Maher, O. (2020). "Reconstructing 3D buildings from aerial LiDAR with AI: details." [Online]. Available: https://medium.com/geoai/reconstructing-3d-buildings-from-aerial-lidar-with-ai-details-6a81cb3079c0

Kulkarni, V. (2017). *Generating Synthetic Mobility Traffic Using RNNs*. DOI: 10.1145/3149808.3149809

Kumar, V., & Patel, A. (2023). Advances in AI-Driven Traffic Management Systems: A Comprehensive Review. *Transportation Research Part B: Methodological*, 158, 334–355.

Kumawat, H., Singh, D. P., Yadav, K. K., Khardia, N., Dhayal, S., Sharma, S., Sharma, K., & Kumawat, A. (2023). Response of fertility levels and liquid Biofertilizers on soil chemical properties and nutrient uptake under wheat (Triticum aestivum L.) crop. *Environment and Ecology*, 41(4), 2248–2256. DOI: 10.60151/envec/ZBLX2641

Law, S., Seresinhe, C. I., Shen, Y., & Gutierrez-Roig, M. (2020). Street-Frontage-Net: Urban Image Classification Using Deep Convolutional Neural Networks. *International Journal of Geographical Information Science*, 34(4), 681–707. DOI: 10.1080/13658816.2018.1555832

Law, S., Shen, Y., & Seresinhe, C. (2017). "An application of convolutional neural network in street image classification," *Proc. 1st Work. Artif. Intell. Deep Learn. Geogr. Knowl. Discov. -. GeoAI*, 17, 5–9. DOI: 10.1145/3149808.3149810

LeCun, Y., Bengio, Y., & Hinton, G. (2015). Deep Learning. *Nature*, 521(7553), 436–444. DOI: 10.1038/nature14539 PMID: 26017442

Lee, I.-S. (2021). A Study on Geospatial Information Role in Digital Twin. *Journal of the Korea Academia-Industrial Cooperation Society*, 22(3), 268–278. DOI: 10.5762/KAIS.2021.22.10.268

Lee, K., & Tsou, M.-H. (2020). Machine learning techniques for geospatial data classification: A comparative study. In *Proceedings of the ACM SIGSPATIAL International Conference on Advances in Geographic Information Systems* (pp. 82-91). ACM. https://doi.org/DOI: 10.1145/3382324.3382364

Leszczynski, A., & Crampton, J. (2016). Introduction: Spatial big data and everyday life. *Big Data & Society*, 3(2), 3. DOI: 10.1177/2053951716661366

Li, H., & Bao, J. (2021). Uncertainties in the surface layer physics parameterizations. *Uncertainties in Numerical Weather Prediction*, 229-236. DOI: 10.1016/B978-0-12-815491-5.00008-2

Li, W. (2022). GeoAI in social science. Handbook of Spatial Analysis in the Social Sciences, 291-304.

Liao, X., Liao, G., & Xiao, L. (2023). Rapeseed Storage Quality Detection Using Hyperspectral Image Technology—An Application for Future Smart Cities. *Journal of Testing and Evaluation*, 51(3), 1740–1752. DOI: 10.1520/JTE20220073

Li, H., Hu, B., Li, Q., & Jing, L. (2021). CNN-based individual tree species classification using high-resolution satellite imagery and airborne LiDAR data. *Forests*, 12(12), 1697. DOI: 10.3390/f12121697

Li, S., Dragicevic, S., Castro, F. A., Sester, M., Winter, S., Coltekin, A., Pettit, C., Jiang, B., Haworth, J., Stein, A., & Cheng, T. (2016). Geospatial big data handling theory and methods: A review and research challenges. *ISPRS Journal of Photogrammetry and Remote Sensing*, 115, 119–133. DOI: 10.1016/j.isprsjprs.2015.10.012

Liu P, Biljecki F (2022) A review of spatially-explicit GeoAI applications in urban geography. Int J Appl Earth Obs Geoinf. . jag. 2022. 102936DOI: 10. 1016/j

Liu Y, Liu X, Gao S, Gong L, Kang C, Zhi Y, Chi G, Shi L (2015) Social sensing: A new approach to understanding our socioeconomic environments. Ann Assoc Am Geogr 105(3):512–530. . 2015. 10187 73DOI: 10. 1080/ 00045 608

Liu, C., Fan, C., & Mostafavi, A. (2024). Graph attention networks unveil determinants of intra- and inter-city health disparity. *Urban Informatics*, 3(1), 18. DOI: 10.1007/s44212-024-00049-5

Liu, K., Chen, J., Li, R., Peng, T., Ji, K., & Gao, Y. (2022). Nonlinear effects of community built environment on car usage behavior: A machine learning approach. *Sustainability (Basel)*, 14(11), 6722. DOI: 10.3390/su14116722

Liu, P., & Biljecki, F. (2022). A review of spatially-explicit GeoAI applications in Urban Geography. *International Journal of Applied Earth Observation and Geoinformation*, 112, 102936. DOI: 10.1016/j.jag.2022.102936

Liu, P., Zhang, Y., & Biljecki, F. (2024). Explainable spatially explicit geospatial artificial intelligence in urban analytics. *Environment and Planning. B, Urban Analytics and City Science*, 51(5), 1104–1123. DOI: 10.1177/23998083231204689

Liu, Y., Hu, J., & Fang, Y. (2021). Real-time flood monitoring using deep learning techniques and multi-source remote sensing data. *Remote Sensing*, 13(1), 45.

Liu, Y., Liu, X., Gao, S., Gong, L., Kang, C., Zhi, Y., Chi, G., & Shi, L. (2015). Social Sensing: A New Approach to Understanding Our Socioeconomic Environments. *Annals of the Association of American Geographers*, 105(3), 512–530. DOI: 10.1080/00045608.2015.1018773

Li, W. (2020). GeoAI: Where machine learning and big data converge in GIScience. *Journal of Spatial Information Science*, (20), 71–77. DOI: 10.5311/JOSIS.2020.20.658

Li, W. (2021). *GeoAI and deep learning*. The International Encyclopedia of Geography. DOI: 10.1002/9781118786352.wbieg2083

Li, W., Batty, M., & Goodchild, M. F. (2020). Real-time GIS for smart cities. *International Journal of Geographical Information Science*, 34(2), 311–324. DOI: 10.1080/13658816.2019.1673397

Li, W., & Hsu, C.-Y. (2022). GeoAI for large-scale image analysis and machine vision: Recent progress of artificial intelligence in geography. *ISPRS International Journal of Geo-Information*, 11(7), 385. DOI: 10.3390/ijgi11070385

Li, W., Raskin, R., & Goodchild, M. F. (2012). Semantic similarity measurement based on knowledge mining: An artificial neural net approach. *International Journal of Geographical Information Science*, 26(8), 1415–1435. DOI: 10.1080/13658816.2011.635595

Li, W., Shao, H., Wang, S., Zhou, X., & Wu, S. (2016). A2CI: A cloud-based, service-oriented geospatial cyberinfrastructure to support atmospheric research. In *Cloud Computing in Ocean and Atmospheric Sciences* (pp. 137–161). Academic Press. DOI: 10.1016/B978-0-12-803192-6.00009-8

Li, Z., Xu, Q., & Tan, X. (2020). Flood depth estimation from synthetic aperture radar imagery using machine learning. *Remote Sensing*, 12(3), 531.

Lv, Z., Shang, W.-L., & Guizani, M. (2022). Impact of Digital Twins and Metaverse on Cities: History, Current Situation, and Application Perspectives. *Applied Sciences (Basel, Switzerland)*, 12(24), 12820. DOI: 10.3390/app122412820

Mahadevan, R., Bhat, V., & Rao, S. (2023). Bangalore's Integrated Traffic Management System: A smart city initiative for urban mobility. *Journal of Urban Technology*, 30(2), 150–169.

Marasinghe, R., Yigitcanlar, T., Mayere, S., Washington, T., & Limb, M. (2024). Towards Responsible Urban Geospatial AI: Insights From the White and Grey Literatures. *Journal of Geovisualization and Spatial Analysis*, 8(2), 24. DOI: 10.1007/s41651-024-00184-2

Marcus, G. (2018). Deep learning: A critical appraisal. 1–27. arXiv preprint arXiv:1801.00631.

Marsal-Llacuna, M.-L. (2020). The people's smart city dashboard (PSCD): Delivering on community-led governance with blockchain. *Technological Forecasting and Social Change*, 158, 120150. DOI: 10.1016/j.techfore.2020.120150

Marzouk, M., & Othman, A. (2020). Planning utility infrastructure requirements for smart cities using the integration between BIM and GIS. *Sustainable Cities and Society*, 57, 102120. DOI: 10.1016/j.scs.2020.102120

Masik, G., Sagan, I., & Scott, J. W. (2021). Smart City strategies and new urban development policies in the Polish context. *Cities (London, England)*, 108, 102970. DOI: 10.1016/j.cities.2020.102970

MHUA. (2024). *Open Data Platform: India Smart Cities*. Smart Cities Mission Data Portal. https://smartcities.data.gov.in/

Miller, H. J. (2004). Tobler's first law and spatial analysis. *Annals of the Association of American Geographers*, 94(2), 284–289. DOI: 10.1111/j.1467-8306.2004.09402005.x

Miller, H. J., & Goodchild, M. F. (2015). Data-driven geography. *GeoJournal*, 80(4), 449–461. DOI: 10.1007/s10708-014-9602-6

Miller, P., Johnson, R., & Hart, D. (2022). GEOINT and artificial intelligence in defense: The integration of AI for real-time national security operations. *Military Technology and Intelligence*, 10(1), 47–59.

Minh, Q. N., Nguyen, V.-H., Quy, V. K., Ngoc, L. A., Chehri, A., & Jeon, G. (2022). Edge Computing for IoT-Enabled Smart Grid: The Future of Energy. *Energies*, 15(17), 6140. DOI: 10.3390/en15176140

Mishra, P., & Singh, G. (2023). Enabling Technologies for Sustainable Smart City. In P. Mishra & G. Singh, *Sustainable Smart Cities* (pp. 59–73). Springer International Publishing. DOI: 10.1007/978-3-031-33354-5_3

Mkumbo, N. J., Mussa, K. R., Mariki, E. E., & Mjemah, I. C. (2022). The use of the DRASTIC-LU/LC model for assessing groundwater vulnerability to nitrate contamination in Morogoro municipality, Tanzania. *Earth (Basel, Switzerland)*, 3(4), 1161–1184. DOI: 10.3390/earth3040067

Mohammadi, B. (2022). Application of machine learning and remote sensing in hydrology. *Sustainability (Basel)*, 14(13), 7586. DOI: 10.3390/su14137586

Mokander, J., & Floridi, L. (2021). Ethics-based auditing to develop trustworthy ai. *Minds and Machines*, 31(2), 323–327. DOI: 10.1007/s11023-021-09557-8

Mollick, T., Azam, M. G., & Karim, S. (2023). Geospatial-based machine learning techniques for land use and land cover mapping using a high-resolution unmanned aerial vehicle image. *Remote Sensing Applications: Society and Environment*, 29, 100859. DOI: 10.1016/j.rsase.2022.100859

Moreno-Álvarez, S., Paoletti, M. E., Sanchez-Fernandez, A. J., Rico-Gallego, J. A., Han, L., & Haut, J. M. (2024). Federated learning meets remote sensing. *Expert Systems with Applications*, 255, 124583. DOI: 10.1016/j.eswa.2024.124583

Mosavi, A., Ozturk, P., & Chau, K. W. (2018). Flood prediction using machine learning models: Literature review. *Water (Basel)*, 10(11), 1536. DOI: 10.3390/w10111536

Mousavi, S. S., Schukat, M., & Howley, E. (2020). Traffic light control using deep policy-gradient and value-function-based reinforcement learning. *Engineering Applications of Artificial Intelligence*, 85, 565–574.

Mueller, J. P. (2006). Mining Google web services: building applications with the Google API. John Wiley & Sons. Jahromi, M. N., Jahromi, M. N., Zolghadr-Asli, B., Pourghasemi, H. R., & Alavipanah, S. K. (2021). Google Earth Engine and its application in forest sciences. *Spatial Modeling in Forest Resources Management: Rural Livelihood and Sustainable Development*, 629-649. Aszkowski, P., Ptak, B., Kraft, M., Pieczyński, D., & Drapikowski, P. (2023). Deepness: Deep neural remote sensing plugin for QGIS. *SoftwareX*, 23, 101495.

Murshed, S. M., Al-Hyari, A. M., Wendel, J., & Ansart, L. (2018). Design and Implementation of a 4D Web Application for Analytical Visualization of Smart City Applications. *ISPRS International Journal of Geo-Information*, 7(7), 276. DOI: 10.3390/ijgi7070276

Musamih, A., Dirir, A., Yaqoob, I., Salah, K., Jayaraman, R., & Puthal, D. (2024). NFTs in Smart Cities: Vision, Applications, and Challenges. *IEEE Consumer Electronics Magazine*, 13(2), 9–23. DOI: 10.1109/MCE.2022.3217660

N, D., & J, N. (2023). Review on malware classification with a hybrid deep learning. *Journal of IoT and Machine Learning*, 18-21. DOI: 10.48001/joitml.2023.1118-21

Nagaraj, A. (2023). Integration of AI and IoT-cloud. *The Role of AI in Enhancing IoT-Cloud Applications*, 116-165. DOI: 10.2174/9789815165708123010008

Nagaraj, A. (2023). Internet of things (IoT) with AI. *The Role of AI in Enhancing IoT-Cloud Applications*, 21-72. DOI: 10.2174/9789815165708123010006

Nagavi, J. C., Shukla, B. K., Bhati, A., Rai, A., & Verma, S. (2024). Harnessing Geospatial Technology for Sustainable Development: A Multifaceted Analysis of Current Practices and Future Prospects. In Sharma, C., Shukla, A. K., Pathak, S., & Singh, V. P. (Eds.), *Sustainable Development and Geospatial Technology* (pp. 147–170). Springer Nature Switzerland., DOI: 10.1007/978-3-031-65683-5_8

Nagendra, S. M. S., Khare, M., & Goyal, R. (2018). Urban air quality management in India: A review. *Atmospheric Environment*, 172, 209–225.

Nayak, D., Surve, N., & Shrivastava, P. (2022). Assessing land use and land cover changes in south Gujarat. *Ecology. Environmental Conservation*, 28(04), 2110–2115. DOI: 10.53550/EEC.2022.v28i04.070

Nguyen, T., & Brown, M. (2021). Geospatial intelligence and artificial intelligence: The future of predictive analysis. *Journal of Geospatial Technologies*, 15(2), 22–38.

Nguyen, T., & Brown, R. (2021). Geospatial intelligence in environmental conservation: Monitoring ecosystems with advanced geospatial data and AI. *Environmental Monitoring & Management*, 22(4), 75–89.

Nguyen, T., & Brown, R. (2021). GIS and geospatial data systems: Foundations and modern applications. *Geospatial Technology Journal*, 15(1), 32–49.

Nguyen, T., & Brown, R. (2021). Improving precision in geospatial data analysis through AI and deep learning models. *International Journal of Geospatial Intelligence*, 16(1), 48–62.

Nguyen, T., & Kim, J. (2022). Advances in spatial data integration and analytics: A survey of current research. In *Proceedings of the International Conference on Geographic Information Science* (pp. 1-10). https://doi.org/DOI: 10.1109/GIScience54041.2022.00001

Nguyen, T., Zhang, Q., & Lee, M. (2022). AI-powered object and pattern recognition in geospatial intelligence for security applications. *Geospatial Security Review*, 18(3), 123–137.

Nieuwenhuijsen, M. J. (2015). *Exposure assessment in environmental epidemiology* (2nd ed.). Oxford University Press. DOI: 10.1093/med/9780199378784.001.0001

Nizzoli, L., Avvenuti, M., Tesconi, M., & Cresci, S. (2020). Geo semantic-parsing: AI-powered geoparsing by traversing semantic knowledge graphs. *Decision Support Systems*, 136, 113346. DOI: 10.1016/j.dss.2020.113346

Nowak, B. (2021). Precision agriculture: Where do we stand? A review of the adoption of precision agriculture technologies on Field crops farms in developed countries. *Agricultural Research*, 10(4), 515–522. DOI: 10.1007/s40003-021-00539-x

Nuckols, J. R., Ward, M. H., & Jarup, L. (2004). Using geographic information systems for exposure assessment in environmental epidemiology studies. *Environmental Health Perspectives*, 112(9), 1007–1015. DOI: 10.1289/ehp.6738 PMID: 15198921

Ogryzek, M., Adamska-Kmieć, D., & Klimach, A. (2020). Sustainable Transport: An Efficient Transportation Network—Case Study. *Sustainability (Basel)*, 12(19), 8274. DOI: 10.3390/su12198274

Organic farming: The way to sustainable agriculture and environmental protection. (2022). *International Journal of Biology, Pharmacy and Allied Sciences, 11*(1 (SPECIAL ISSUE)). DOI: 10.31032/IJBPAS/2022/11.1.1004

Osupile, K., Yahya, A., & Samikannu, R. (2022). A review on agriculture monitoring systems using Internet of things (IoT). *2022 International Conference on Applied Artificial Intelligence and Computing (ICAAIC), 7*, 1565-1572. DOI: 10.1109/ICAAIC53929.2022.9792979

Ozcelik, A. E. (2024). Blockchain-oriented geospatial architecture model for real-time land registration. *Survey Review*, 56(394), 1–17. DOI: 10.1080/00396265.2022.2156755

Panahi, M., Rezaie, F., & Shirzadi, A. (2022). Flood susceptibility mapping using hybrid machine learning algorithms: A case study from Iran. *Geocarto International*, 37(3), 734–749.

Pandya, S., Srivastava, G., Jhaveri, R., Babu, M. R., Bhattacharya, S., Maddikunta, P. K. R., Mastorakis, S., Piran, M., & Gadekallu, T. R. (2023). Federated learning for smart cities: A comprehensive survey. *Sustainable Energy Technologies and Assessments*, 55, 102987. DOI: 10.1016/j.seta.2022.102987

Pang, J., Huang, Y., Xie, Z., Li, J., & Cai, Z. (2021). Collaborative city digital twin for the COVID-19 pandemic: A federated learning solution. *Tsinghua Science and Technology*, 26(5), 759–771. DOI: 10.26599/TST.2021.9010026

Papageorgiou, M., Diakaki, C., & Aboudolas, K. (2022). Real-time traffic signal control for urban networks: The potential of AI-based adaptive systems. *IEEE Transactions on Intelligent Transportation Systems*, 23(2), 512–523.

Parikh, N. (2024). Unveiling the role of AI product managers: Shaping the future. DOI: 10.36227/techrxiv.172504030.01820212/v1

Park, J., & Yoo, S. (2023). Evolution of the smart city: Three extensions to governance, sustainability, and decent urbanisation from an ICT-based urban solution. *International Journal of Urban Sciences*, 27(sup1), 10–28. DOI: 10.1080/12265934.2022.2110143

Patel, M., Mehta, A., & Chauhan, N. C. (2021). Design of Smart Dashboard based on IoT & Fog Computing for Smart Cities. *2021 5th International Conference on Trends in Electronics and Informatics (ICOEI)*, 458–462. DOI: 10.1109/ICOEI51242.2021.9452744

Pérez Del Hoyo, R., Visvizi, A., & Mora, H. (2021). Inclusiveness, safety, resilience, and sustainability in the smart city context. In *Smart Cities and the un SDGs* (pp. 15–28). Elsevier., DOI: 10.1016/B978-0-323-85151-0.00002-6

Petrocchi, E., Tiribelli, S., Paolanti, M., Giovanola, B., Frontoni, E., & Pierdicca, R. (2024). GeomEthics: Ethical considerations about using artificial intelligence in Geomatics. *Lecture Notes in Computer Science*, 14366, 282–293. DOI: 10.1007/978-3-031-51026-7_25

Petrov, S., Dimitrov, S., & Ihtimanski, I. (2024). Integrated application of geospatial technologies for digital twining of urbanized territories for microscale urban planning. In Michaelides, S. C., Hadjimitsis, D. G., Danezis, C., Kyriakides, N., Christofe, A., Themistocleous, K., & Schreier, G. (Eds.), *Tenth International Conference on Remote Sensing and Geoinformation of the Environment (RSCy2024)* (p. 4). SPIE. DOI: 10.1117/12.3034288

Pierdicca, R., & Paolanti, M. (2022). GeoAI: A review of artificial intelligence approaches for the interpretation of complex geomatics data. *Geoscientific Instrumentation, Methods and Data Systems*, 11(1), 195–218. DOI: 10.5194/gi-11-195-2022

Postert, P., Wolf, A. E. M., & Schiewe, J. (2022). Integrating Visualization and Interaction Tools for Enhancing Collaboration in Different Public Participation Settings. *ISPRS International Journal of Geo-Information*, 11(3), 156. DOI: 10.3390/ijgi11030156

Pourebrahim, N., Thill, J.-C., Sultana, S., & Mohanty, S. (2018). "Enhancing trip distribution prediction with twitter data: Comparison of neural network and gravity models," *Proc. 2nd ACM SIGSPATIAL Int. Work. AI Geogr. Knowl. Discov. GeoAI 2018*. DOI: 10.1145/3281548.3281555

Pourghasemi, H. R., Rahmati, O., & Gokceoglu, C. (2017). Modeling the impact of human activities and geomorphology on flood susceptibility. *Geocarto International*, 32(3), 244–259.

Pulvirenti, L., Chini, M., Pierdicca, N., & Guerriero, L. (2016). An algorithm for operational flood mapping from synthetic aperture radar (SAR) data using fuzzy logic. *Remote Sensing*, 8(7), 565.

Q. Li, (2017). "Visual Landmark Sequence-based Indoor Localization," no. 1.

Qudus, T., Ade, S., Modibbo, M. A., Aleem, K. F., Lawal, M. A., & Musa, L. A. (2024). Advancements and Innovations in Object-Oriented Feature Extraction Algorithms for Nigeria-Sat2 Data: A Comprehensive Review. *International Journal of Research Publication and Reviews*, 5, 815–829.

Radočaj, D., & Jurišić, M. (2022). GIS-based cropland suitability prediction using machine learning: A novel approach to sustainable agricultural production. *Agronomy (Basel)*, 12(9), 2210. DOI: 10.3390/agronomy12092210

Radočaj, D., Jurišić, M., Gašparović, M., Plaščak, I., & Antonić, O. (2021). Cropland suitability assessment using satellite-based biophysical vegetation properties and machine learning. *Agronomy (Basel)*, 11(8), 1620. DOI: 10.3390/agronomy11081620

Rahmati, O., Pourghasemi, H. R., & Melesse, A. M. (2019). Application of GIS and remote sensing techniques in flood risk management: A case study in Ilam Province, Iran. *Environmental Earth Sciences*, 78(1), 40.

Rai, P. K. (2013). Forest and land use mapping using Remote Sensing and Geographical Information System: A case study on model system. *Environmental Skeptics and Critics*, 2(3), 97.

Ramu, S. P., Boopalan, P., Pham, Q.-V., Maddikunta, P. K. R., Huynh-The, T., Alazab, M., Nguyen, T. T., & Gadekallu, T. R. (2022). Federated learning enabled digital twins for smart cities: Concepts, recent advances, and future directions. *Sustainable Cities and Society*, 79, 103663. DOI: 10.1016/j.scs.2021.103663

Rana, A. (2022). Land use and land cover change mapping: A Spatio temporal and correlational analysis of Ramganjmandi Tehsil, Kota, Rajasthan, India. *Ecology. Environmental Conservation*, •••, 1384–1389. DOI: 10.53550/EEC.2022.v28i03.040

Rana, A., & Sharma, R. (2020). Drinking water quality assessment and predictive mapping: Impact of Kota stone mining in Ramganjmandi Tehsil, Rajasthan, India. *Nature Environment and Pollution Technology*, 19(3), 1219–1225. DOI: 10.46488/NEPT.2020.v19i03.036

Ranganathan, A., Ramachandran, R., & Patel, V. (2021). Challenges and opportunities in implementing AI-based traffic management systems in Indian smart cities. *International Journal of Urban Sciences*, 25(4), 549–565.

Rani, S., Mishra, R. K., Usman, M., Kataria, A., Kumar, P., Bhambri, P., & Mishra, A. K. (2021). Amalgamation of Advanced Technologies for Sustainable Development of Smart City Environment: A Review. *IEEE Access : Practical Innovations, Open Solutions*, 9, 150060–150087. DOI: 10.1109/ACCESS.2021.3125527

Reichstein, M., Camps-Valls, G., Stevens, B., Jung, M., Denzler, J., Carvalhais, N., & Prabhat, . (2019). Deep learning and process understanding for data-driven Earth system science. *Nature*, 566(7743), 195–204. DOI: 10.1038/s41586-019-0912-1 PMID: 30760912

Ren, Y., Chen, H., Han, Y., Cheng, T., Zhang, Y., & Chen, G. (2020). A Hybrid Integrated Deep Learning Model for the Prediction of Citywide Spatio-Temporal Flow Volumes. *International Journal of Geographical Information Science*, 34(4), 802–823. DOI: 10.1080/13658816.2019.1652303

Reutov, V., Mottaeva, A., Varzin, V., Jallal, M. A. K., Burkaltseva, D., Shepelin, G., Blazhevich, O., Faskhutdinov, A., Trofimova, A., Niyazbekova, S., & Babin, M. (2023). Smart city development in the context of sustainable development and environmental solutions. *E3S Web of Conferences, 402*, 09020. DOI: 10.1051/e3sconf/202340209020

R, G., & P, D. L. (1693-1696). R, G. R., & P, L. (2024). Ethical considerations in artificial intelligence development. *International Journal of Research Publication and Reviews*, 5(6), 1693–1696. Advance online publication. DOI: 10.55248/gengpi.5.0624.1453

Ribeiro, M. P., De Melo, K., Chen, D., & Valente, R. A. (2024). Prioritization of New Green Infrastructures Aimed at Protecting Urban Biodiversity. *IGARSS 2024 - 2024 IEEE International Geoscience and Remote Sensing Symposium*, 5447–5452. DOI: 10.1109/IGARSS53475.2024.10640597

Rinaldi, F. M., & Nielsen, S. B. (2024). Artificial intelligence. *Artificial Intelligence and Evaluation*, 287-308. DOI: 10.4324/9781003512493-14

Rojas, E., Bastidas, V., & Cabrera, C. (2020). Cities-Board: A Framework to Automate the Development of Smart Cities Dashboards. *IEEE Internet of Things Journal*, 7(10), 10128–10136. DOI: 10.1109/JIOT.2020.3002581

Rolnick, D., Veit, A., Belongie, S., & Shavit, N. (2017). Deep learning is robust to massive label noise. arXiv preprint arXiv:1705.10694.

Rosa, L., Silva, F., & Analide, C. (2021). Mobile Networks and Internet of Things Infrastructures to Characterize Smart Human Mobility. *Smart Cities*, 4(2), 894–918. DOI: 10.3390/smartcities4020046

Roussel, C., & Böhm, K. (2023). Geospatial XAI: A Review. *ISPRS International Journal of Geo-Information*, 12(9), 355. DOI: 10.3390/ijgi12090355

Roy, S. S., & Ray, R. (2024). An Integrated Approach of Multi-Criteria Decision Analysis and Spatial Interaction on Urbanization using Geospatial Techniques—A Case Study of Barasat Sub-division, North 24 Parganas District, West Bengal, India. *Journal of Interdisciplinary and Multidisciplinary Research*, 19(8), 102–116.

Rustamov, J., Rustamov, Z., & Zaki, N. (2023). Green space quality analysis using machine learning approaches. *Sustainability (Basel)*, 15(10), 7782. DOI: 10.3390/su15107782

Rustamov, Z., Rustamov, J., Zaki, N., Turaev, S., Sultana, M. S., Tan, J. Y., & Balakrishnan, V. (2023). Enhancing cardiovascular disease prediction: A domain knowledge-based feature selection and stacked ensemble machine learning approach. DOI: 10.21203/rs.3.rs-3068941/v1

Saha, R., Misra, S., & Deb, P. K. (2021). FogFL: Fog-Assisted Federated Learning for Resource-Constrained IoT Devices. *IEEE Internet of Things Journal*, 8(10), 8456–8463. DOI: 10.1109/JIOT.2020.3046509

Sameen, M. I., & Pradhan, B. (2019). Assessing landslide susceptibility using machine learning and geospatial data in the high mountain environment. *Geomorphology*, 327, 11–22.

Sampson, C. C., Smith, A. M., Bates, P. D., Neal, J. C., Trigg, M. A., & Brewer, P. A. (2015). A high-resolution global flood hazard model. *Water Resources Research*, 51(9), 7358–7381. DOI: 10.1002/2015WR016954 PMID: 27594719

Sánchez, O., Castañeda, K., Vidal-Méndez, S., Carrasco-Beltrán, D., & Lozano-Ramírez, N. E. (2024). Exploring the influence of linear infrastructure projects 4.0 technologies to promote sustainable development in smart cities. *Results in Engineering*, 23, 102824. DOI: 10.1016/j.rineng.2024.102824

Sarker, I. H. (2024). AI-Enabled Cybersecurity for IoT and Smart City Applications. In I. H. Sarker, *AI-Driven Cybersecurity and Threat Intelligence* (pp. 121–136). Springer Nature Switzerland. DOI: 10.1007/978-3-031-54497-2_7

Schumann, G. (2011). The need for precise global information in flood risk management. *International Journal of Remote Sensing*, 32(22), 5963–5968.

Schumann, G., & Di Baldassarre, G. (2010). The direct use of radar satellites for continuous monitoring of flood inundation. *Remote Sensing of Environment*, 115(1), 2880–2890.

Scott, G. J., England, M. R., Starms, W. A., Marcum, R. A., & Davis, C. H. (2017). Training Deep Convolutional Neural Networks for Land-Cover Classification of High-Resolution Imagery. *IEEE Geoscience and Remote Sensing Letters*, 14(4), 549–553. DOI: 10.1109/LGRS.2017.2657778

Shahat Osman, A. M., & Elragal, A. (2021). Smart Cities and Big Data Analytics: A Data-Driven Decision-Making Use Case. *Smart Cities*, 4(1), 286–313. DOI: 10.3390/smartcities4010018

Shamsuzzoha, A., Nieminen, J., Piya, S., & Rutledge, K. (2021). Smart city for sustainable environment: A comparison of participatory strategies from Helsinki, Singapore and London. *Cities (London, England)*, 114, 103194. DOI: 10.1016/j.cities.2021.103194

Sharma, K., & Shivandu, S. K. (2024). Integrating artificial intelligence and Internet of things (IoT) for enhanced crop monitoring and management in precision agriculture. *Sensors International*, 5, 100292. DOI: 10.1016/j.sintl.2024.100292

Sharma, L. K., & Naik, R. (2024). *Conservation of Saline Wetland Ecosystems: An Initiative towards UN Decade of Ecological Restoration*. Springer Nature Singapore., DOI: 10.1007/978-981-97-5069-6

Silva, A., Zhang, Y., & Lee, M. (2021). AI and GEOINT in environmental conservation: Monitoring ecosystems and biodiversity. *Environmental Monitoring and AI Research*, 19(2), 98–110.

Silva, A., Zhang, Y., & Lee, M. (2021). AI and geospatial intelligence for environmental protection and sustainability. *Environmental Monitoring and AI Research*, 19(1), 102–115.

Silva, A., Zhang, Y., & Lee, M. (2021). The role of geospatial intelligence in smart cities: Integrating IoT for urban sustainability. *Journal of Smart City Innovations*, 9(1), 24–38.

Silva, C., Zhang, Y., & Hogan, M. (2021). Geospatial intelligence for smart city development. *Journal of Urban Technology*, 10(3), 27–42.

Silva, D. S., & Holanda, M. (2022). Applications of geospatial big data in the Internet of Things. *Transactions in GIS*, 26(1), 41–71. DOI: 10.1111/tgis.12846

Singh, V., & Dubey, A. (2012). Land use mapping using remote sensing & GIS techniques in Naina-Gorma Basin, part of Rewa district, MP, India. *International Journal of Emerging Technology and Advanced Engineering*, 2(11), 151–156.

Smékalová, L., & Kučera, F. (2020). Smart City Projects in the Small-Sized Municipalities: Contribution of the Cohesion Policy. *Scientific Papers of the University of Pardubice, Series D. Faculty of Economics and Administration*, 28(2). Advance online publication. DOI: 10.46585/sp28021067

Smith, A., Patel, D., & Williams, H. (2021). Geospatial intelligence in military operations and national security. *Military Intelligence Journal*, 13(1), 5–20.

Smith, J., & Kline, G. (2020). Urban planning and smart cities: The role of geospatial intelligence in modernizing infrastructure. *Journal of Urban Development*, 13(2), 56–68.

Smith, J., Tan, R., & Chen, H. (2021). Ethical considerations in geospatial intelligence: Data privacy, surveillance, and biases in AI algorithms. *Geospatial Ethics Review*, 11(3), 101–113.

Smith, R., Patel, K., & Williams, S. (2021). Advances in satellite imaging: Radar and infrared technology in geospatial intelligence. *Remote Sensing and Geospatial Intelligence Review*, 6(3), 60–78.

Song, H., Srinivasan, D., & Choy, M. C. (2017). Hybrid cooperative traffic signal control using a novel multi-agent reinforcement learning. *IEEE Transactions on Intelligent Transportation Systems*, 18(8), 2143–2157.

Song, Y., Kalacska, M., Gašparović, M., Yao, J., & Najibi, N. (2023). Advances in geocomputation and geospatial artificial intelligence (GeoAI) for mapping. [Elsevier.]. *International Journal of Applied Earth Observation and Geoinformation*, 120, 103300. DOI: 10.1016/j.jag.2023.103300

Sonnenschein, T., Scheider, S., de Wit, G. A., Tonne, C. C., & Vermeulen, R. (2022). Agent-based modeling of urban exposome interventions: Prospects, model architectures, and methodological challenges. *Exposome*, 2(1), osac009. DOI: 10.1093/exposome/osac009 PMID: 37811475

Srivastava, S., Vargas-Muñoz, J. E., Swinkels, D., & Tuia, D. (2018)."Multilabel Building Functions Classification from Ground Pictures using Convolutional Neural Networks," *Proc. 2nd ACM SIGSPATIAL Int. Work. AI Geogr. Knowl. Discov.*, pp. 43–46. DOI: 10.1145/3281548.3281559

Statista. (2024). *Internet of Things (IoT) connected devices installed base worldwide from 2015 to 2025.* https://www.statista.com/statistics/471264/iot-number-of-connected-devices-worldwide/

Stuss, M., & Fularski, A. (2024). Ethical considerations of using artificial intelligence (AI) in recruitment processes. *Edukacja Ekonomistów i Menedżerów*, 71(1). Advance online publication. DOI: 10.33119/EEIM.2024.71.4

Sudarmadji, S., & Santoso, S. (2023, April). The role of nanofluid and ultrasonic vibration in coolant radiator. In *AIP Conference Proceedings* (Vol. 2531, No. 1). AIP Publishing.

Sun, W., Guo, L., & Zhang, Y. (2022). Climate-informed flood risk prediction using LSTM networks and geospatial analysis. *Water Resources Research, 58*(4), e2022WR031301.

Sun, T., Di, Z., & Wang, Y. (2018).. . *Combining Satellite Imagery and GPS Data for Road Extraction*, 3281550(c), 4–7.

T. O. (2018). Satellite-based and D. U. U. Convolutional. "Towards Operational Satellite-Based Damage-Mapping Using U-Net Convolutional Network: A Case Study of 2011 Tohoku".

Tahar, A., Mendy, G., & Ouya, S. (2024). Efficient and Optimized Geospatial Data Representation in Blockchain-Based Land Administration. In X.-S. Yang, S. Sherratt, N. Dey, & A. Joshi (Eds.), *Proceedings of Ninth International Congress on Information and Communication Technology* (Vol. 1014, pp. 519–535). Springer Nature Singapore. DOI: 10.1007/978-981-97-3562-4_41

Tannous, H. O., Major, M. D., & Furlan, R. (2021). Accessibility of green spaces in a metropolitan network using space syntax to objectively evaluate the spatial locations of parks and promenades in Doha, State of Qatar. *Urban Forestry & Urban Greening*, 58, 126892. DOI: 10.1016/j.ufug.2020.126892

Tao, H., Liu, Y., & Chen, S. (2022). Smart cities and GEOINT integration: Case studies from Singapore and Barcelona. *Urban Systems & Planning Journal*, 15(4), 211–227.

Tao, P., Zhang, Y., & Silva, C. (2022). The impact of geospatial intelligence on disaster response and recovery. *International Journal of Emergency Management*, 11(1), 48–62.

Tao, Q., Garcia, L., & Chen, H. (2022). AI-driven geospatial intelligence for disaster management: Case studies from natural catastrophes. *Journal of Disaster Response and AI Solutions*, 15(3), 53–65.

Tao, Q., Garcia, L., & Chen, H. (2022). AI-driven geospatial intelligence in public health: Real-time tracking of disease outbreaks during the COVID-19 pandemic. *Journal of Public Health Informatics*, 18(4), 55–69.

Tao, Q., Garcia, L., & Chen, H. (2022). Using GEOINT for disaster management and public health: Case studies in real-time geospatial applications. *International Journal of Disaster Response and Public Health*, 20(3), 65–83.

Tarboton, D. G. (1997). A new method for the determination of flow directions and upslope areas in grid digital elevation models. *Water Resources Research*, 33(2), 309–319. DOI: 10.1029/96WR03137

Tareke, B., Filho, P. S., Persello, C., Kuffer, M., Maretto, R. V., Wang, J., Abascal, A., Pillai, P., Singh, B., D'Attoli, J. M., Kabaria, C., Pedrassoli, J., Brito, P., Elias, P., Atenógenes, E., & Santiago, A. R. (2024). User and Data-Centric Artificial Intelligence for Mapping Urban Deprivation in Multiple Cities Across the Globe. *IGARSS 2024 - 2024 IEEE International Geoscience and Remote Sensing Symposium*, 1553–1557. DOI: 10.1109/IGARSS53475.2024.10640428

Tariq, A., Jiango, Y., Li, Q., Gao, J., Lu, L., Soufan, W., Almutairi, K. F., & Habib-ur-Rahman, M. (2023). Modelling, mapping and monitoring of forest cover changes, using support vector machine, kernel logistic regression and naive bayes tree models with optical remote sensing data. *Heliyon*, 9(2), e13212. DOI: 10.1016/j.heliyon.2023.e13212 PMID: 36785833

Taromideh, F., Fazloula, R., Choubin, B., Emadi, A., & Berndtsson, R. (2022). Urban flood-risk assessment: Integration of decision-making and machine learning. *Sustainability (Basel)*, 14(8), 4483. DOI: 10.3390/su14084483

Tehrany, M. S., Pradhan, B., & Jebur, M. N. (2015). Flood susceptibility mapping using a novel ensemble weights-of-evidence model. *Geocarto International*, 30(6), 660–685.

Teja, K. R., Liu, C. M., & Chopra, S. R. (2023). Water Assessment Using Geospatial and Data Science Tools. *2023 International Conference on Emerging Smart Computing and Informatics (ESCI)*, 1–6. IEEE. DOI: 10.1109/ESCI56872.2023.10099538

Tetteh, A. T., Moomen, A.-W., Yevugah, L. L., & Tengnibuor, A. (2024). Geospatial approach to pluvial flood-risk and vulnerability assessment in Sunyani Municipality. *Heliyon*, 10(18), e38013. DOI: 10.1016/j.heliyon.2024.e38013 PMID: 39381211

Tewary, A., Upadhyay, C., & Singh, A. (2023). *Emerging role of AI, ML and IoT in modern sustainable energy management* (Vol. 2). IoT and Analytics in Renewable Energy Systems., DOI: 10.1201/9781003374121-23

Tharayil, S. M., Krishnapriya, M. A., & Alomari, N. K. (2024). How multimodal AI and IoT are shaping the future of intelligence. *Internet of Things and Big Data Analytics for a Green Environment*, 138-167. DOI: 10.1201/9781032656830-8

Thomas, J. P., & Frantz, R. (2023). AI in traffic management: A global review of implementations, challenges, and future directions. *Journal of Urban Planning and Development*, 149(1), 04022035.

Tian, S., Kang, L., Xing, X., Tian, J., Fan, C., & Zhang, Y. (2022). A Relation-Augmented Embedded Graph Attention Network for Remote Sensing Object Detection. *IEEE Transactions on Geoscience and Remote Sensing*, 60, 1–18. DOI: 10.1109/TGRS.2021.3073269

Tohidi, N., & Rustamov, R. B. (2020). A review of the machine learning in gis for megacities application. *Geographic Information Systems in Geospatial Intelligence*, 29-53.

Trisovic, A., Lau, M. K., Pasquier, T., & Crosas, M. (2022). A large-scale study on research code quality and execution. *Scientific Data*, 9(1), 60. DOI: 10.1038/s41597-022-01143-6 PMID: 35190569

Tuia, D., Kellenberger, B., Beery, S., Costelloe, B. R., Zuffi, S., Risse, B., Mathis, A., Mathis, M. W., van Langevelde, F., Burghardt, T., Kays, R., Klinck, H., Wikelski, M., Couzin, I. D., van Horn, G., Crofoot, M. C., Stewart, C. V., & Berger-Wolf, T. (2022). Perspectives in machine learning for wildlife conservation. *Nature Communications*, 13(1), 792. Advance online publication. DOI: 10.1038/s41467-022-27980-y PMID: 35140206

Uddin, K., & Matin, M. A. (2021). Potential flood hazard zonation and flood shelter suitability mapping for disaster risk mitigation in Bangladesh using geospatial technology. *Progress in Disaster Science*, 11, 100185. DOI: 10.1016/j.pdisas.2021.100185

Vahidnia, M. H. (2024). Empowering geoportals HCI with task-oriented chatbots through NLP and deep transfer learning. *Big Earth Data*, •••, 1–41. DOI: 10.1080/20964471.2024.2403166

Van Liempt, I. (2011). From dutch dispersal to ethnic enclaves in the UK: The relationship between segregation and integration examined through the eyes of somalis. *Urban Studies (Edinburgh, Scotland)*, 48(16), 3385–3398. DOI: 10.1177/0042098010397401

Verma, J. P., Sharma, N., Krishnan, S., Gautam, S., & Balas, V. E. (Eds.). (2024). *Green Computing for Sustainable Smart Cities: A Data Analytics Applications Perspective*. CRC PRESS.

Vlahogianni, E. I., Karlaftis, M. G., & Golias, J. C. (2015). Short-term traffic forecasting: Where we are and where we're going. *Transportation Research Part C, Emerging Technologies*, 43, 3–19. DOI: 10.1016/j.trc.2014.01.005

Vopham, T., Hart, J. E., Laden, F., & Chiang, Y. Y. (2018). "Emerging trends in geospatial artificial intelligence (geoAI): Potential applications for environmental epidemiology," *Environ. Heal. A Glob.Access Sci. Source*, 17(1), 1–6.

VoPham, T., Hart, J. E., Laden, F., & Chiang, Y.-Y.VoPham. (2018). Emerging trends in geospatial artificial intelligence (geoAI). *Environmental Health*, 17(1), 40. Advance online publication. DOI: 10.1186/s12940-018-0386-x PMID: 29665858

Walks, A. (2020). On the meaning and measurement of the ghetto as a form of segregation. In *Handbook of Urban Segregation*. Edward Elgar Publishing. DOI: 10.4337/9781788115605.00032

Wang, J., & Biljecki, F. (2022). Unsupervised machine learning in urban studies: A systematic review of applications. *Cities (London, England)*, 129, 103925. DOI: 10.1016/j.cities.2022.103925

Wang, S., Cai, W., Tao, Y., Sun, Q. C., Wong, P. P. Y., Thongking, W., & Huang, X. (2023). Nexus of heat-vulnerable chronic diseases and heatwave mediated through tri-environmental interactions: A nationwide fine-grained study in Australia. *Journal of Environmental Management*, 325, 116663. DOI: 10.1016/j.jenvman.2022.116663 PMID: 36343399

Wang, S., Liu, Y., Lam, J., & Kwan, M.-P. (2021). The effects of the built environment on the general health, physical activity and obesity of adults in Queensland, Australia. *Spatial and Spatio-temporal Epidemiology*, 39, 100456. DOI: 10.1016/j.sste.2021.100456 PMID: 34774262

Wang, Y., & Liu, W. (2019). Real-time geospatial data processing using edge computing. In *Proceedings of the IEEE International Conference on Big Data* (pp. 1025–1034). IEEE. DOI: 10.1109/BigData47090.2019.9006068

Wang, Y., & Zhang, Q. (2021). Real-Time Traffic Flow Prediction with Machine Learning and IoT: A Survey. *IEEE Access : Practical Innovations, Open Solutions*, 9, 214674–214688.

Wei, A., Bi, P., Guo, J., Lu, S., & Li, D. (2021). Modified drastic model for groundwater vulnerability to nitrate contamination in the Dagujia river basin, China. *Water Science and Technology: Water Supply*, 21(4), 1793–1805. DOI: 10.2166/ws.2021.018

Wei, B., Guo, X., Wu, Z., Zhao, J., & Zou, Q. (2023). Construct Fine-Grained Geospatial Knowledge Graph. In El Abbadi, A., Dobbie, G., Feng, Z., Chen, L., Tao, X., Shao, Y., & Yin, H. (Eds.), *Database Systems for Advanced Applications. DASFAA 2023 International Workshops* (Vol. 13922, pp. 267–282). Springer Nature Switzerland., DOI: 10.1007/978-3-031-35415-1_19

Wen, R., & Li, S. (2022). Spatial Decision Support Systems with Automated Machine Learning: A Review. *ISPRS International Journal of Geo-Information*, 12(1), 12. DOI: 10.3390/ijgi12010012

Wheeler, S. (2013). *Planning for Sustainability: Creating Livable, Equitable and Ecological Communities* (2nd ed.). Routledge., DOI: 10.4324/9780203134559

White, R. (2020). Geospatial intelligence in environmental monitoring: Tracking biodiversity and combating climate change. *Environmental Science and GeoData*, 19(2), 102–115.

Wolniak, . (2023). Wolniak. (2023). Smart mobility in smart city – Copenhagen and Barcelona comparision. *Scientific Papers of Silesian University of Technology Organization and Management Series*, 2023(172). Advance online publication. DOI: 10.29119/1641-3466.2023.172.41

Wu, Z., Liu, Y., Fang, S., Shen, W., Li, X., Mao, Z., & Wu, S. (2024). Integration of Geographic Features and Bathymetric Inversion in the Yangtze River's Nantong Channel Using Gradient Boosting Machine Algorithm with ZY-1E Satellite and Multibeam Data. *Geomatica*, 100027(2), 100027. Advance online publication. DOI: 10.1016/j.geomat.2024.100027

Xing, J., & Sieber, R. (2021) Integrating XAI and GeoAI. In: GIScience 2021 Short Paper Proceedings, UC Santa Barbara: Center for Spatial Studies. DOI: 10.25436/ E2301473

Xu, Q. (2021). Evaluation of Rural Tourism Spatial Pattern Based on Multifactor-Weighted Neural Network Algorithm Model in Big Data Era. *Scientific Programming*, 2021, 1–11. DOI: 10.1155/2021/8108287

Xu, Y., Pan, L., Du, C., Li, J., Jing, N., & Wu, J. (2018). "Vision-based UAVs Aerial Image Localization: A Survey," *Proc. 2nd ACM SIGSPATIAL Int. Work. AI Geogr. Knowl. Discov.*, pp. 9–18, 2018. DOI: 10.1145/3281548.3281556

Yabe, T., Jones, N. K. W., Rao, P. S. C., Gonzalez, M. C., & Ukkusuri, S. V. (2022). Mobile phone location data for disasters: A review from natural hazards and epidemics. *Computers, Environment and Urban Systems*, 94, 101777. DOI: 10.1016/j.compenvurbsys.2022.101777

Yadav, A., & Sagi, S. (2024). Management Cases Studies and Technical Use Cases on Web 3. In Darwish, D. (Ed.), (pp. 267–287). Advances in Web Technologies and Engineering. IGI Global., DOI: 10.4018/979-8-3693-1532-3.ch013

Yan, B., Janowicz, K., Mai, G., & Gao, S. (2017). From itdl to place2vec: Reasoning about place type similarity and relatedness by learning embeddings from augmented spatial contexts. In *Proceedings of the 25th ACM SIGSPATIAL International Conference on Advances in Geographic Information Systems*, page 35. ACM. DOI: 10.1145/3139958.3140054

Yang, B., Wang, S., Li, S., Zhou, B., Zhao, F., Ali, F., & He, H. (2022). Research and application of UAV-based hyperspectral remote sensing for smart city construction. *Cognitive Robotics*, 2, 255–266. DOI: 10.1016/j.cogr.2022.12.002

Yang, C. (2024). Application and assessment of GIS technology in flash flood risk management. *Sustainable Environment*, 9(1), 26. DOI: 10.22158/se.v9n1p26

Yap, J., Tay, R., & Ong, K. (2020). Smart cities and urban transport: Opportunities and challenges in the deployment of artificial intelligence for real-time traffic management. *Journal of Transport and Land Use*, 13(1), 25–45.

Yin, J., Dong, J., Hamm, N. A. S., Li, Z., Wang, J., Xing, H., & Fu, P. (2021). Integrating remote sensing and geospatial big data for urban land use mapping: A review. *International Journal of Applied Earth Observation and Geoinformation*, 103, 102514. DOI: 10.1016/j.jag.2021.102514

Yohana, A. R., Makoba, E. E., Mussa, K. R., & Mjemah, I. C. (2023). Evaluation of groundwater potential using aquifer characteristics in Urambo district, Tabora region, Tanzania. *Earth (Basel, Switzerland)*, 4(4), 776–805. DOI: 10.3390/earth4040042

York, P., & Bamberger, M. (2024). The applications of big data to strengthen evaluation. *Artificial Intelligence and Evaluation*, 37-55. DOI: 10.4324/9781003512493-3

Youssef, A. M., Pradhan, B., & Pourghasemi, H. R. (2023). Urban flood susceptibility mapping using hybrid AI models and GIS: Case study of Cairo, Egypt. *Geocarto International*, 38(1), 95–114.

Yuan, M., Buttenfield, B. P., Gahegan, M. N., & Miller, H. (2004). Geospatial data mining and knowledge discovery. In *A Research Agenda for Geographic Information Science* (p. 24). CRC Press. DOI: 10.1201/9781420038330-14

Zhang, Q., Nguyen, T., & Brown, R. (2021). Enhancing agriculture through AI and GEOINT: Precision farming for

Zhang, F., Wu, L., Zhu, D., & Liu, Y. (2019). Social Sensing from Street-Level Imagery: A Case Study in Learning Spatio-Temporal Urban Mobility Patterns. *ISPRS Journal of Photogrammetry and Remote Sensing*, 153, 48–58. DOI: 10.1016/j.isprsjprs.2019.04.017

Zhang, F., Zhou, B., Liu, L., Liu, Y., Fung, H. H., Lin, H., & Ratti, C. (2018). Measuring Human Perceptions of a Large-Scale Urban Region Using Machine Learning. *Landscape and Urban Planning*, 180, 148–160. DOI: 10.1016/j.landurbplan.2018.08.020

Zhang, J., Huang, Y., & Ma, H. (2018). Application of ensemble methods in flood prediction: A case study in the Yangtze River Basin, China. *Water (Basel)*, 10(12), 1801.

Zhang, P., Yi, W., Song, Y., Thomson, G., Wu, P., & Aghamohammadi, N. (2024). Geospatial learning for large-scale transport infrastructure depth prediction. *International Journal of Applied Earth Observation and Geoinformation*, 132, 103986. DOI: 10.1016/j.jag.2024.103986

Zhang, Q. (2021). Geospatial intelligence in precision agriculture: Supporting food security through satellite imagery and AI. *Agricultural Geospatial Intelligence*, 12(3), 67–80.

Zhang, Q., Li, J., & Xu, W. (2020). Machine learning applications in GEOINT: Tracking environmental changes and land use. *Geospatial Technology Journal*, 8(5), 73–89.

Zhang, Q., Nguyen, T., & Brown, R. (2020). Geospatial intelligence in precision agriculture: Enhancing food security and agricultural efficiency. *Agricultural Geointelligence Review*, 6(4), 149–163.

Zhang, W. (1988). Shift-invariant pattern recognition neural network and its optical architecture. *Proceedings of Annual Conference of* the Japan Society of Applied Physics.

Zhang, X., Wang, Q., & Li, H. (2023). Mapping flood propagation using graph neural networks and geospatial data. *Environmental Modelling & Software*, 160, 105349.

Zhang, X., & Zhao, X. (2022). Machine learning approach for spatial modeling of ride sourcing demand. *Journal of Transport Geography*, 100, 103310. DOI: 10.1016/j.jtrangeo.2022.103310

Zhang, Y., & Cheng, T. (2020). Graph Deep Learning Model for Network-Based Predictive Hotspot Mapping of Sparse Spatio-Temporal Events. *Computers, Environment and Urban Systems*, 79, 101403. DOI: 10.1016/j.compenvurbsys.2019.101403

Zhang, Y., Liu, Y., & Zheng, Y. (2021). Predictive traffic management using machine learning: A case study of large metropolitan areas in China. *Journal of Transportation Systems Engineering and Information Technology*, 21(4), 234–246.

Zhang, Y., & Raubal, M. (2022). Street-level traffic flow and context sensing analysis through semantic integration of multi-source geospatial data. *Transactions in GIS*, 26(8), 3330–3348. DOI: 10.1111/tgis.13005

Zhang, Y., Wei, C., He, Z., & Yu, W. (2024). GeoGPT: An assistant for understanding and processing geospatial tasks. *International Journal of Applied Earth Observation and Geoinformation*, 131, 103976. DOI: 10.1016/j.jag.2024.103976

Zhang, Z., Qian, Z., Zhong, T., Chen, M., Zhang, K., Yang, Y., Zhu, R., Zhang, F., Zhang, H., Zhou, F., Yu, J., Zhang, B., Lü, G., & Yan, J. (2022). Vectorized rooftop area data for 90 cities in China. *Scientific Data*, 9(1), 66. DOI: 10.1038/s41597-022-01168-x PMID: 35236863

Zhang, Z., Wen, F., Sun, Z., Guo, X., He, T., & Lee, C. (2022). Artificial intelligence-enabled sensing technologies in the 5G/internet of things era: From virtual reality/augmented reality to the digital twin. *Advanced Intelligent Systems*, 4(7), 2100228. DOI: 10.1002/aisy.202100228

Zhao, L., Song, Y., Zhang, C., Liu, Y., Wang, P., Lin, T., Deng, M., & Li, H. (2019). T-GCN: A Temporal Graph Convolutional Network for Traffic Prediction. *IEEE Transactions on Intelligent Transportation Systems*, 21(9), 3848–3858. DOI: 10.1109/TITS.2019.2935152

Zhao, W., Du, S., & Qi, J. (2019). CNN-based deep learning method for mapping flood areas. *Water (Basel)*, 11(10), 2060.

Zheng, C., Yuan, J., Zhu, L., Zhang, Y., & Shao, Q. (2020). From digital to sustainable: A scientometric review of smart city literature between 1990 and 2019. *Journal of Cleaner Production*, 258, 120689. DOI: 10.1016/j.jclepro.2020.120689

Zheng, Z., Shen, W., Li, Y., Qin, Y., & Wang, L. (2020). Spatial equity of park green space using KD2SFCA and web map API: A case study of zhengzhou, China. *Applied Geography (Sevenoaks, England)*, 123, 102310. DOI: 10.1016/j.apgeog.2020.102310

Zhou, Q., Zhu, M., Qiao, Y., Zhang, X., & Chen, J. (2021). Achieving resilience through smart cities? Evidence from China. *Habitat International*, 111, 102348. DOI: 10.1016/j.habitatint.2021.102348

Zhu, A. X., Lu, G., Liu, J., Qin, C. Z., & Zhou, C. (2018). Spatial Prediction Based on Third Law of Geography. *Annals of GIS*, 24(4), 225–240. DOI: 10.1080/19475683.2018.1534890

Zhu, D., Zhang, F., Wang, S., Wang, Y., Cheng, X., Huang, Z., & Liu, Y. (2020). Understanding Place Characteristics in Geographic Contexts through Graph Convolutional Neural Networks. *Annals of the American Association of Geographers*, 110(2), 408–420. DOI: 10.1080/24694452.2019.1694403

Zhu, R., Zhang, X., Kondor, D., Santi, P., & Ratti, C. (2020). Understanding spatio-temporal heterogeneity of bike-sharing and scooter-sharing mobility. *Computers, Environment and Urban Systems*, 81, 101483. DOI: 10.1016/j.compenvurbsys.2020.101483

Zhu, X. X., Tuia, D., Mou, L., Xia, G.-S., Zhang, L., Xu, F., & Fraundorfer, F. (2017). Deep Learning in Remote Sensing: A Comprehensive Review and List of Resources. *IEEE Geoscience and Remote Sensing Magazine*, 5(4), 8–36. DOI: 10.1109/MGRS.2017.2762307

Zhu, X., & Li, S. (2020). Integrating geospatial big data and artificial intelligence for smart city applications: Challenges and opportunities. *International Journal of Applied Earth Observation and Geoinformation*, 88, 102035. DOI: 10.1016/j.jag.2019.102035

Zou, X., Yan, Y., Hao, X., Hu, Y., Wen, H., Liu, E., Zhang, J., Li, Y., Li, T., Zheng, Y., & Liang, Y. (2025). Deep learning for cross-domain data fusion in urban computing: Taxonomy, advances, and outlook. *Information Fusion*, 113, 102606. DOI: 10.1016/j.inffus.2024.102606

About the Contributors

Dina Darwish, Vice dean, faculty of computer science and information technology, Ahram Canadian university, Egypt. I obtained my Ph.D from Cairo university at 2009, engineering faculty. I am professor since 2020, also my special domain is artificial intelligence and I have many publications in the following topics, including: Artificial intelligence, wireless ad hoc networks, Internet of things, Big data analytics, Blockchain applications, Cloud computing applications, Web 3 technologies, Chatbots development and others. My email is: dina.g.darwish@gmail.com

Houssem Chemingui, is an Assistant Professor in Information Systems at Brest Business School. With a PhD in Computer Science from Université Paris 1 Panthéon Sorbonne, his research focuses on streamlining product line configuration processes. He's published extensively on topics like Artificial Intelligence, Data Analytics, and Software Configuration. Beyond academia, he's contributed to development projects and consultancy initiatives, showcasing his commitment to advancing technology and education. Dr. Chemingui's dedication to research and teaching has earned him recognition as a prominent figure in the field of Computer Science.

***

Pankaj Bhambri is affiliated with the Department of Information Technology at Guru Nanak Dev Engineering College in Ludhiana. Additionally, he fulfills the role of the Convener for his Departmental Board of Studies. He possesses nearly two decades of teaching experience. His research work has been published in esteemed worldwide and national journals, as well as conference proceedings. Dr. Bhambri has garnered extensive experience in the realm of academic publishing, having served as an editor for a multitude of books in collaboration with esteemed publishing houses such as CRC Press, Elsevier, Scrivener, and Bentham Science. In addition to his editorial roles, he has demonstrated his scholarly prowess by authoring numerous books and contributing chapters to distinguished publishers within the academic community. Dr. Bhambri has been honored with several prestigious accolades, including the ISTE Best Teacher Award in 2023 and 2022, the I2OR National Award in 2020, the Green ThinkerZ Top 100 International Distinguished Educators award in 2020, the I2OR Outstanding Educator Award in 2019, the SAA Distinguished Alumni Award in 2012, the CIPS Rashtriya Rattan Award in 2008, the LCHC Best Teacher Award in 2007, and numerous other commendations from various government and non-profit organizations. He has provided guidance and oversight for numerous research projects and dissertations at the postgraduate and Ph.D. levels. He successfully organized a diverse range of educational programmes, securing financial backing from esteemed institutions such as the AICTE, the TEQIP, among others. Dr. Bhambri's areas of interest encompass machine learning, bioinformatics, wireless sensor networks, and network security.

Rupanshi Bhatnagar is working as a Senior Analyst at EY India, specializing in the Business Technology department. She holds a degree in Electronics and Communication from Bharati Vidyapeeth's College of Engineering, New Delhi, where she was honored as the Best Student of the Year. Rupanshi excels in combining technical expertise with analytical skills, notably leading projects in machine learning, including a significant one on diagnosing Parkinson's disease. Beyond academics, she has been the President of EduTech, a technical subchapter of EduMinerva, and the General Secretary of the IOSC Student's Club. Her commitment to leveraging technology for impactful solutions, along with her leadership roles, showcases her dedication to advancing both her field and community.

Veena Christy completed her graduation in Commerce from the renowned Madras Christian College, affiliated to University of Madras. Her urge for higher studies led her to complete her MBA with laurels. She was awarded the University rank in her PG program which was also affiliated to University of Madras. Lured by passion for research, she completed her doctoral studies in organisational behaviour and has many Scopus-indexed publications to her credit. With ten years of teaching experience in different institutions of repute and six years of research expertise, she is also the recipient of Innovative Researcher and Dedicated Academician Award conferred by The Innovative Scientific Research Professional Institute, Chennai. She is an Associate member of Madras Management Association (MMA) (Mem No: AM863) and Affiliate member of Association of Behavior Analysis International (Mem No: 102469). Presently, She is serving as Assistant Professor in SRM University, KTR Campus, Chennai.

Monica Gupta, a distinguished scholar in Electronics and Communication Engineering, earned her Ph.D. from Delhi Technological University in 2022, highlighting a relentless commitment to academic excellence. With over 17 years of teaching experience, she currently serves as an Associate Professor at Bharati Vidyapeeth's College of Engineering, New Delhi. A prolific researcher, Dr. Gupta has made substantial contributions in Low Power Memory Design, Image and Video Processing, VLSI, IoT, Artificial Intelligence and Machine Learning, evident in her numerous publications in reputable journals and conferences. Beyond academia, she has initiated the IOSC-BVP student club, in collaboration with Intel OneAPI Student ambassador, addressing the skill gap between graduates and industry expectations. Inaugurated on November 7, 2022, the club focuses on skill enhancement programs, organizing national-level events to deepen students' understanding of crucial areas like Artificial Intelligence, Internet of Things, Machine Learning, and Website Development. Dr. Monica Gupta's impact extends beyond the classroom, shaping the future of engineering by bridging academic knowledge with industry needs. Her dynamic leadership and commitment to excellence make her a trailblazer in the landscape of Electronics and Communication Engineering education.

Arbia Hlali is an Assistant Professor at the Taibah University, Saudi Arabia. She has international experience in teaching and researching and earlier worked at University of Puerto Rico at Rio Piedras, San Juan, Puerto Rico, and prior to that at Mediterranean University, Mediterranean Institute of Tunisia, MIT Nabeul, Tunisia. Her research interests include Transport Economics, Maritime Transport, Transport Management, Logistics, Container Shipping, Transportation Science, Containers, Port Management, Business Model Innovation, Advanced Econometrics, Standard Statistical Analysis, Shipping, Total Quality Management, Food Quality Management, Transportation, Supply Chain, and Green Supply Chain. She is a prolific author and has several publications in renowned journals and books.

Amit Sai Jitta has research interests in the area of Computer Science. His passion lies in on Artificial Intelligence, especially its its application in finance domain. He is keenly interested on how machine learning can help improve financial decision-making and predictions. His research goal is to create AI solution that tackle real-world problems and make things more efficient across various industries. Outside of coding and research, he loves collaborating with co-researchers on various use-cases and projects. He truly believes that AI can drive positive change, and is excited to be part of its future, especially in transforming financial forecasting.

Vijaya Kittu Manda A multi-dimensional personality, Dr. Vijaya Kittu Manda has nearly 13+ years of experience in Business Management and Technology. He is a Researcher at PBMEIT, India and works in capital markets, financial planning, and investing. He is an Advocate, a technocrat, an academician, a book writer, and a stock market enthusiast. He has 11 University Postgraduate Degrees in various disciplines. He is a Ph.D. in Management (Finance). His thesis was on Mutual Funds and their Market Competition. His thesis won the prestigious NSE-IEA Best Thesis Award in 2023. He is currently pursuing his second Ph.D. in Computer Science with focus on Blockchain. He contributed over 750 articles to various magazines. He is the Chief Editor for a Management Book Series, is a Peer Review, is a Certified Peer Review Supervisor, is a Session Chair and an Advisor for various academic and industrial conferences. He writes Research Papers and Case Studies, Book Chapters, and sits on Editorial Boards of various publishers. He is a guest speaker for colleges and universities.

R.Parvathi is a Professor of School of Computing Science and Engineering at VIT University, Chennai since 2011. She received the Doctoral degree in the field of spatial data mining in the same year. Her teaching experience in the area of computer science includes more than two decades and her research interests include data mining, big data and computational biology.

R. Rajeshkanna is an accomplished academic with over 16 years of experience in the field of computer science and information technology. He holds a Ph.D. from Bharathiar University, Coimbatore, and an M.Phil from Vinayaka Missions University, Salem. He also earned his MCA from Periyar University, Salem. Currently, Dr. Rajeshkanna serves as an Assistant Professor in the Department of Computer Science at CHRIST (Deemed to be University), Bangalore, a position he has held since July 2022. Prior to this, he was the Professor and Head of the Department of Information Technology at Dr. N.G.P Arts & Science College (Autonomous), Coimbatore. His tenure at this institution, spanning from January 2020 to June 2022, followed a series of progressive roles, including Associate Professor and Head of the Department, and earlier, as an Assistant Professor. Dr. Rajeshkanna has made significant academic contributions through his research and scholarly activities. He has published 35 papers in reputed international journals with high impact factors and has contributed 16 book chapters through IGI Global Publications.

Rachna Rana is working as Associate Professor in the Department of Computer Science & Engineering at Ludhiana Group of Colleges in Ludhiana. She possesses nearly two decades of teaching experience. Her areas of interest includes machine learning, wireless sensor networks, and network security.

Mohana Priya T is Assistant Professor at CHRIST(Deemed to be University), Bangalore, India. She Holds a PhD degree in Computer Science in Machine learning at Bharathiar University, Coimbatore, India. Her research areas are Machine Learing and medical image analysis. She is acted as Coordinator for curriculum development cell and Software development cell.She published 10 Journal articles in various international peer reviewed journals, 2 Book publications and 15 presentations in various international level conference held in india and abroad.. She can be contacted at email: drtmohanapriya @gmail.com

Theodore Tarnanidis is a marketing scholar, adjunct professor at the International Hellenic University and researcher in Applications of D.Sc. and MCDA. Theodore has six years experience as a marketing and decision making practitioner. He made his post-doc research in the area of sustainable entrepreneurship from the University of Macedonia. He obtained a Ph.D. from the University of London Met., UK. He received his M.B.A from Liverpool University, UK and is a graduate from the University of Macedonia (Business Administration) and Alexander Technological Educational Institute (Marketing). His research focuses on International Marketing, Multi-cultural Marketing, Marketing Management, Consumer Science by Means of Rank-Coded Data, Preference Measurement Techniques, Quantitative Methods & Structural Equation Modelling. His work has been published in various internationally renowned scientific conferences (Academy of Marketing, European Marketing Academy, PROMETHEE Days, Hellenic Operational Research) and in journals (Journal of Business Ethics, World Review of Entrepreneurship, Management and Sustainable Development, Journal of Retailing and Consumer Services, Current Issues in Tourism, Marketing Science & Inspirations, Management Science Letters). Email: tarnanidis@ihu.gr

Pattabiraman Venkatasubbu obtained his Bachelor's from Madras University and Master's degree from Bharathidasan University. He completed his PhD from Bharathiar University, India. He has a total Professional experience of more than 16 years working in various prestigious institutions. He has published more than 30 papers in various National and International peer reviewed journals and conferences. He visited various countries namely few China, Singapore, Malaysia, Thailand and South Africa etc. for presenting his research contributions as well as to giving key note address. He is currently an Associate Professor and Program-Chair for Master's Programme at VIT University-Chennai Campus, India. His teaching and research expertise covers a wide range of subject area including Data Structures, Knowledge Discovery and Data mining, Database echnologies, Big Data Analytics, Networks and Information security etc.

Yashwant Waykar is a dedicated educator and researcher at Dr. Babasaheb Ambedkar Marathwada University, specializing in Software Engineering, Machine Learning, Artificial Intelligence, and Learning Management Systems (LMS). With over 14 years of experience and a Ph.D., Dr. Waykar has authored 40+ publications in prestigious journals indexed in SCOPUS, UGC CARE, and Web of Science. Their research has earned accolades and includes several book chapters and four patents (3 granted, 1 filed) in LMS and AI. Dr. Waykar's teaching in AI, Web Development, and Software Testing is highly regarded, fostering a dynamic learning environment. They have completed three research projects funded by UGC and Dr. B.A.M.U. and have mentored 3 M.Phil and 2 Ph.D. students, guiding them in Machine Learning and LMS research.

Sucheta S. Yambal has been an Assistant Professor in the Department of Management Science of Dr. Babasaheb Ambedkar Marathwada University in Chhatrapati Sambhaji Nagar, Maharashtra. Her total teaching experience is 19 years, and she has about 16 years of research experience. She has taken on several significant duties as a member of the Board of Directors, organized multiple workshops, and given guest lectures at numerous universities. Dr. Yambal has supervised four M.Phil. and four Ph.D. research students at Dr. Babasaheb Ambedkar Marathwada University to date. She has also reviewed Ph.D. theses for Tilak Maharastra Vidyapeeth, Pune. More than thirty national and international research papers have been published by her at prestigious institutes' international conferences, such as IIT Delhi and IIM Raipur. Her research has also been published in prestigious journals, such as SCOPUS and UGC- Care Listed. In November 2019, Dr. Yambal scored highly in the e-business course on the Nptel platform during an exam administered by IIT Kharagpur. She is deemed a "Women of Substance" 2018 Award which was organized at MGM, Chhtrapati Sambhaji Nagar. She had a stellar academic record from elementary school to the Dr. Babasaheb Ambedkar Marathwada University Ph.D. admission exam, where she received the best possible mark. Her areas of interest include employment, AI/ML, automation in industries, database management systems and data mining, modern and inclusive education, and employability.

Index

Symbols

3d modeling 198, 246
4d modeling 242, 259, 261, 269

A

AI in Traffic Management 219, 220, 225, 238
ai models 30, 54, 55, 56, 57, 58, 59, 62, 63, 78, 80, 160, 163, 164, 167, 168, 169, 175, 178, 224, 225, 231, 232, 233, 234, 235, 240, 248, 256, 258, 284, 286, 287, 297
Artificial Intelligence 1, 2, 3, 4, 5, 10, 16, 17, 18, 19, 20, 21, 22, 23, 24, 25, 27, 28, 29, 32, 39, 44, 46, 49, 50, 51, 52, 53, 54, 62, 64, 66, 68, 69, 70, 76, 77, 78, 80, 84, 85, 86, 87, 88, 91, 92, 93, 96, 98, 99, 100, 101, 102, 103, 121, 123, 129, 130, 131, 134, 135, 136, 137, 141, 142, 155, 156, 157, 158, 161, 162, 166, 170, 178, 180, 181, 206, 210, 211, 212, 213, 214, 217, 218, 219, 222, 229, 230, 233, 235, 236, 237, 238, 241, 243, 245, 256, 262, 266, 271, 272, 273, 276, 281, 282, 283, 284, 285, 286, 287, 288, 289, 291, 292, 293, 297, 298, 300, 301, 302, 305, 307, 310, 311, 312, 313, 315, 316, 319
Autonomous Vehicles 35, 77, 79, 81, 108, 136, 181, 206, 221, 224, 229, 231, 235, 236, 237, 254, 301

B

Big Data 5, 8, 10, 16, 17, 18, 19, 22, 24, 30, 32, 33, 42, 43, 44, 45, 65, 70, 101, 102, 103, 110, 115, 119, 120, 121, 122, 123, 124, 125, 127, 129, 130, 133, 135, 161, 169, 172, 200, 202, 209, 213, 218, 237, 242, 243, 244, 250, 257, 265, 268, 300, 314, 316, 317, 320
Blockchain 114, 121, 126, 129, 131, 134, 251, 259, 260, 266, 270

C

Computer Vision 29, 101, 124, 161, 162, 246, 300, 312, 315, 318

D

Deep Learning 1, 2, 5, 8, 10, 11, 15, 16, 22, 25, 27, 28, 29, 30, 34, 35, 36, 37, 62, 65, 67, 70, 78, 101, 124, 158, 159, 161, 162, 164, 166, 168, 170, 174, 176, 180, 192, 194, 195, 224, 227, 230, 245, 262, 269, 282, 286, 293, 297, 298, 299, 300, 302, 311, 312, 314, 315, 316, 318, 319, 320, 321, 322, 323
digital twins 17, 113, 124, 130, 133, 208, 240, 241, 245, 248, 249, 251, 258, 266, 269, 308, 310
disaster management 27, 28, 29, 30, 34, 38, 39, 41, 47, 64, 73, 75, 77, 81, 83, 85, 86, 89, 94, 95, 107, 110, 112, 113, 129, 161, 162, 165, 166, 167, 168, 169, 171, 172, 178, 181, 183, 184, 186, 192, 194, 196, 197, 199, 202, 203, 204, 206, 207, 209, 271, 273, 278, 286, 288, 290, 292, 293, 294, 295, 305

E

Environmental Monitoring 9, 10, 27, 28, 30, 33, 38, 39, 41, 42, 44, 45, 49, 57, 58, 61, 73, 74, 76, 84, 96, 108, 112, 136, 148, 149, 150, 161, 162, 163, 166, 167, 169, 171, 172, 181, 183, 185, 186, 192, 193, 196, 197, 199, 201, 203, 204, 206, 207, 209, 220, 241, 243, 252, 261, 270, 287, 290, 293, 294, 295, 302, 304, 307, 308, 316, 317

G

GeoAI 1, 2, 3, 4, 5, 6, 8, 10, 11, 12, 14, 16, 17, 18, 19, 20, 21, 22, 23, 24, 25,

101, 161, 170, 171, 245, 246, 256, 297, 298, 300, 310, 311, 312, 313, 314, 315, 316, 319, 320, 322

Geographic Information Systems 1, 2, 3, 4, 5, 25, 27, 28, 31, 37, 69, 74, 75, 76, 77, 80, 101, 109, 113, 142, 150, 182, 185, 200, 240, 248, 272, 275, 278, 281, 301, 303, 320, 322

geospatial ai 1, 2, 4, 10, 16, 17, 50, 56, 57, 58, 59, 63, 124, 128, 136, 155, 156, 157, 159, 162, 163, 164, 165, 177, 178, 179, 181, 203, 204, 205, 206, 208, 214, 240, 241, 243, 245, 248, 256, 260, 261, 264, 266, 271, 297, 302, 307, 308, 316

Geospatial Analysis 5, 9, 19, 27, 28, 29, 30, 31, 32, 34, 35, 36, 37, 38, 39, 40, 41, 42, 43, 44, 45, 46, 48, 49, 50, 64, 65, 70, 71, 74, 110, 117, 120, 122, 123, 160, 169, 171, 173, 175, 184, 194, 199, 206, 214, 256, 293, 297

Geospatial Challenges 172, 298

Geospatial Techniques 56, 109, 133, 158, 201

geospatial technologies 50, 54, 55, 56, 57, 60, 61, 63, 64, 65, 76, 80, 88, 96, 103, 104, 106, 107, 109, 110, 111, 123, 127, 131, 132, 136, 156, 170, 172, 181, 185, 187, 188, 189, 190, 191, 192, 199, 201, 209, 239, 240, 241, 242, 243, 244, 245, 249, 250, 251, 252, 253, 254, 255, 256, 260, 293, 317

Geospatial Technology 25, 73, 74, 75, 76, 77, 80, 81, 82, 83, 84, 85, 87, 89, 90, 91, 94, 95, 96, 98, 106, 110, 129, 131, 132, 134, 181, 184, 188, 190, 191, 192, 199, 200, 201, 203, 214, 292, 293, 295, 305, 306, 307, 308, 316

GIS 1, 2, 3, 4, 5, 9, 10, 18, 21, 25, 27, 28, 30, 31, 32, 33, 34, 37, 38, 44, 54, 55, 65, 67, 68, 69, 70, 73, 74, 77, 89, 90, 91, 94, 95, 98, 103, 109, 112, 113, 117, 119, 121, 130, 131, 133, 135, 136, 140, 141, 142, 143, 148, 149, 150, 151, 152, 159, 160, 171, 172, 175, 180, 182, 183, 184, 185, 187, 188, 189, 190, 191, 192, 199, 200, 201, 202, 203, 214, 240, 244, 245, 248, 249, 250, 254, 256, 260, 268, 269, 270, 272, 275, 278, 279, 281, 293, 297, 301, 302, 303, 304, 305, 314, 319, 322, 323

GPS 14, 15, 23, 32, 34, 42, 55, 73, 75, 76, 77, 103, 110, 114, 115, 116, 136, 161, 169, 177, 178, 180, 181, 182, 183, 185, 186, 187, 188, 189, 190, 192, 194, 199, 200, 201, 203, 204, 205, 206, 208, 209, 214, 218, 219, 223, 227, 230, 233, 234, 235, 243, 244, 248, 263, 275, 289, 298, 300, 302

I

Industrial Growth 188, 191, 192, 209, 214

Internet Of Things 2, 17, 24, 103, 114, 121, 122, 133, 134, 136, 178, 182, 183, 184, 206, 210, 212, 213, 218, 223, 233, 237, 243, 244, 255, 267, 284, 289

iot 33, 35, 43, 47, 49, 67, 103, 107, 110, 114, 115, 119, 120, 121, 122, 126, 127, 130, 131, 132, 133, 134, 136, 161, 177, 178, 179, 181, 182, 183, 184, 190, 192, 193, 194, 195, 199, 201, 203, 204, 205, 208, 210, 211, 212, 213, 214, 218, 220, 223, 227, 233, 234, 236, 238, 240, 241, 243, 244, 245, 248, 249, 250, 253, 255, 258, 259, 260, 265, 267, 269, 284, 289, 294, 302, 317

IoT Sensors 35, 49, 115, 127, 183, 204, 205, 208, 214, 234, 248, 250, 258

L

LiDAR scans 169, 177, 195, 196, 197, 207

M

Machine Learning 1, 2, 3, 4, 5, 11, 22, 24, 25, 27, 28, 31, 32, 33, 35, 39, 40, 41, 42, 44, 46, 47, 48, 49, 50, 51, 52, 53, 54, 55, 58, 59, 64, 66, 67, 68, 69, 70, 76, 77, 78, 101, 116, 119, 121, 123, 124, 128, 130, 140, 145, 146, 148,

149, 150, 151, 155, 156, 157, 158, 159, 160, 161, 162, 166, 167, 168, 169, 170, 171, 172, 174, 175, 178, 180, 184, 192, 193, 194, 195, 199, 200, 202, 207, 209, 210, 211, 212, 214, 217, 219, 220, 222, 224, 226, 227, 228, 229, 230, 232, 233, 234, 238, 245, 249, 253, 262, 272, 273, 274, 276, 282, 284, 285, 286, 295, 297, 298, 301, 302, 304, 305, 306, 311, 314, 320, 321, 322

Machine Learning Algorithms 31, 47, 53, 54, 67, 128, 145, 148, 150, 157, 159, 169, 170, 171, 174, 180, 194, 202, 209, 217, 219, 220, 226, 227, 230, 232, 233, 234, 249, 253, 284, 285, 301

P

precision agriculture 39, 77, 80, 81, 85, 86, 107, 167, 171, 183, 186, 197, 204, 206, 207, 209, 210, 211, 212, 213, 289, 293, 295

Predictive Analytics 27, 40, 41, 42, 51, 52, 58, 65, 70, 90, 103, 122, 125, 127, 135, 178, 221, 235, 246, 258, 263, 281, 286, 291, 302, 306, 307, 308

R

Real-Time Traffic Prediction 236

Remote Sensing 2, 9, 28, 30, 31, 32, 35, 37, 38, 39, 40, 43, 47, 48, 50, 54, 58, 65, 67, 70, 71, 73, 74, 75, 76, 80, 89, 101, 103, 109, 113, 115, 124, 130, 132, 133, 134, 135, 136, 139, 141, 149, 150, 152, 155, 156, 157, 158, 159, 160, 161, 162, 163, 164, 165, 166, 167, 169, 174, 175, 180, 185, 187, 188, 189, 190, 192, 195, 196, 199, 200, 201, 203, 206, 207, 243, 251, 253, 256, 262, 264, 265, 268, 272, 275, 278, 281, 290, 294, 300, 316, 317, 318, 319, 321, 322, 323

Resource Management 19, 31, 32, 49, 54, 56, 58, 63, 65, 74, 80, 85, 86, 88, 94, 95, 96, 97, 100, 113, 129, 135, 173, 179, 184, 186, 187, 189, 191, 194, 199, 203, 208, 209, 242, 282, 289, 293, 307

S

smart cities 3, 54, 55, 79, 80, 82, 103, 104, 105, 106, 107, 109, 110, 111, 112, 113, 114, 115, 116, 121, 122, 123, 124, 125, 126, 127, 129, 130, 131, 132, 133, 135, 136, 167, 181, 183, 190, 193, 199, 201, 203, 206, 208, 209, 211, 218, 219, 223, 237, 238, 239, 240, 241, 242, 243, 244, 245, 248, 249, 252, 253, 255, 256, 259, 260, 262, 263, 265, 266, 267, 268, 269, 270, 272, 279, 282, 285, 288, 289, 292, 293, 294, 307, 310, 316

Smart City Technologies 221, 222, 223, 227, 245, 263

spatial data analysis 82, 215

Sustainable Development Goals 54, 57, 60, 107, 124

T

Traffic Flow Optimization 193, 217

U

Urban Ecosystems 193

Urban Infrastructure 49, 104, 106, 107, 121, 129, 131, 135, 183, 218, 219, 257, 288

Urban Planning 1, 3, 10, 11, 17, 21, 27, 28, 29, 30, 31, 33, 34, 35, 38, 40, 41, 42, 44, 45, 47, 49, 52, 54, 55, 60, 61, 62, 64, 65, 73, 74, 75, 76, 80, 83, 94, 95, 97, 101, 105, 107, 112, 114, 115, 127, 132, 136, 160, 161, 163, 166, 167, 169, 172, 181, 184, 187, 190, 192, 196, 199, 200, 206, 207, 217, 220, 237, 238, 240, 241, 242, 243, 244, 248, 249, 250, 251, 256, 258, 259, 261, 263, 269, 270, 271, 272, 275, 278, 279, 281, 283, 287, 288, 289, 291, 293, 294, 302, 304, 305, 311, 322

W

Web Mapping 120